Bridged Free Radicals

Bridged Free Radicals

LEONARD KAPLAN

Department of Chemistry
The University of Chicago
Chicago, Illinois

1972

MARCEL DEKKER, INC.   New York

PREFACE

While the possibility of bridging in free radicals is often considered or invoked in the research literature and for instructional purposes, the work which is frequently cited and the large amount of potentially relevant work which is usually ignored have not been discussed in a thorough and critical way. Hence, this book.

Although my primary goal has been to create a source for reference and teaching in an important area of free radical chemistry, its implementation has been guided without compromise by my belief that no harm should be done. I have tried to guide the reader, but have not made choices for him. I have tried to present accurately the work which I discuss, but have given no special weight to the views of the authors involved. I have tried to present the experimental observations in such a way as to enable the reader, without consulting the original literature, not just to formulate his own general conclusions but also to decide for himself whether the reaction mixture really was shown to contain 30% isopropyl chloride. Consequently, I have willfully obliged the reader to make his way through a "comprehensive" and somewhat critical book. My hope is that he will arrive at informed judgments with confidence and with the secure belief that he has not been misled.

My decision to base the primary organization of the book on the structure of the bridged(?) radical, rather than on the nature of the experiments performed or on the type of arguments employed, has led unavoidably to some repetition and referrals to other parts of the book. The result, however, is increased ease of obtaining information on specific points.

Lest the deceptively narrow title obscure the generality of the coverage, I should point out that I have discussed or tabulated in detail many areas of solution and gas phase kinetics, structural and mechanistic organic chemistry, free radical reactions, and stereochemistry.

iii

My thanks to Barbara Kaplan for reading the manuscript and to George Zacharias for drawing the structures.

Leonard Kaplan

CONTENTS

LIST OF TABLES

Bridged Free Radicals

Chapter 1

DEFINITIONS

It is convenient to define a "bridge" in terms of one of three attributes which may, but need not, be interrelated: structure, energy, and ability to control stereochemistry. Each of these classes of definition will be illustrated with β-X-ethyl radicals and discussed in general terms with reference to meaningfulness and utility. In subsequent sections, work relevant to "bridging" will be reviewed in the light of these definitions.

It is important that experiments be designed and performed with an eye toward a generally applicable definition and that they be interpreted within the framework of that definition. Otherwise, an investigation may be of value only in a very narrow sense, namely, a concept of "bridging" implicitly defined in terms of the results of the particular experiments performed.

I. STRUCTURE

While potentially the least meaningless, a definition based on the structure of the radical is also the least useful, because of the lack of techniques for the determination of the structure of transient intermediates in complex systems.

Definitions may be constructed in order to permit a response to questions of the type: "Is the radical bridged?" or "How much is it bridged?". One geared to the former [e.g. (a) and (b) below] could yield a clear answer, but that answer would be strongly dependent on the arbitrary factors that comprise the definition. One constructed around the latter [e.g. (c) below] would be better in that the question approaches reality closer. However, caution would be necessary in applying it because it pretends to an ability to determine details of structure to a degree that may be unattainable at present.

(a) Definition: X-C̶-C̶· is bridged if X is equidistant from the two carbon atoms.

Definitions similar to this have been applied to radicals having X = Br:

". . . the Br [in (I)] may be centrally located

(I)

between C_1 and C_2 . . . ." [1]

". . . an intermediate in which the bromine atom is centrally located between the ethylenic carbon atoms . . . ." [2]

". . . an intermediate radical (I) in which a Br atom is midway between the (originally) olefinic carbons." [3]

While clear and reasonable when applied to radicals which, if bridged, would have a plane of symmetry not containing the two carbon atoms (e.g. XCH_2CH_2·), this definition is arbitrary, unrealistic, and narrow otherwise [e.g. (I) with Y ≠ H]. In the latter circumstance (b) may be preferred.

(b) Definition: X-C̶-C̶· is bridged if the two carbon atoms are symmetric [4-9] with respect to X.

Although this definition is more reasonable than (a) for molecules lacking the possibility of attaining a structure having a plane of symmetry not containing the two carbon atoms, it is wide open to the caution given in the opening paragraph of this section; the experimental application of such a definition would be a nightmare. Also note that this is not a definition based on structure.

(c) Definition: X-C̶-C̶· is bridged to the extent that the larger X-C-C angle is smaller than would be expected from conventional ideas of structure.

The value of this definition clearly hinges on its last part. While its meaning might potentially be vague and variable, the definition may actually be viable in view of the fact that "conventional ideas of structure" may not leave one's expectation much leeway (Tables 1 - 3).

TABLE 1

Structure of Molecules Containing a $ICH_2-\overset{|}{\underset{|}{C}}-$ Group

Compound	I–C–C angle	Method	Ref.
ICH_2CH_3	112.02 ± 0.01	Microwave	29
ICH_2CH_3	$112.2 \ \pm 0.3$	Microwave	37
$ICH_2CH_2SSSCH_2CH_2I$	109	X-ray	27
$ICH_2CH_2SSSCH_2CH_2I$	~100 [a]	X-ray	28
$ICH_2C \equiv CH$	111 ± 3	Electron diffraction	20

[a] However, see Ref. 27.

TABLE 2

Structure of Molecules Containing a $BrCH_2-\overset{|}{\underset{|}{C}}-$ Group

Compound	Br–C–C angle	Method	Ref.
$BrCH_2CH_3$	111	Microwave/infrared	30
$BrCH_2CH_3$	111.03	Microwave	31
$BrCH_2CH_3$	110.5	Microwave	24
$BrCH_2CH_3$	109 ± 2	Electron diffraction	23
$BrCH_2CHBrCH_3$	110 ± 1	Electron diffraction	25
$BrCH_2C \equiv CH$	112 ± 2	Electron diffraction	20
$BrCH_2C \equiv CH$	112.0	Microwave	164
$BrCH_2C \equiv CCH_2Br$	111.5	Electron diffraction	26
$BrCH_2CONH_2$	$113 \pm \sim 4$	X-ray	36

This definition also can, with a greater degree of arbitrariness, be formulated in terms of interatom distances.

II. ENERGY

Given that what happens and how are determined by relative free energies, it is reasonable to attempt to define what is purported to be an extraordinary molecule in terms of extraordinary energetics.

TABLE 3

Structure of Molecules Containing a $ClCH_2-\overset{|}{\underset{|}{C}}-$ Group

Compound	Cl-C-C angle	Method	Ref.
$ClCH_2CH_3$	110.5	Microwave	62
$ClCH_2CH_3$	111.03	Microwave	32
$ClCH_2CH_3$	110.5	Microwave	22
$ClCH_2CH_3$	111.5 ± 2	Electron diffraction	23
$ClCH_2CH_3$	111	Infrared/microwave	30
$ClCH_2CH_2CH_3$	111 ± 2	Electron diffraction	18
$ClCH_2C(CH_3)_3$	111 ± 2	Electron diffraction	17
$ClCH_2CH_2Cl$	109	Infrared/microwave	30
$ClCH_2CH_2OH$	111 ± 2	Electron diffraction	14
$ClCH_2CHOHCH_2Cl$	111 ± 1.5	Electron diffraction	15
$ClCH_2C\equiv CH$	111 ± 2	Electron diffraction	20
$ClCH_2C\equiv CH$	111.8 ± 0.3	Microwave	21
$ClCH_2C\equiv CCH_3$	111.5	Microwave	19
$ClCH_2C\equiv CCH_2Cl$	111 ± 2	Electron diffraction	16
$ClCH_2CONH_2$	∼ 115	X-ray	10
$ClCH_2CONH_2$	∼ 113	Electron diffraction	11
$ClCH_2CONH_2$	∼ 108	X-ray	12
$ClCH_2CONH_2$	∼ 116	X-ray	13

(a) Definition: $X-\overset{|}{\underset{|}{C}}-\overset{|}{\underset{|}{C}}\cdot$ is bridged to the extent that its free energy of formation is more negative than would be expected from conventional ideas.

This definition has been employed, implicitly or explicitly and with different bases of expectation, on at least three occasions:

" . . . stabilization by interaction of the Br with C_1 as illustrated by the following contributing structures (II - VI)" [1]

(II)   (III)   (IV)   (V)   (VI)

". . . an interaction between the filled 3p-atomic orbital of the substituent Cl atom and a half-filled 2pz-orbital of a trivalent (sp^2) carbon atom . . . ." [33]

" . . . the barrier for rotation about the C_2C_3 bond [in a β-Br radical] may be higher than expected because of interaction between the C_3 - Br and the trivalent carbon, e.g., the radical may have a bridged structure." [34]

There are problems with the quoted definitions as well as with one worded as in II(a). It is improper to speak only under special circumstances of "interaction" between parts of a molecule. To the degree that the word has any meaning, all parts of all molecules "interact." Assuming that a satisfactory definition which permits quantification of the concept can be formulated, the least one must do is speak of the sign and magnitude of the "interaction." However, the best that this could lead to is a definition of the type II(a), where one's expectation would be based on the approximately calculated energy of a hypothetical molecule which lacks a particular "interaction." This is what is implied in the first two quoted statements; the conceptual, computational, and practical problems resulting from so approaching a transient intermediate may be grasped by reflecting on the results of certain approaches to defining the "resonance energy" of stable molecules, on which any necessary experiments could be performed conveniently. It would be better if one's expectation were based on the energy of a real molecule, as is done in the last quoted statement; however, to be more practical and general, one may not wish to choose one rotational form of the radical itself as the model compound. Perhaps the least distasteful approach would be to compare ΔG_1 with ΔG_2 after appropriate

$$X-\underset{|}{\overset{|}{C}}-\underset{|}{\overset{|}{C}}-H \;\rightarrow\; X-\underset{|}{\overset{|}{C}}-\underset{|}{\overset{|}{C}}\cdot \;+\; H\cdot \tag{1}$$

$$H-\underset{|}{\overset{|}{C}}-\underset{|}{\overset{|}{C}}-H \;\rightarrow\; H-\underset{|}{\overset{|}{C}}-\underset{|}{\overset{|}{C}}\cdot \;+\; H\cdot \tag{2}$$

empirical correction [35] has been made for the contribution of the reactants to any departure of $\Delta G_1 - \Delta G_2$ from zero.

Further discussion, including a fatally damaging criticism, of a definition based on energy is deferred until Sec. IVA.

III. CONTROL OF STEREOCHEMISTRY

The most commonly used, although rarely explicitly stated, definition of bridging is the ability of the "bridging" group to control the direction of attack on the radical. For example:

". . . the evidence for bridged bromoalkyl radicals dictates that such species be incorporated into the mechanistic scheme . . . [whenever] steric control operates in HBr additions. Experiments . . . clearly indicate that a bromine atom adjacent to a radical site can effectively control the stereochemistry at the radical-bearing carbon." [40]

(a) Definition: A radical is bridged to the extent that, given the possibility of forming two diasteriomeric [41] or enantiomeric products, it produces one preferentially. Or, equivalently, a radical is bridged to the extent that, given the possibility of its formation by loss of either of two groups, it is produced preferentially by loss of one particular group.

In this way a bridged molecule is defined in terms of its chemical behavior, rather than in terms of its structure or energy, i.e., the molecule is defined not in terms of its own properties but in terms of the relative free energies of transition states.

IV. INTERFACING BETWEEN DEFINITIONS

To what degree does classification of a radical according to one definition imply that it is also classified according to another?

A. Energy - Structure

The semantics of definition II(a) and the penultimate paragraph of Sec. II demand that an abnormal energy be accompanied by an abnormal geometry. Abnormality in energy is defined as a deviation from expectation, the expectation being based on what one would expect the energy of a "normal" structure to be, based on a real model molecule. Energy and structure are inextricably linked. However, an abnormal energy does not require a particular type of abnormality in structure. A molecule strongly bridged according to definition II(a) need not be bridged according to definitions I(a), (b), or (c). An "origin" of an abnormal energy which could lead to an abnormality in structure of a type different from that required by definitions I(a) or (c) has been mentioned:

"Resonance stabilization of the β radical by the neighboring bromine atoms is another possible explanation of the enhanced stability

$$CH_3CH_2\overset{\cdot}{C}HC\overset{H}{\underset{H}{\diagup}}\!\!-\!\!Br \longleftrightarrow CH_3CH_2CH\!=\!C\overset{H}{\underset{H}{\diagdown}}\;Br \cdot$$

of the β radical." [38]

"We ascribe the stabilization of these radicals to hyperconjugative effects represented by the valence bond structures

$$\left[R_3M \overset{H}{\underset{H}{\diagdown}}\!\!-\!\!\overset{\cdot}{C} \;-\; \overset{\cdot}{C}H_2 \longleftrightarrow R_3M \overset{H}{\underset{H}{\diagdown}}C\!=\!CH_2 \right] \qquad [39]$$

"It would seem possible to explain the enhanced reactivity [of the

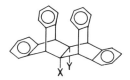

X = Br, Y = H compound compared to the X = Y = H compound, toward photobromination] by the assumption of a syn hyperconjugative electron delocalization in the syn-bromo radical and in the transition state leading to it." [194]

B. Structure - Energy

Bridging according to definitions I(a) or (c) may or may not be accompanied by an unusually low free energy depending on how the energy lowering is defined, i.e., whether the free energy must be lower than expected depends upon the basis of the expectation. If an energy-based definition which has been rejected is employed and the exact energy of structure (VII) could be computed,

$$X - \overset{|}{\underset{|}{C}} - \overset{|}{\underset{|}{C}} \cdot \qquad \overset{\cdot}{\underset{/}{\underset{}{\diagup}}}\overset{X}{\underset{}{C\!\!-\!\!C}}\overset{/}{\underset{\diagdown}{}}$$

(VII)   (VIII)

then, if the actual structure is (VIII), the free energy of the real structure (VIII) must be lower than the calculated free energy of (VII) and the bridged [I(a), (c)] structure must have a free energy which is

lower than would be expected. However, there would be no such requirement if definition II(a) were implemented as suggested in the penultimate
paragraph of Sec. II but without making the specified empirical correction.
Since there is no a priori requirement that $\Delta G_3 \leq \Delta G_2$, there would be no
requirement that $\Delta G_4 \leq \Delta G_2$.

$$X-\underset{|}{\overset{|}{C}}-\underset{|}{\overset{|}{C}}-H \;\;\rightarrow\;\; X-\underset{|}{\overset{|}{C}}-\underset{|}{\overset{|}{C}} \cdot \;\; [unbridged,\; I(a)\; or\; (c)\,] \;+\;\; H\cdot \qquad (3)$$

$$X-\underset{|}{\overset{|}{C}}-\underset{|}{\overset{|}{C}}-H \;\;\rightarrow\;\; X-\underset{|}{\overset{|}{C}}-\underset{|}{\overset{|}{C}} \cdot \;\; [bridged,\; I(a)\; or\; (c)\,] \;\;+\;\; H\cdot \qquad (4)$$

C. Control of Stereochemistry - Structure

There is no requirement whatever that a molecule bridged according
to definition III(a) be bridged according to definition I(a), (b), or (c). If no
additional connotations are attached to the results of the application of
definition III(a), there is no problem in classifying groups according to the
degree to which they bridge. However, if it is being used as a criterion of
a particular structural feature, e.g., conformance to definition I(a) or (c),
then one must worry about whether another structural (or kinetic) feature
or "effect," perhaps not even directly involving the bridging [III(a)] group,
might be capable of bringing about the observed control of stereochemistry,
i.e., the observed difference in energies of transition states. See also
Secs. 1;IVD and 4;III. *

D. Geometry - Control of Stereochemistry

In principle there can be no requirement that a radical bridged according to definition I(a), (b), or (c) also be bridged according to III(a). The
relative rates at which a molecule undergoes competing reactions depend
on the free energies of the corresponding transition states, not on the
structure of the reactant. The absence of such a requirement, however,
does not preclude the possibility that a strong case may be made for the

---

*Note: The number before the semi-colon refers to the chapter;
numbers and letters after the semi-colon refer to the section(s) within
that chapter.

expectation that a radical of particular structure [e.g. (VIII)] would exhibit a strong preference for formation of a product of particular stereochemistry. There is, however, no sound experimental or theoretical basis for such an expectation.

E. Energy and Control of Stereochemistry

These factors are related to each other at most to the extent that each is related to structure.

V. APPLICATION OF DEFINITIONS

A. Structure

In addition to direct approaches [533], there are indirect ones, as implied in Secs. IVA and C.

If energy were to be used as an indicator of geometry [I(a), (b), (c)], it would be necessary to rebut the type of argument exemplified by the quotes in Sec. IVA. In most cases this would be a nearly impossible task to accomplish without doing experiments on a significantly different model system.

Use of control of stereochemistry to reveal structure has been widespread. For this to be done properly, it would be necessary for all of the other possible structural and kinetic features alluded to in IVC to be eliminated as causative factors.

Finally, formation of products in which X has migrated from one carbon atom to another would support, but not require, its having been bridged in the radical.

B. Energy

In addition to direct methods for the determination of the relative energies of transient radicals which will not be reviewed, there is one very common indirect method: determination of the relative energies of transition states for the formation or destruction of the radicals. Such an approach, although very common, is treacherous. Minimal requirements for its validity are:

(i) Demonstration that the transition state strongly resembles the radical [42-49].

(ii) Demonstration that bridging in the transition state is not masked by a rate-retarding "effect."

(iii) Demonstration of the absence of a stabilizing "effect" on the transition state which may be peculiar to the transition state and which is absent in the radical. In view of the frequency with which this particular problem has been ignored some hypothetical examples will be given.

In the decomposition of RCO_3R' it is possible, even with the transition state being far along the reaction coordinate and with the (usually unsubstantiated) assumption of two-bond ("concerted") cleavage, that there is significant charge separation in the transition state, e.g., as exemplified by the resonance structure $R^+ CO_2^- OR'$. Thus, the transition state may be of unusually low energy because of the ability of a substituent on R to "bridge" to a (partially) cationic, rather than to a radical, center. (There would be several other ambiguities in the interpretation of a rate enhancement. Our intent here is only to illustrate a particular one.) In addition to the extensive literature on polar effects on radical reactions in general, there have been several observations relevant to the question of polar effects on those for which neighboring group participation has been invoked. (these results have not been presented in detail sufficient for the reader to judge their validity; our purpose here is only to establish a cause for concern and not to prove a case, which could not be done by analogy anyway):

(i) The logarithms of the first-order (through 85% reaction) rate

$$(X = H, Cl, F, CH_3, OCH_3)$$

$$(IX)$$

constants for the decomposition of $\sim 2.9 \times 10^{-5}$ M peroxides (IX) (95-99% pure) in the presence of α, γ-bis (biphenylene)-β-phenylallyl in benzene at 50° were reported to give "a reasonably linear plot vs. σ^+; $\rho = -1.16$." [52]

(ii) "The relative rates of decomposition of peroxides . . . (X) in chlorobenzene at 70° correlate with σ^+—constants with a ρ of -1.8 . . .

The rates of decomposition of compounds . . . (XI) correlate with σ—constants with a ρ of +0.7." [53]

(X)

(X = H, $\underline{m}$ - NO$_2$, $\underline{p}$ - NO$_2$)

(XI)

(X = H, $\underline{p}$ - NO$_2$, $\underline{p}$ - CH$_3$O)

(iii) A Hammett plot (correlation with σ) of the first-order rate constants for the decomposition of 0.030 M solutions of peresters (XII) in the presence of 0.2 M styrene in chlorobenzene at 50.2° yielded ρ = -1.3 to -1.4 [54].

(XII)

(X = H, CH$_3$O, Cl, NO$_2$)

R· may abstract X from XCH$_2$CH$_2$Y much faster than from XCH$_2$CH$_3$, not because the unusual stability of the developing · CH$_2$CH$_2$Y is reflected in the transition state, but because of partial coordination ("bridging"?) of R· to Y as X is being abstracted. Possibilities of this type have been discussed in the context of reaction of Cl· with alkylbenzenes [50] and of · CCl$_3$ with phenylalkenes [51].

Finally, all problems (e.g. IVA) associated with the interpretation of direct determinations of energy apply to that of indirect ones.

C. Control of Stereochemistry

The experimental approach is self-evident. Of the many appropriate caveats, one in particular will be stated explicitly because of the frequency with which it has gone unheeded: It is necessary to show that the overall observed stereochemistry faithfully portrays the stereochemistry of the elementary act of interest.

Chapter 2

SILYL-SUBSTITUTED RADICALS

I. β-SILYLALKYL RADICALS

A. Structure

Electron spin resonance (ESR) spectra attributed to β-silylalkyl radicals on the basis of method(s) of generation and qualitative consistency with the assigned structures are summarized in Table 4.

The observable spin density on silicon may be discussed in terms of "hyperconjugation" or "spin polarization" "effects" (Sec. 1; IVA*), among others, obviating the need to consider a bridged structure, or it may be discussed in terms of bridging. The unexceptional α coupling constants make it difficult to consider that the "hyperconjugation effect" is a large one (in terms of spin density distribution, "stabilization" being a separate question), or that the radical is bridged [1; I(a)] or strongly bridged [1; I(c)], unless the spin density at the α-proton would be grossly insensitive to the changes in electron density distribution which would result from a bridged structure. The small (β-H coupling constant of $CH_3CH_2\cdot$ = 26.87 G) β-H coupling constant of $(CH_3CH_2)_3SiCH_2CH_2\cdot$ was discussed [39] in terms of details of internal rotation distinct from the requirements of bridging [1; I(a)-(c)] on structure. We consider the α-H coupling constant to be much less ambiguous than the β- and much more ambiguous than the ^{13}C coupling constant as an indicator of bridging.

In summary, there is no evidence that points toward bridged [1; I(b)-(c)] structures for the β-silylalkyl radicals in Table 4; at most, there is

_____

*The number before the semi-colon refers to the chapter; numbers and letters after the semi-colon refer to the section(s) within that chapter.

TABLE 4

ESR Spectra Attributed to $R_3M\text{-}\overset{|}{C}\text{-}\overset{|}{C}\text{-}$

Radical	Method of generation	Temperature	Coupling constants (G)			Ref.
			α	β	M	
$H_3CCH_2CH_2\cdot$	b	$-110°$	22.14	30.33		473
$H_3SiCH_2CH_2\cdot$	b	$-70°$	21.39	17.68		473
$(CH_3CH_2)_3SiCH_2CH_2\cdot$	a	$-102°$	20.97	17.87	37.4	39
$(CH_3CH_2)_3SiCH_2CH_2\cdot$	b	$-119°$	21.01	17.67		473
$PhSiH_2CH_2CH_2\cdot$	b	$-65°$	21.20	17.72		473
$Ph_2SiHCH_2CH_2\cdot$	b	$-87°$	21.08	17.79		473
$Ph_3SiCH_2CH_2\cdot$	b	$-60°$	21.24	18.64		473
$(CH_3CH_2)_3SiCH_2CHCH_3\cdot$	b	$-101°$	21.10	16.74		473
$H_3GeCH_2CH_2\cdot$	b	$-80°$	20.96	15.84		473
$(CH_3CH_2)_3GeCH_2CH_2\cdot$	a	$-107(-118)°$	20.55(20.70)	16.71(16.42)		39(473)
$(CH_3)_3SnCH_2CH_2\cdot$	c	$-133°$	19.74	15.75		473
$(CH_3CH_2)_3SnCH_2CH_2\cdot$	a	$-96°$	19.77	16.12		39
$(\underline{n}\text{-}C_4H_9)_3SnCH_2CH_2\cdot$	b	$-148°$	19.64	15.67		473

[a] Photolysis of $(CH_3CH_2)_4M$ and $(Bu^tO)_2$ in cyclopropane.

[b] Photolysis of $-\overset{|}{M}\text{-}H$, $(Bu^tO)_2$, and $\overset{\diagdown}{\diagup}C=C\overset{\diagup}{\diagdown}$.

[c] Photolysis of $Me_3SnSnMe_3$ and ethylene.

evidence which may not be inconsistent with them. The $-\overset{|}{\underset{|}{Si}}-CH_2CH_2\cdot$ radicals are not bridged [1; I(a)] with equivalent geminal hydrogens.

B. Energy

Results of the chlorination of silylalkanes and alkanes [66] by SO_2Cl_2 are given in Tables 5 and 5A, respectively. In view of the lack of adequate proofs of structure of the products of the former reactions and the lack of demonstration for any reaction that the relative amounts of products are equal to the relative rates of abstraction of the corresponding hydrogens, we reserve judgment on whether these results are relevant to the question of bridging. If they are assumed to be so and are taken at face value, they indicate no extraordinary stabilization (relative rate constants being an exponential, and sensitive, function of relative energies of transition states) of the transition states leading to β-, γ-, or δ-silylalkyl radicals.

C. Control of Stereochemistry

The addition of $SiHCl_3$ to 1-methylcyclohexene produced 1-methyl-2-trichlorosilyl cyclohexane (Table 6) [2]. Our assessment of these data is the same as that (2; IB) of those in Table 5. Should the reported cis/trans ratios be correct and should they reflect relative rates of attack on a 1-methyl-2-trichlorosilylcyclohexyl radical, the radical could be considered to be moderately bridged [1; III(a)]. If this is the case, it would still not be possible to conclude anything about the structure of the radical, because of the other possible "causes" of trans-addition.

II. β-SILYLSILICOETHYL RADICALS

Structure

Photolysis of $(CH_3)_3SiSi(CH_3)_3$ as per Table 4 produced a solution whose ESR spectrum (Table 7) was considered to be that of $(CH_3)_3SiSi(CH_3)_2-CH_2\cdot$ on the basis of its qualitative consistency with the assigned structure [39]. The contents of Table 7 provide no cause to consider a bridged structure [1; I(a)-(c)].

TABLE 5

Reaction of Silylalkanes with SO_2Cl_2 [58]

Reactant	Reaction conditions	Product identification	Reported products, relative amounts[a]				Control experiments
			α-Cl	β-Cl	γ-Cl	δ-Cl	
$CH_3CH_2CH_2SiCl_3$	b	d	0.19	0.67	0.39		e
$CH_3CH_2CH_2SiCl_3$	c	f	0.19±0.01	0.57±0.02	0.27±0.01		h,i
$CH_3CH_2CH_2SiMeCl_2$	c	f	0.38±0.01	0.92±0.02	0.29±0.01		g,h,i
$CH_3CH_2CH_2X$			See Table 50				
$CH_3CH_2CH_2CH_2SiCl_3$	c	f	0.09±0.01	0.49±0.03	1.1±0.06	0.25±0.01	h,i
$CH_3CH_2CH_2CH_2SiMeCl_2$	c	f	0.37±0.03	0.84±0.04	1.2±0.05	0.32±0.01	h,i
$CH_3CH_2CH_2CH_2X$			See Table 51				

[a] For all rows except the first "the values listed are relative to one of the three aliphatic hydrogens of toluene which is assigned unit." The values reported in the first row are relative yields of α, β, and γ products, divided by 2, 2, and 3, respectively, and normalized so that the α-Cl entry is 0.19.

[b] Silane (3.78 moles) reacted with 4 moles of sulfuryl chloride and 4 mmoles of benzoyl peroxide. Reported products formed in 90% yield; conversion = 74% [57].

[c] Silane, SO_2Cl_2, and $PhCH_3$ refluxed in CCl_4. Conversion to products ≤ 20% [58].

[d] Identifications based on acceptable Cl analyses and indirect inferences drawn from chemistry involving KOH.

^{e}None.

^{f}Identifications based on NMR spectra (spectrum of $CH_3CH_2CH_2CHClSiCl_3$ reported to have a doublet at $\tau = 8.79$) except for that of $CH_3CH_2CH_2CHClSiCl_3$ which rests on its being different from what are reported to be the β-, γ-, and δ-chlorides.

^{g}Changing the relative amount of SO_2Cl_2 by a factor of 5 or the concentration of all reagents by a factor of 2 resulted in less than a 15% change in the relative amount of any product.

^{h}"VPC analysis also showed the extent of dichlorination to be below 5% relative to monochlorination even in the worst cases."

^{i} ". . . the vapor phase chromatograms showed the expected product peaks as well as the starting materials but no other peaks."

TABLE 5A

Reaction of Alkanes with SO_2Cl_2 ^{a}

Alkane	Product ratio ^{b}
n–Heptane	3.7
n–Octane	3.4
n–Decane	3.4
n–Dodecane	3.3

^{a}SO_2Cl_2 (18 mmoles) and 0.36 mmole of Bz_2O_2 in 12 ml of alkane at 85°.

^{b}Ratio of 2° to 1° chlorides, corrected for the relative numbers of methylene and methyl hydrogens in the alkane.

TABLE 6

Reaction of 1-Methylcyclohexene with $SiHCl_3$

Reaction conditions	% Yield of 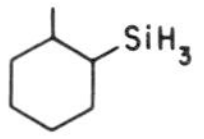	Reported ratio of isomers[a]
1. 9 mole % Ac_2O_2, 45-60° 2. $LiAlH_4$	80	cis/trans = 5.7
1. $h\nu$, 49-59° 2. $LiAlH_4$	39	cis/trans = 8.1

[a] Cis vs. trans assigned on the basis of the von Auwers-Skita rule and by comparison of trends of physical properties with those for the 1,2-dimethylcyclohexanes. No control experiments were reported.

TABLE 7

$$XSi(CH_3)_3 + (Bu^tO)_2 \xrightarrow[\Delta \text{ solution}]{h\nu} \text{ESR spectrum}$$

Assigned structure	Temperature	α Coupling constant, G
$(CH_3)_3SiSi(CH_3)_2CH_2\cdot$	- 95°	20.15
$(CH_3)_3SiCH_2\cdot$	-126°	20.88

III.   O-SILYLMETHYLOL RADICALS

Structure

Included in Tables 8 and 8A are coupling constants obtained from the ESR spectra of radicals of this type whose structures were assigned on the basis of their qualitative consistency with the spectra. Employing considerations very similar to those outlined in Sec. 2; IA, we find no reason to consider these radicals to be bridged [1; I(a)-(c)].

TABLE 8

Spectra Attributed to $R_3\overset{|}{M}O\overset{|}{C}\cdot$ Radicals

Assigned structure	Temp.	Coupling constant, G			Method of generation	Ref.
		α	β	Si		
$(CH_3)_3COCH_2\cdot$	-102°	17.29			a	39
$(CH_3)_3SiOCH_2\cdot$	-109°	19.08		1.5	a	39,476
$(CH_3)_2(CH_3O)SiOCH_2\cdot$	-90°	19.27			a	39
$(CH_3O)_3SiOCH_2\cdot$	-22°	19.69			a	39
$(CH_3CH_2O)_2SiHO\overset{\cdot}{C}HCH_3$		17.76	23.35		d	476
$CH_3CH_2O\overset{\cdot}{C}HCH_3$		13.8	21.9		b	55
$CH_3CH_2O\overset{\cdot}{C}HCH_3$		13.8	21.5		b	478
$(CH_3)_2CHO\overset{\cdot}{C}(CH_3)_2$	room		20.2		b	55
$(CH_3CH_2)_3SiO\overset{\cdot}{C}(CH_3)_2$	-40°		20.3		c	56
$(CH_3O)_3GeOCH_2\cdot$	-58°	19.13			a	39
$(CH_3CH_2CH_2CH_2)_3SnOCH_2\cdot$	-40°	16.30			a	39

[a] Photolysis of R_3MOCH_3 and $(Bu^tO)_2$ in cyclopropane solution.

[b] R· produced by reaction of RH with H_2O_2/titanous ion in acid solution.

[c] Photolysis of CH_3COCH_3, $(Bu^tO)_2$, and Et_3SiH.

[d] Photolysis of $(CH_3CH_2O)_3SiH$ and $(Bu^tO)_2$.

TABLE 8A

ESR Spectra Attributed to ROĊPh$_2$ Radicals

Assigned structure	Method of generation	Temp.	Coupling constant, G				g-Value	Ref.
			Ortho	Meta	Para	Hydroxy		
Ph$_2$ĊOSiPh$_3$ [b]	a	127°	3.16 ± 0.03	1.23	3.48	–	2.00305	492
Ph$_2$ĊOH	c	25°	3.20 – 3.23	1.23 – 1.24	3.64 – 3.69	2.78 – 3.42	2.00298 – 2.00300	491

[a] Reaction of Ph$_2$CO with Ph$_3$SiH.

[b] "... the e.s.r. spectrum drastically changed its structure when benzophenone-d$_{10}$ was used." We believe that the observed spectrum is not necessarily inconsistent with Ph$_2$ĊOH as the carrier.

[c] Photoirradiation of Ph$_2$CO in methanol, ethanol, 2-propanol, or tetrahydrofuran.

IV. β-SILYLVINYL RADICALS

Control of Stereochemistry

Results of the addition of $SiHCl_3$ to terminal alkynes are given in Table 9. In view of the lack of control experiments for all alkynes except $(CH_3)_3CC\equiv CH$, for which there are indications that the products may not be stable under conditions somewhat similar to those of the reaction, we must reserve judgment on whether these results are relevant to the question of bridging.

The photochemical reaction of Me_3SiH with 24-246% $CF_3C\equiv CCF_3$ yielded 45-77% $Me_3Si(CF_3)C=CHCF_3$ (100% trans) and 15-50% $Me_3Si(CF_3)$-$CHCH(CF_3)SiMe_3$ [63]. It was shown that the photochemical reaction of trans-$Me_3Si(CF_3)C=CHCF_3$ with 610 mole % Me_3SiH yielded 10% $Me_3Si(CF_3)CHCH(CF_3)SiMe_3$ and 89% recovered trans-olefin; no cis-olefin was detected. Since the fate of any cis-olefin under the reaction conditions remains unknown, it is not possible to interpret the formation of only the trans-olefin as an indication that the radical is bridged [1; III(a)].

Reaction (5) has been reported to proceed as shown [64]:

$$(CH_3)_3SiSi(CH_3)_2H + (CH_3)_3Si\underset{\underset{CH_3}{\overset{\displaystyle\longrightarrow}{\vert}}}{\overset{\overset{\displaystyle\longrightarrow}{CH_3}}{\overset{\vert}{Si}}}\!-C\equiv CH \xrightarrow[135°]{(Bu^tO)_2} \underset{H}{\overset{Me_5Si_2}{\diagdown}}C=C\underset{Si_2Me_5}{\overset{H}{\diagup}} +$$

$$\qquad\quad\text{(XIII)}\qquad\qquad\qquad\text{(XIV)}\qquad\qquad\qquad\qquad\text{(XV)}$$

$$\underset{H}{\overset{Me_5Si_2}{\diagdown}}C=C\underset{H}{\overset{Si_2Me_5}{\diagup}} + Me_5Si_2C\equiv CSi_2Me_5 \qquad\qquad (5)$$

$$\qquad\text{(XVI)}\qquad\qquad\qquad\text{(XVII)}$$

Data are given in Table 10.

V. β-SILYLOXYSILICOETHYL RADICALS

Structure

Photolysis of $(CH_3)_3SiOSi(CH_3)_3$ at −108° as per Table 4 produced a solution whose ESR spectrum was considered to be that of

TABLE 9

Addition of $SiHCl_3$ to Acetylenes

Alkyne	Reaction conditions	Reported % yield of $RCH=CHSiCl_3$ [a]	$RCH=CHSiMe_3$ % yield [b]	Product cis/trans	Controls	Ref.
1-Pentyne	d	35	27	3.8^g	i	59
1-Hexyne	d	36	28	3.3^g	n	59
1-Heptyne	d	43	32	3.0^g	j	59
3-Methyl-1-butyne	d	37	23	2.6^g	n	59
3,3-Dimethyl-1-butyne	d	39	28	$\sim 0^g$	k, m	59
3,3-Dimethyl-1-butyne	e	68		0.20^g	k, l, m	59
5-Chloro-1-pentyne	f	54		2.8^h	n	60
4-Chloro-1-butyne	c	31-33		$2.3-4.0^h$	n	61

[a] Product identification based on C & H analyses and conversion to $RCH=CHSiMe_3$.

[b] Based on alkyne.

[c] Alkyne (0.4 mole) refluxed in 1.2 moles of cyclohexane with 0.8 mole of $SiHCl_3$ and 7 mole % Bz_2O_2.

[d] Alkyne heated in cyclohexane at 60-70° for 20 h with 2 equivalents of $SiHCl_3$ and 7 mole % Bz_2O_2. Product then refluxed in ether with 1.7 equivalents of MeMgBr.

e Same as d except 1 equivalent of SiHCl$_3$ used.

f Alkyne (0.68 mole) refluxed in 5.6 moles of cyclohexane with 1.48 moles of SiHCl$_3$ and 8 mole % Bz$_2$O$_2$.

g RCH=CHSiMe$_3$.

h RCH=CHSiCl$_3$.

i 1-Trimethylsilyl-1-pentene (79% cis) refluxed for 10 h with 5 mole % I$_2$ and 5 mole % Mg to yield product which was 71% cis; when refluxed for 9 h with 5 mole % I$_2$ and 2.5 mole % Bz$_2$O$_2$, it produced 59% cis product; when refluxed for 48 h with 100 mole % MeI and 5 mole % Bz$_2$O$_2$, it produced 77% cis product.

j The cis/trans ratios of samples of reaction mixture taken at 4, 10, and 20 h and treated as in d were "essentially constant."

k cis-1-Trimethylsilyl-3,3-dimethyl-1-butene was heated in cyclohexane at 80° for 20 h in the presence of 1 equivalent of SiHCl$_3$ and 7 mole % Bz$_2$O$_2$; treatment of the reaction mixture with MeMgBr yielded 1-trimethylsilyl-3,3-dimethyl-1-butene (39% trans) and "no appreciable di-adduct." The trans-isomer, under similar treatment, "remained unchanged." The cis-isomer, under the same conditions except for the absence of SiHCl$_3$, yielded olefin which was 8% trans.

l Samples withdrawn after 2.5 and 18 h yielded RCH=CHSiMe$_3$ cis/trans ratios of 0.52 and 0.32, respectively.

m 1-Trichlorosilyl-3,3-dimethyl-1-butene (88% cis) was refluxed in cyclohexane for 20 h in the presence of 100 mole % SiHCl$_3$ and 0.7 mole % Bz$_2$O$_2$; treatment of the reaction mixture with MeMgBr yielded 45% olefin (13% cis) and 55% "diadduct." The trans isomer, under similar treatment, "remained unchanged." The 88% cis-isomer, under the same conditions except for the absence of SiHCl$_3$ and the presence of 8 mole % Bz$_2$O$_2$

or under the same conditions except for the absence of Bz_2O_2, underwent "a very small change in isomer distribution."

n None.

TABLE 10

Reaction of Me_5Si_2H with $Me_5Si_2C \equiv CH$ at 135°

Mole % $(Bu^tO)_2$	Products [a]		
	(XV)	(XVI)	(XVII)
25	28	–	8
7	19	19	4
4	12	17	2

[a] "Structures of (XV) . . . , (XVI) . . . , and (XVII) . . . were determined by physical constants, elemental analyses, GLC, hydrogenation, and IR and NMR spectra after separation from the mixture." No control experiments were reported.

$(CH_3)_3SiOSi(CH_3)_2CH_2\cdot$ on the basis of its qualitative consistency with the assigned structure [39]. The observed α-H coupling constant of 21.09 G provides no cause to consider a bridged structure [1; I(a)-(c)].

VI. γ-, δ-, and ϵ-SILYLALKYL RADICALS

Energy

See Sec. 2, IB.

VII.   MISCELLANEOUS

The opinion has been expressed that "the addition of $SiHCl_3$ . . . to norbornadiene . . . yields products whose stereochemistry can be rationalized by assuming formation of precursive bridged intermediates." [65]

VIII.   SUMMARY

There is no evidence which points toward 1, 2-, 1, 3-, 1, 4-, or 1, 5-bridged radicals [1; II(a), III(a)]. There is some evidence (2;IA, IIA, IIIA, V) whose extremes of interpretation vary from inconclusive to indicative of an at-most-weakly 1, 2- or 1, 3-bridged [1; I(b) and (c)] structure. $-\overset{|}{\underset{|}{Si}}CH_2CH_2\cdot$ are not bridged [1; I(a)] with equivalent geminal hydrogens.

Chapter 3

GERMYL-SUBSTITUTED RADICALS

I. β-GERMYLALKYL RADICALS

Structure
See Sections 2;IA and Table 4.

There is no evidence which points toward bridged [1; I(b), (c)] * structures for $(CH_3CH_2)_3GeCH_2CH_2 \cdot$ or $H_3GeCH_2CH_2 \cdot$; at most, there is evidence which may not be inconsistent with them.   The radicals are not bridged [1; I(a)] with equivalent geminal hydrogens.

II. O-GERMYLMETHYLOL RADICALS

Structure
See Sections 2;III and 4;II and Table 8.

We find no reason to consider $(CH_3O)_3GeOCH_2 \cdot$ to be bridged [1; I(a)-(c)].

III. β-GERMYLVINYL RADICALS

Control of Stereochemistry
Results of the addition of germanes to acetylenes are in Table 11. In view of the lack of adequate control experiments in all cases, we must reserve judgment on whether these results are relevant to the question of bridging.

---

*The number before the semi-colon refers to the chapter; numbers and letters after the semi-colon refer to the section(s) within that chapter.

TABLE 11

Addition of Germanes to Acetylenes

Alkyne	Germane	Reaction conditions	% Yield of -GeCH=CHR	Trans/cis	Controls	Ref.
HC≡CC(CH₃)=CH₂	$(CH_3CH_2)_2ClGeH$	a	10.5^c	0^c	b	67
HC≡CCH₂Cl	$(CH_3CH_2)_2ClGeH$	a	42^d	0.68	e	68
HC≡CCH₂Cl	$(CH_3CH_2)_2ClGeH$	f	44^g	0.76	e	68
HC≡CCH₂Cl	$(CH_3CH_2)Cl_2GeH$	h	81	0.65	i	68
HC≡CCH₂Cl	$(CH_3CH_2)Cl_2GeH$	j	54	0.54	i	68
CF₃C≡CCF₃	$(CH_3CH_2)_3GeH$	k	29^m	115^m	l	63

[a] Neat; AIBN initiator; 80°.

[b] Reaction totally blocked by galvinoxyl.

[c] Basis of assignment of structure not mentioned.

[d] In addition, formation of $H_2C=C(CH_2Cl)GeEt_2Cl$, $Et_2ClGeCH_2(Cl)C=CH_2 + Et_2ClGeCH=CClCH_3$, and reduced alkyne in yields of 9, 34, and 13%, respectively, was reported.

[e] Except for the reported formation of a very small amount of $H_2C=C(CH_2Cl)GeEt_2Cl$, the reaction was inhibited by galvinoxyl.

[f] Neat; hν; 80°.

[g] Same as d, except: 7% instead of 9%.

[h] Temperature rose spontaneously to 120°; reaction mixture then kept at 80° for 1 h; neat.

[i] Reaction totally inhibited by galvinoxyl.   Pure cis and trans product stable in refluxing hexane.   Irradiation of either cis or trans product for 24 h at 50° in hexane produced a mixture wherein trans/cis = 5.7.

[j] MeCN solution; 80°; 5 h.

[k] hν; 7 days.

[l] No reaction occurred in the dark.

[m] Assignment of structure rests solely upon NMR spectra: The isomer with $J(F-C-C=C-H) = 2.5$ cps and $\delta(=C-H) = 6.59$ was considered to be cis and that with $J < 1$ cps and $\delta = 5.82$ was considered to be trans.

IV. SUMMARY

There is no evidence which points toward 1, 2-bridged radicals [1; II(a); III(a)]. There is some evidence (3; I, II) whose extremes of interpretation vary from inconclusive to indicative of an at-most-weakly 1, 2- or 1, 3-bridged [1;I(b) and (c)] structure. $-\overset{|}{\text{Ge}}\text{CH}_2\text{CH}_2\cdot$ are not bridged [1; I(a)] with equivalent geminal hydrogens.

Chapter 4

STANNYL-SUBSTITUTED RADICALS

I. β-STANNYLETHYL RADICALS

Structure

See Sec. 2, IA and Table 4. There is no evidence which points toward bridged [1; I(b) and (c)]* structures for $R_3SnCH_2CH_2\cdot$; at most, there is evidence which may not be inconsistent with them. The radicals are not bridged [1; I(a)] with equivalent geminal hydrogens.

II. O-STANNYLMETHYLOL RADICALS

Structure

See Sec. 2; III and Table 8. Whether $Bu_3SnOCH_2\cdot$ has a particularly low α coupling constant (16. 30 G) depends on whether the compound used for comparison is $Me_3SiOCH_2\cdot$ (19. 08 G) or $Me_3COCH_2\cdot$ (17. 29 G). We do not believe that there is information at hand sufficient to permit meaningful discussion of these data; there are many "competing factors" "operating," and it would be premature and reckless to attempt to draw conclusions about structure. Not only are the differences in coupling constants undramatic but, even if their possible "explanations" could be "narrowed down" to those importing a significant spin density on Sn, we would at best still be left with the problems outlined in Sec. 2; IA.

---

* The number before the semi-colon refers to the chapter; numbers and letters after the semi-colon refer to the section(s) within that chapter.

III. β-STANNYLVINYL RADICALS

Control of Stereochemistry

Results of the addition of R_3SnH to terminal acetylenes are given in Table 12. In view of the well-known ability of R_3SnH to isomerize olefins and the lack of adequate control experiments for all cases except those described in Ref. 69, we doubt that the other results are relevant to the question of bridging. In the reaction of 1-hexyne with Et_3SnH there are positive indications (footnotes a and b to Table 12) that cis- is converted to trans- $Et_3SnCH=CHR'$ under the reaction conditions. In particular, the fact that cis/trans increases dramatically as the percent conversion decreases (cis/trans > 10 at 2% conversion) may indicate that $Et_3Sn\overset{\cdot}{C}H=CR'$ reacts with Et_3SnH to give much more cis- than trans-$Et_3SnCH=CHR'$. If this is the case, then the radical could be considered to be strongly bridged [1, III(a)], although it would still not be possible to conclude anything about its structure [1, IV(c) and (d)]. However, in view of the low $D(Et_3Sn-H)$, it would be less improper, other things being equal, for this reaction than for others to ascribe to the reactant properties revealed by transition states [42-49].

It was found that reaction (6) gave cis-trans/trans-trans = 0.67, but

$$\underset{C\equiv CH}{\overset{C\equiv CH}{\bigcirc}} + 2\ Et_3SnH \xrightarrow[\text{reflux}]{\text{PhH}} \underset{CH=CHSnEt_3}{\overset{CH=CHSnEt_3}{\bigcirc}} \qquad (6)$$

no control experiments were reported [72].

In reaction (7) cis-α/trans-α was reported to be > 10, 6.8, 6.5,

$$Me_3SnH + MeC\equiv C-CO_2Et \rightarrow \underset{H}{\overset{Me}{\diagup}}C=C\underset{CO_2Et}{\overset{SnMe_3}{\diagup}} + \underset{Me}{\overset{H}{\diagup}}C=C\underset{CO_2Et}{\overset{SnMe_3}{\diagup}}$$

$$\underline{\text{Cis-}\alpha} \qquad\qquad \underline{\text{Trans-}\alpha}$$

$$+ \quad \underset{Me_3Sn}{\overset{Me}{\diagup}}C=C\underset{CO_2Et}{\overset{H}{\diagup}} \qquad (7)$$

$$\underline{\text{Cis-}\beta}$$

TABLE 12

Addition of R_3SnH to Terminal Acetylenes

Acetylene	R	Reaction conditions	% Conversion of R_3SnH	Reported % yield of $R_3SnCH=CHR'$	Reported product cis/trans	Controls	Ref.
1-Hexyne	Et	c	2.0^a		$>10^p$	b	69
1-Hexyne	Et	c	57.5^a		3.1^p	b	69
1-Hexyne	Et	d	55		1.1^p	e	70
1-Hexyne	Et	d	85	72^f	0.47^p	e	70
1-Hexyne	Et	g			0.47^p	e	70
1-Hexyne	Me	h	85		0.42^i	e	70
1-Hexyne	Ph	j	>80		5.7^k	e	70
PhC≡CH	Me	j	75		2.4^l	e	70
PhC≡CH	Et	h	75		2.3^m	e	70
PhC≡CH	Ph	n	90		0.39^o	e	70
$CH_3C≡CH$	Me	q		29	0.41^p	r	71
$Me_2C(OH)C≡CH$	Et	s		28	2^t	e	485
cyclopentyl C≡CH / OH	Et	s		21	2^t	e	485

[a] Percent conversions of 20.5, 28.0, 37.5, and 46.5 were reported to correspond to cis/trans = 7.7, 6.5, 5.0, and 3.8, respectively.

b "Rates of formation of the α-adduct $[H_2C=C(R')SnR_3]$ as well as that of the β-adducts $[R'CH=CHSnR_3]$ appeared to be practically solvent-independent." [butyronitrile ($\epsilon = 20.3$) vs. decane ($\epsilon = 1.99$)]. Introduction of 2.4 mole %

· into a reaction mixture after about 50% conversion of R_3SnH resulted in the sudden acquisition of near-zero slopes by plots of $[R_3SnH]$, $[\underline{cis}\text{-Et}_3SnCH=CHC_4H_9]$, or $[\underline{trans}\text{-Et}_3SnCH=CHC_4H_9]$ vs. time.

Introduction of 2.4 mole % AIBN at a similar stage of the reaction resulted in significant increases in the rates of disappearance of R_3SnH and appearance of $\underline{trans}\text{-Et}_3SnCH=CHC_4H_9$ and a significant decrease in the rate of appearance of $\underline{cis}\text{-Et}_3SnCH=CHC_4H_9$ (its rate of appearance was negative for a while; at about 56 min of reaction cis/trans $\cong$ 6; AIBN was introduced at about 100 min; at about 108 min cis/trans $\cong$ 0.37).

c R_3SnH/alkyne $\approx$ 1/2, o-xylene, 49.9°.

d R_3SnH/alkyne = 0.91; 50°.

e None.

f Isolated.

g R_3SnH/alkyne = 0.91; 20°.

h 60°.

i Assignments of structure based solely on NMR spectra. cis: $\delta = 5.76$ (d of t; J = 12.5, 1.0 cps), 6.42 (d of t; J = 12.5, 7.2 cps). trans: $\delta = 5.92$ (broad).

j 50°.

k Assignments of structure based solely on NMR spectra.   cis: $\delta = 6.11$ (d of t; $J = 12.0$, 0.8 cps), 6.25 (d of t; $J = 12.0$, 6.8 cps).   trans: $\delta = 6.25$ (broad).

l Assignments of structure based solely on NMR spectra.   cis: $\delta = 6.15$ (d, $J = 13.6$ cps), 7.55 (d, $J = 13.6$ cps).   trans: $\delta = 6.82$ (broad).

m Assignments of structure based solely on NMR spectra.   cis: $\delta = 6.15$ (d, $J = 13.7$ cps), 7.55 (d, $J = 13.7$ cps).   trans: $\delta = 6.83$ (broad).

n $20°$ and $60°$.

o Amount of trans-isomer deduced from amount of PhC≡CH reacted and amount of cis-isomer produced. Assignment of structure of cis-isomer based solely on NMR spectrum: $\delta = 6.43$ (d, $J = 13.4$ cps), 7.80 (d, $J = 13.4$ cps).

p Assignment of structure based on moderately sound evidence.

q $100°$, 18 h.

r The refluxing of propenyltrimethyltin (cis/trans = 71/29) with 0.8 mole % Bz_2O_2 in CCl_4 resulted in quantitative recovery of material whose cis/trans = 71/29 and 65/35 after 4 and 72 h, respectively.   A similar experiment with 1.4 mole % AIBN resulted in "nearly quantitative" recovery of 65/35 material after 65 h.   Irradiation for 10 min in hexane of cis/trans = 71/29, 100/0, or 0/100 material resulted in cis/trans = 25/75, a ratio which did not change upon irradiation for 20 h.

s $R_3SnH/alkyne = 0.9$; $100°$.

t Basis of identification of product: none.

5.8, and 5.4 at % conversions of Me_3SnH of 5.0, 48.5, 58.5, 65.5, and 77.5, respectively, while α/β remained almost constant at ~ 2 [69]. These results may be subject to the same interpretation as was given those of the reaction between 1-hexyne and Et_3SnH.

IV. MISCELLANEOUS

The following opinion has been expressed:

"The addition of trimethyltin hydride to 1,3-butadiene gave a predominance of 1,4-addition along with 7.1% of 1,2-addition. The predominance of the cis product of the 1,4-addition is unusual because free-radical addition to this diene usually gives predominantly trans products . . . [The major feature of] a scheme which accounts for these results is . . . the formation of the bridged radical (XVIII) in which a d orbital of the tin atom serves to delocalize the odd electron while two others form bonds with the carbons at the ends of the four-carbon chain." [65]

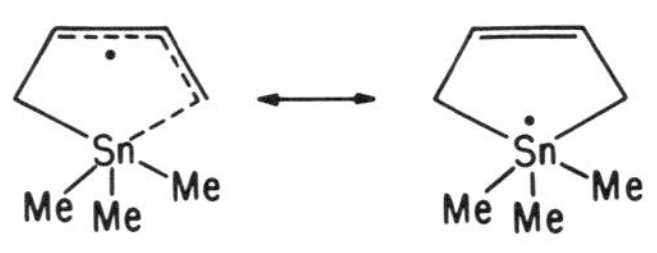

(XVIII)

V. SUMMARY

There is some evidence (4;I, II) whose extremes of interpretation vary from inconclusive to indicative of a weakly 1,2-bridged [1; I(b) and (c)] structure and there are some strong hints (4;III) that $Et_3SnCH=\overset{\bullet}{C}C_4H_9$ is strongly bridged [1; III(a)]. $-\overset{|}{S}nCH_2CH_2\cdot$ are not bridged [1; I(a)] with equivalent geminal hydrogens.

Chapter 5

CHLORO-SUBSTITUTED RADICALS

I. β-CHLOROALKYL RADICALS

A. Structure

Electron spin resonance spectra, which have been assigned to β-chloroalkyl and other radicals on the basis of qualitative consistency with the structures, are summarized in Table 13.

The "normal" α-H coupling constant and the very small Cl coupling constant attributed to $HOCHCH_2Cl$ provide no cause to consider the radical to be bridged [1; I(a)-(c)].*

The claim [74] that the spectrum of $HOC(CH_2Cl)_2$ was obtained has recently been challenged [75] on the grounds that the β-H coupling constant, Cl coupling constant, and g-factor of 19.6 G, 0.5 G, and 2.00413, respectively, attributed [74] to $HOC(CH_2Cl)_2$ are very close to the values (19.3, 19.7) G, 0.42 G, and 2.0043, respectively, attributed [75] to $\cdot CH_2COCH_2Cl$.

We agree that $HOC(CH_2Cl)_2$, unless it is bridged [1; I(a)-(c)], is unlikely to have a g-factor of 2.00413, while such a value is to be expected (see Table 74) for an acetonyl radical.

The α-H coupling constants attributed to $HC(CH_2Cl)_2$ and $\cdot CH_2CH_2Cl$ are "normal." The significant Cl coupling constants may be discussed in terms of the usual "hyperconjugation" or "spin polarization" "effects," obviating the need to consider a bridged structure, or they may be

---

*The number before the semi-colon refers to the chapter number; numbers and letters after the semi-colon refer to the section(s) within that chapter.

37

TABLE 13

ESR Spectra Attributed to β-Halogen–Containing and Other Radicals

Radical	Method of generation	Coupling constants (G)				g	Temperature	Ref.
		C_1-H	C_2-H	C_2-F	Cl			
$HO\dot{C}HCH_3$	a	16.0	22.2					55
$HO\dot{C}HCH_3$	l	15.09	22.06			2.00316	28.5	166
$HO\dot{C}HCH_3$	e	15.24	22.14			2.00319	57	76
$HO\dot{C}HCH_3$	m	15.16	22.59			2.00315	27.5	166
$HO\dot{C}HCH_3$	e	15.50	22.27			2.00321	-33	76
$HO\dot{C}HCH_3$	p	15.8	22.9				-88	527
$HO\dot{C}HCH_2CH_3$	e	15.06	21.40			2.00319	26	76
$HO\dot{C}HCH_2CH_3$	e	15.20	21.92, 21.03			2.00318	-40	76
$HO\dot{C}HCH_2CH_3$	a	14.7	21.1				22	78
$HO\dot{C}HCH_2OH$	f	17.6	9.4				22	78
$HO\dot{C}HCH_2OH$	a	15.4 ± 0.4	20.6 ± 0.4					73
$HO\dot{C}HCH_2OH$	e	17.54	9.94			2.00308	26	76
$HO\dot{C}HCH(CH_3)_2$	a	12.7	20.0					501
$HO\dot{C}HCH_2OCH_2CH_2OH$	a	18.9	22.2				25	77
$HO\dot{C}HCH_2Cl$	a	15.0	11.1				22	78
$HO\dot{C}HCH_2Cl$	a	15.4 ± 0.4	20.6 ± 0.4		<1			73
$HO\dot{C}(CH_3)_2$	e		19.66			2.00317	26	76

Radical						g		
HOĊ(CH₃)₂	j		20.0 ± 0.5				-196	500
HOĊ(CH₃)₂	e		19.57			2.00309	-22.5	76
HOĊ(CH₃)₂	e		20				-196	502
HOĊ(CH₃)₂	a		19.5				22	78
HOĊ(CH₃)₂	a		20					501
HOĊ(CH₃)₂	g		19.1 ± 0.4			2.003	-100	79
HOĊ(CH₃)₂	k		20				-196	503
HOĊ(CH₃)₂	n		19.90			2.00315	~-15	523
HOĊ(CH₃)₂	o		19.48			2.00307	27.5	166
HOĊ(CH₃)₂	m		19.96			2.00313	28	166
HOĊ(CH₂OH)₂	f		10.1				22	78
HOĊ(CH₂Cl)₂ᶜ	b		19.6		0.5	2.00413		74
HĊ(CH₂Cl)₂	d	21.3	11.4		14.2			75
·CH₂CH₂OH	e	22.00	31.67			2.00247	-71	76
·CH₂CH₂OH	f	22.2	27.6				22	78
·CH₂CH₂OCH₃	h	22.08	31.45				-41	473
·CH₂CH₂OCH₂CH₃	f	21.7	27.9				25	77
·CH₂CH₂OCF₃	i	22.73	28.72		2.00		-74	473
·CH₂CF₃	d	23.9		29.9			~-60	75
·CH₂CH₂F	d	22.3	26.9	47.1			~-60	75
·CH₂CH₂Cl	d	21.5	11.5	17.4			~-60	75

[a] $R\cdot$ generated by mixing in a flow system a solution containing H_2SO_4 and $TiCl_3$ with a solution containing H_2SO_4, RH, and H_2O_2.

[b] Reaction of $HOCH(CH_2Cl)_2$ with Ti^{3+}/H_2O_2.

[c] See Section 5;IA.

[d] "... [the radicals] were mostly prepared from the corresponding alkyl bromides by reaction with triethylsilyl radicals, in a few cases hydrogen atoms were abstracted from the parent alkane by t-butoxy radicals."

[e] $R\cdot$ generated by photolysis of RH containing H_2O_2.

[f] $R\cdot$ generated by mixing in a flow system a solution containing $FeSO_4$-EDTA with a solution containing RH and H_2O_2.

[g] Acetone γ-irradiated at -196°.

[h] Photolysis of $CH_3OCH_2CH_2CO_2\!\!+_2$ in cyclopropane.

[i] Photolysis of CF_3OOCF_3 in ethylene.

[j] $R\cdot$ generated by photolysis at 77°K of RH in water in the presence of ceric perchlorate and $HClO_4$.

[k] $R\cdot$ generated by X-irradiation at 77°K of RH.

[l] $R\cdot$ generated by photolysis of CH_3COCH_3/RH (95/5).

[m] $R\cdot$ generated by photolysis of H_2O/CH_3COCH_3/RH (85/10/5).

[n] Electron irradiation of $(CH_3)_2CHOH$ in H_2O.

[o] $R\cdot$ generated by photolysis of CH_3COCH_3/RH (98/2).

[p] $R\cdot$ generated by photolysis of uranyl perchlorate/RH.

discussed in terms of bridging. On the basis of a common crude treatment of the internal-rotational-angular dependence of the β-H coupling
constants, it was stated [75] that the β-H coupling constants attributed to
these radicals are about 2 G lower than the minimum possible value
(13.5 G) which could result from the treatment. It was suggested that
· CH_2CH_2Cl "is essentially locked in a conformation with chlorine eclipsing
the odd electron orbital" and that there may be "some deformation of the
chloromethyl group with chlorine moving towards a bridging position and
the β-protons moving towards the nodal plane of the π-orbital . . ."
Although this description may be correct, we feel that it is premature
because we do not believe that the 13.5-G figure rests on a basis sufficiently sound to enable a 2-G deviation from it to be considered significant
and to merit a special explanation. It should also be noted that β-H coupling
constants which are "too small" have been reported for $HO\dot{C}HCH_2OH$ and
$HO\dot{C}(CH_2OH)_2$ (Table 13). The data presently at hand are not inconsistent
with slightly bridged [1; I(c)] structures for · CH_2CH_2Cl and $H\dot{C}(CH_2Cl)_2$
and might, with some difficulty ("normal" α-H coupling constants), be
shown to be consistent with such structures. On the other hand, there is no
strong indication that either molecule is bridged [1; I(a)-(c)]; · CH_2CH_2Cl
is not bridged [1; I(a)] with equivalent geminal hydrogens.

B. 1,2-Cl Migrations

    A contention that a radical is bridged is supported by the observa-
tion of products whose structures indicate that the supposedly bridging
group has migrated from one terminus of the alleged bridge to another.
Of course, the observation of such an intramolecular migration requires
only that a bridged structure correspond to an extremum, not necessarily
a minimum, on an energy surface. Also, observation of a given product
does not in itself prove that the product was formed via one particular
route, e.g., an intramolecular 1,2-shift.

    There have been many observations related to the possibility of 1,2-Cl
shifts in radicals. Our purpose is only to indicate that a claim for the
existence of 1,2-Cl-bridged radicals is not diminished by the lack of supporting evidence in the form of observed 1,2-Cl shifts. We will not be

concerned with the mechanisms of the reactions which will be listed, out-
side of what is relevant to their being free-radical. Indeed the mechanisms of many are unknown and can plausibly be formulated so as not to involve an intramolecular 1, 2-Cl shift; also, there is evidence that some of
them do not proceed via intramolecular 1, 2-Cl shifts.

Observations and data have been tabulated (only those products possibly indicative of a Cl shift) according to the following possible means of
obtaining β-chloroalkyl radicals: radiolysis or photolysis of alkyl chlorides
(Table 14), addition of HBr to chloroolefins (Table 15), addition of thiols to
chloroolefins (Table 16), dimerization of chloroolefins (Table 17), reaction
of diazo compounds with halides (Table 18), reaction of $BrCCl_3$ and CCl_4
with chloroolefins (Table 19), and reaction of CH_3CO_3H with chloroalkanes
(Table 20). In addition, it has been reported that irradiation of $CCl_3CBr=CH_2$
with 546-mμ light in the presence of 1, 1.8, and 4.6 mole % Br_2* at 30,
40, and 50° gave $CCl_2=CClCH_2Br$ as the "sole product" and that
photolysis of $CCl_3CBr=CH_2$ at 60° gave $CCl_2=CClCH_2Br$ in yields of up to
41% [511].

C. Energy

1. "Direct" Tests of Bridging

Based on results of studies [258] of the system $[Cl_2C=CHCl + Cl \cdot \rightleftarrows$
$C_2HCl_4 \cdot]$, $D(C_2HCl_4-H)$ has been estimated [257, 258] to be 96 ± 3 kcal/
mole. Similarly, based on results of studies [259, 260] of the system
$[Cl_2C=CCl_2 + Cl \cdot \rightleftarrows CCl_3CCl_2 \cdot]$, $D(CCl_3CCl_2-H)$ has been estimated
[257, 259, 260, 261] to be 95 ± 3 kcal/mole. In view of the value of
$D(CH_3CH_2-H)$, 98 ± 1 kcal/mole [248, 250, 262], and the effect of α-Cl
substitution on $D(R-H)$ [$D(CH_3-H)$ = 104 ± 1 kcal/mole [247, 248, 263,
264], $D(CCl_3-H)$ = 95.7 ± 1.5 kcal/mole [231], the quoted values of
$D(C_2HCl_4-H)$ and $D(CCl_3CCl_2-H)$, taken at face value, indicate that
$C_2HCl_4 \cdot$ and $CCl_3CCl_2 \cdot$ are not bridged [1; II(a)].

---

*No reaction occurred in the absence of Br_2.

TABLE 14

Radiolysis and Photolysis of Alkyl Chlorides

Reactant	Reported product	Reaction conditions	Yield of product	Controls	Ref.
$CH_3CH_2CH_2Cl$	$CH_3CHClCH_3$	γ-rays	$G = 70$[a]	b	80
$CH_3CH_2CH_2Cl$	$CH_3CHClCH_3$	$h\nu$	?		80
$CH_3CH_2CH_2Cl$	$CH_3CHClCH_3$	γ-rays, gas phase	$G = 6.1$[n]		81
$CH_3CH_2CH_2Cl$	$CH_3CHClCH_3$	o	$G = 220$		81
$CH_3CH_2CH_2Cl$	$CH_3CHClCH_3$	p	"tenths of a %"		81
$CH_3CH_2CH_2Cl$	$CH_3CHClCH_3$	q	$G =$ "several hundred"	s	81
$CH_3CH_2CH_2Cl$	$CH_3CHClCH_3$	γ-rays, 22°	$G = 56 \pm 5$	u	82
$CH_3CH_2CH_2Cl$	$CH_3CHClCH_3$	t	$G = 80 \pm 4$		82
$CH_3CH_2CH_2CH_2Cl$	$CH_3CH_2CHClCH_3$	e	$G = 0.7$		81
$CH_3CH_2CH_2CH_2Cl$	$CH_3CH_2CHClCH_3$	f	$G = 0.8$		81
$CH_3CH_2CH_2CH_2Cl$	$CH_3CH_2CHClCH_3$	g	$G = 1.0$		81
$CH_3CH_2CH_2CH_2Cl$	$CH_3CH_2CHClCH_3$	h	$G = 2.7$[j]		81
$CH_3CH_2CH_2CH_2Cl$	$CH_3CH_2CHClCH_3$	i	$G = 60\text{–}120$[j]		81
$CH_3CH_2CH_2CH_2Cl$	$CH_3CH_2CHClCH_3$	k	$G = 1.9$[m]	r	81
$CH_3CH_2CH_2CH_2Cl$	$CH_3CH_2CHClCH_3$	l	$G \cong 90$[m]	r	81
$CH_3CH_2CH_2CH_2Cl$	$CH_3CH_2CHClCH_3$[w]	v	$G = 60$		83
$(CH_3)_2CHCH_2Cl$	$(CH_3)_3CCl$	c	$G \cong 20$[d]		81

a Dose = 18×10^{20} eV g^{-1} .

b Addition of 1% I_2 reduced G($CH_3CHClCH_3$) to 0.4, independent of dose and reduced G($CH_3CHClCH_2Cl$) from 0.59 to a "level too low or too poorly resolved for quantitative evaluation," independent of dose at a dose rate of 1.4×10^{18} eV g^{-1} min^{-1} . Addition of 5 mole % C_2H_4 was reported to result in the production of $CH_3CH_2CH_2CHClCH_3$ and $CH_3CH_2CH_2CH(CH_3)CH_2Cl$ with G-values of 2.5 and 3.6, respectively, at a dose of 6.6 x 10^{20} eV g^{-1} at 1.4×10^{18} eV g^{-1} min^{-1} .

c γ-rays, 1.5 mole % HCl, 25° .

d Dose = 5×10^{19} eV g^{-1} at 3.6×10^{18} eV g^{-1} min^{-1} .

e γ-rays, -196° , polycrystalline solid.

f γ-rays, -196° , glass.

g γ-rays, -196° , polycrystalline solid or glass, 0.4-1.6 mole % HCl.

h γ-rays, gas phase, 25° , dose = 3.2×10^{20} eV g^{-1} at 2.5×10^{18} eV g^{-1} min^{-1} , 0 mole % HCl.

i As in h, except 260 mole % HCl.

j G($CH_3CH_2CHClCH_3$) is a monotonic function of the mole % HCl.

k γ-rays, liquid, 25° .

l γ-rays, liquid, 25° , 1.7 mole % HCl.

m G($CH_3CH_2CHClCH_3$) is a linear function of the mole % HCl.

n Dose = 0.75×10^{20} eV g^{-1} .

o γ-rays, gas phase, ~ 75 mole % HCl.

p $_{h\nu}$, gas phase, 1.5 mole % HCl.

q γ-rays, liquid, 160°, 1.2 mole % HBr.

r "In the presence of 0.25 or 0.5 mole % I_2 or 0.1 mole % diphenylpicrylhydrazyl, G ($\underline{sec}$-C_4H_9Cl) . . . was 0.4."

s "Relatively negligible isomerization occurred . . . without irradiation."

t γ-rays, 22°, 0.1-0.3 mole % HCl.

u Addition of 0.1-0.2 weight % diphenylamine or 2,6-di-$\underline{t}$-butyl-$\underline{p}$-cresol resulted in G-factors of 10-40 and 40-51, respectively.

v γ-rays, ~ 25°, 3.2 x 10^5 r hr^{-1}.

w Identified from VPC retention time; $(CH_3)_2CHCH_2Cl$ not excluded as a possible structure.

TABLE 15

Reaction of HBr with Chloroolefins

Olefin	Reaction conditions	Reported product	Reported % yield of product	Controls	Ref.
$CH_2=CHCCl_2F$[a]	hν	$BrCH_2CHClCHClF$[b]	18	c	84
$CCl_3CH=CH_2$[w]	hν	$BrCH_2CHClCHCl_2$[d]	87		84
$CCl_3CH=CH_2$[w]	e	$BrCH_2CHClCHCl_2$[d]	77		85
$CCl_3CH=CH_2$[w]	x	$BrCH_2CHClCHCl_2$[d]	77	y	89
$CHCl_2CH=CH_2$	g	$CH_2ClCHClCH_2Br$[h]	35		86
$CHCl_2CH=CH_2$	ad	$ClCH_2CHClCH_2Br$[h]	46		90
$CCl_3C(CH_3)=CH_2$	i	$CHCl_2C(CH_3)ClCH_2Br$[j]	59		87
$CCl_3C(CH_3)=CH_2$	i	$CHCl_2C(CH_3)ClCH_2Cl$[l]	11 (crude)		87
$CCl_3CCl=CH_2$[m]	n	$CHCl_2CCl_2CH_2Cl$[o]	11		88
$CCl_3CCl=CH_2$[m]	n	$CHCl_2CCl_2CH_2Br$[r]	30		88
$CH_3CCl_2CH=CH_2$	z	$CH_3CHClCHClCH_2Br$[aa]	73		90
$CHCl_2C(CH_3)=CH_2$	ae	$ClCH_2C(CH_3)ClCH_2Cl$[ac]	<15	ah	90
$CHCl_2C(CH_3)=CH_2$	ae	$ClCH_2C(CH_3)ClCH_2Br$[af]	<46, >42	ah	90
$CH_3CCl_2CBr=CH_2$[ai]	am	$BrCH_2CCl=CClCH_3$[an]	90		91

[a] Basis of assignment of structure: (1) prepared in 59% yield from $ClCH_2CH_2CCl_2F$; (2) picrate of iso-thiuronium derivative gave an acceptable N analysis.

[b] Basis of assignment of structure: (1) acceptable C, H, and F analyses; (2) picrate of isothiuronium derivative gave acceptable C, H, and N analyses; (3) reacted with "alkali" at 0° in ethyl cellosolve to give 54% $CHFClCCl=CH_2$.

[c] There was a "need for irradiation with uv light." The reaction was inhibited by small amounts of hydroquinone or acetone.

[d] Basis of assignment of structure: reaction with alcoholic KOH at 0° to give what was reported to be $CH_2=CClCHCl_2$.[f]

[e] 1.2 mole % Bz_2O_2 in CCl_4, 50°.

[f] Basis of assignment of structure: (1) "differs in physical constants from the four known of the six possible trichloropropenes" [the two unknown ones were $CH_2=CClCHCl_2$ and $CHCl=CHCHCl_2$]; (2) ozonolysis gave $CHCl_2COOH$.

[g] 0.7 mole % Bz_2O_2, 60–70°.

[h] Assignment of structure OK.

[i] 1.1 mole % Bz_2O_2 in CCl_4, 70°.

[j] Basis of assignment of structure: (1) acceptable C, H, Br, and Cl analyses; (2) reaction with KOH in $MeOCH_2CH_2OH$ at 0–5° gave 57% of what was reported to be $CCl_2=C(CH_3)CH_2Br$.[k]

[k] Basis of assignment of structure: (1) acceptable C, H, and Cl analyses; (2) diethylamine derivative, which had an acceptable Cl analysis, was the same as that prepared from $CCl_3C(CH_3)=CH_2$.

l Basis of assignment of structure: (1) acceptable C and H analyses; (2) boiling point, index of refraction, and density agreed with those of one of the photochlorination products (which had an acceptable Cl analysis) of $(CH_3)_3CCl$.

m Basis of assignment of structure: (1) Prepared in 50% yield by reaction of $CCl_3CHClCH_2Cl$ with KOH/EtOH; (2) reacted with Cl_2 to give 62% of material which reacted with KOH/EtOH to give 58% of material which gave Cl_3CCO_2H upon ozonolysis.

n 0.3 mole % Bz_2O_2 at 60–70°.

o Basis of assignment of structure: (1) acceptable C and H analyses; (2) ir spectrum identical to that of a product[p] of the chlorination of $CHCl_2CCl=CH_2$ and "differs sharply" from that of what was reported to be $CCl_3CHClCH_2Cl$.[q]

p Basis of assignment of structure: melting point, index of refraction, and density agreed with those of a compound whose assignment of structure is based on (1) an acceptable Cl analysis; (2) its method of preparation (from Cl_2 and $CHCl=CClCH_2Cl$); (3) agreement of its boiling point, refractive index, and density with those of a product of the reaction of SO_2Cl_2 and $ClCH_2CCl_2CH_2Cl$.

q The account [Chem. Abstr., 46, 1957c(1952)] of its preparation from Cl_2 and $CCl_2=CHCH_2Cl$ describes it as "probably" this compound.

r Basis of assignment of structure: (1) acceptable halogen, C, and H analyses; (2) same as product of reaction of Cl_2 and what was reported to be $BrCH_2CCl=CHCl$[s]; (3) ir spectrum "differed noticeably" from that of what was reported to be $CCl_3CHClCH_2Br$[t]; (4) reacted with KOH in ethanol to give what was reported to be $BrCH_2CCl=CCl_2$.[u]

s Basis of assignment of structure: (1) acceptable C and H analyses; (2) prepared by reaction of ClCH$_2$CCl=CHCl and NaBr in acetone.

t Basis of assignment of structure: (1) acceptable C and H analyses; (2) prepared by reaction of CCl$_2$=CHCH$_2$Br and Cl$_2$.

u Basis of assignment of structure: (1) prepared by reaction of what was reported to be ClCH$_2$CCl=CCl$_2$[v] and NaBr in acetone; (2) boiling point, refractive index, and density agreed with those of material which gave acceptable C and H analyses; (3) Et$_2$NH derivative same as that (which gave an acceptable N analysis) prepared from what was reported to be ClCH$_2$CCl=CCl$_2$.[v]

v Basis of assignment of structure: (1) boiling point, refractive index, and density agreed with those of material which is described in a publication unavailable to us; (2) Et$_2$NH derivative gave an acceptable N analysis.

w Basis of assignment of structure: (1) prepared in 27% yield by reaction of CCl$_3$CH$_2$CH$_2$Cl with KOH in ethanol at 0°; (2) ozonolysis yielded CCl$_3$CHO.

x Bz$_2$O$_2$, 50–70°.

y No reaction occurred in the absence of initiator. Minute amounts of inhibitors completely stopped the reaction.

z 1.6 mole % Bz$_2$O$_2$ in CCl$_4$, 70–75°.

aa Basis of assignment of structure: (1) acceptable C and H analyses; (2) acceptable C, H, and N analyses for picrate of isothiourea derivative; (3) reaction with KOH in ethyl cellosolve yielded 60% of what was reported to be CH$_3$CHClCCl=CH$_2$.[ab]

ab Basis of assignment of structure: (1) boiling point, index of refraction, and density agreed with those of previously reported material[ac]; (2) ozonolysis yielded CH_2O and $CH_3CHClCOOH$.

ac Basis of assignment of structure: none.

ad 0.46 mole % Bz_2O_2 in heptane, $80°$.

ae 0.57 mole % Bz_2O_2 in CCl_4, $80°$.

af Basis of assignment of structure: (1) boiling point, index of refraction, and density agreed with those of a sample which gave acceptable C and H analyses and was prepared from Cl_2 and $CH_2=C(CH_3)CH_2Br$; (2) reaction with Zn in ethanol gave what was reported to be $ClCH_2C(CH_3)=CH_2$.[ag]

ag Basis of assignment of structure: picrate of isothiourea derivative same as that (which gave acceptable C and H analyses) prepared from a "known sample."

ah "In order to exclude the possibility that the rearranged reaction products were formed by allyl isomerization of the original unsaturated compounds with subsequent addition of hydrogen bromide to yield the structures $CH_3CHClCHBrCH_2Cl$ and $CH_2ClC(CH_3)BrCH_2Cl$, respectively, control experiments were carried out in which hydrogen bromide was added to $CH_3CCl=CHCH_2Cl$ and $CHCl=C(CH_3)CH_2Cl$ (allylic isomers of the original unsaturated compounds) in the presence of benzoyl peroxide; these experiments gave negative results."

ai Basis of assignment of structure: (1) prepared in 59% yield by reaction of what was reported to be $BrCH_2CHBrCCl_2CH_3$[aj] with KOH in ethyl cellosolve; (2) acceptable C and H analyses; (3) picrate of isothiourea derivative gave acceptable C, H, and N analyses; (4) reaction with Me_2NH yielded what was reported to be $CH_3CCl=CBrCH_2NMe_2$.[al]

aj Basis of assignment of structure: (1) prepared in 77% yield by reaction of what was reported to be $CH_3CCl_2CH=CH_2$[ak] with Br_2 in HOAc; (2) acceptable C and H analyses.

ak Basis of assignment of structure: (1) prepared in 43% yield by reaction of $BrCH_2CH_2CCl_2CH_3$ with KOH in ethyl cellosolve; (2) acceptable C and H analyses; (3) picrate of thiourea derivative gave acceptable C, H, and N analyses and was the same as that obtained from $CH_3CCl=CHCH_2Cl$.

al Basis of assignment of structure: picrate gave acceptable C, H, and N analyses and was the same as that prepared from what was reported to be $BrCH_2CHBrCCl_2CH_3$.[aj]

am 1.2 mole % Bz_2O_2 in CCl_4.

an Basis of assignment of structure: (1) acceptable C and H analyses; (2) picrate of thiourea derivative gave acceptable N and Cl analyses; (3) picrate of dimethylamino derivative gave acceptable C, H, and N analyses and contained no Br.

TABLE 16

Reaction of Thiols with Chloroolefins

Olefin	Thiol	Reaction conditions	Reported product	Reported % yield of product	Controls	Ref.
$CH_2=C(CH_3)CH_2Cl$	CH_3SH	$h\nu$, 15°, RSH/halide = 0.5	$CH_3SCH_2CCl(CH_3)_2$ [a]	41		92
$CH_2=C(CH_3)CH_2Cl$	CH_3SH	$h\nu$, 15°, RSH/halide = 1.0	$CH_3SCH_2CCl(CH_3)_2$ [a]	31		92
$CH_2=C(CH_3)CH_2Cl$	CH_3SH	$h\nu$, 15°, RSH/halide = 2.0	$CH_3SCH_2CCl(CH_3)_2$ [a]	5		92
$CH_2=C(CH_3)CH_2Cl$	CH_3SH	$h\nu$, 15°, RSH/halide = 6.0	$CH_3SCH_2CCl(CH_3)_2$ [a]	nil		92
$CH_2=CHCH_2Cl$	CH_3SH	$h\nu$, 15°, RSH/halide = 0.8	$CH_3SCH_2CClHCH_3$ [a]	11 (isolated)		92
$CH_2=C(CH_3)CH_2Cl$	CH_3CH_2SH	$h\nu$, 15°, RSH/halide = 1	$CH_3CH_2SCH_2CCl(CH_3)_2$ [a]	22 (isolated)		92
$CCl_3CH=CH_2$	PhSH	$h\nu$, 80–90(110), RSH/halide = 0.5	$CHCl_2CHClCH_2SPh$ [d]	35 (30)		93,98
$CCl_3CBr=CH_2$	PhSH	$h\nu$, 105–110°, RSH/halide = 1.3	$CCl_2=CClCH_2SPh$ [b]	55	c	93
$CCl_3CH=CH_2$	$PhCH_2SH$	$h\nu$, 110–115°, RSH/halide = 0.67	$Cl_2CHCHClCH_2SCH_2Ph$ [h]	11 (crude)		93,98

[a] Identification of product OK.

[b] Basis of assignment of structure: (1) reacted with H_2O_2 in HOAc to give same product as is obtained in

57% yield from reaction of $CCl_2=CClCH_2Br$ with PhSNa followed by reaction with H_2O_2; (2) material before and after reaction with H_2O_2 gave acceptable C and H analyses.

[c] Product formed in 46% yield in the absence of light. "In a specially conducted experiment, in which $CCl_3CBr=CH_2$ had not reacted completely with PhSH, it was found that the unreacted trichlorobromopropene corresponded to $CCl_3CBr=CH_2$ in its structure. Furthermore, it was estimated that authentic $CCl_2=CClCH_2Br$ did not react with PhSH under conditions at which the reaction of PhSH with $CCl_3CBr=CH_2$ took place."

[d] Basis of assignment of structure: (1) acceptable H analysis; (2) boiling point, index of refraction, and density agreed with those of material prepared in 56% yield by reaction of what was reported to be $CHCl_2CHClCH_2Br$[e] with PhSNa; (3) reaction with H_2O_2 in HOAc yielded 80% of material (acceptable C, H, Cl, and S analyses) which was the same as that similarly prepared in 80% yield from the material mentioned in (2); (4) reaction with Zn in MeOH yielded 50% (acceptable C, H, and Cl analyses) of what was reported to be $CHCl=CHCH_2SPh$.[f]

[e] Basis of assignment of structure: see footnote d to Table 15.

[f] Basis of assignment of structure: oxidation yielded 34% of what was reported as trans-$CHCl=CHCH_2SO_2Ph$.[g]

[g] Basis of assignment of structure: (1) prepared by reaction of trans-$CHCl=CHCH_2Cl$ with PhSNa (46% yield) followed by reaction with H_2O_2; (2) acceptable C, H, and Cl analyses.

[h] Basis of assignment of structure: (1) acceptable C, H, and Cl analyses; (2) reaction with H_2O_2 yielded 89% of what was reported to be $Cl_2CHCHClCH_2SO_2CH_2Ph$.[i]

[i] Basis of assignment of structure: (1) prepared by reaction of what was reported to be $Cl_2CHCHClCH_2Br$[e] with $PhCH_2SNa$ followed by H_2O_2; (2) acceptable C, H, and Cl analyses.

TABLE 17

Dimerization of Chloroolefins

Olefin	Reaction conditions	Reported product	Reported % yield of product	Controls	Ref.
$CFCl_2CH=CH_2$	$h\nu$	$ClCH_2CHClCFClCH_2CH=CClF$[a]	41		94
$CFCl_2CCl=CH_2$	$h\nu$	$ClCH_2CCl_2CFClCH_2CCl=CClF$[b]	12		94
$CCl_3CH=CH_2$	c	$CCl_2=CHCH_2CCl_2CHClCH_2Cl$[d]	27	j	95
$CFCl_2CBr=CH_2$ [m]	90 or 140°	$BrCH_2C(CCl_2F)BrCH_2CCl=CClF$[o]	42	k	96
$CFCl_2CBr=CH_2$ [m]	l	$BrCH_2C(CCl_2F)BrCH_2CCl=CClF$[o]	3		96
$CClF_2CBr=CH_2$ [p]	$h\nu$	$BrCH_2C(CF_2Cl)BrCH_2CCl=CF_2$[p]	11		97

[a] Basis of assignment of structure: reaction with 95% H_2SO_4 produced 81% of what was reported to be $ClCH_2CHClCFClCH_2CH_2COOH$. Treatment of the latter with Zn in HOAc at 120° produced 30% of what was reported to be $H_2C=CHCHFCH_2CH_2COOH$ (reaction with H_2SO_4 yielded what was reported to be $CH_3CH_2COCH_2CH_2COOH$[g]) and 27% of what was reported to be $CH_3CH_2COCH_2CH_2COOH$.[g]

[b] Basis of assignment of structure not mentioned.

[c] 0.5–1 mole % Bz_2O_2 in benzene.

[d] Basis of assignment of structure: (1) acceptable C and H analyses; (2) reaction with 95% H_2SO_4 yielded 62% of what was reported to be $ClCH_2CHClCCl_2CH_2CH_2COOH$[e]; (3) reaction with Zn in HOAc yielded 18% of "1,1-dichlorohexadiene" (acceptable C and H analyses), 16% of what was reported to be $Cl_2C=CHCH_2CCl=CHCH_3$,[h] and 24% of what was reported to be $CCl_2=CHCH_2CCl=CHCH_2Cl$.[i]

e Basis of assignment of structure: (1) acceptable C, H, and Cl analyses; (2) reaction with Zn in HOAc yielded 52% of what was reported to be $CH_3CH=CClCH_2CH_2COOH$. f

f Basis of assignment of structure: (1) reaction with 95% H_2SO_4 yielded what was reported to be $CH_3CH_2COCH_2CH_2COOH$g; (2) reaction with H_2 and Pd/$BaSO_4$ yielded 71% $CH_3CH_2CH_2CH_2CH_2COOH$.

g Basis of assignment of structure: 2,4-dinitrophenylhydrazine derivative gave acceptable C, H, and N analyses.

h Basis of assignment of structure: (1) acceptable C and H analyses; (2) reaction with H_2SO_4 at 40° yielded what was reported to be $CCl_2=CHCH_2COCH_2CH_3$ (2,4-dinitrophenylhydrazine derivative gave acceptable C, H, and Cl analyses).

i Basis of assignment of structure: picrate of Et_2NH derivative gave acceptable C, H, and N analyses.

j $CCl_2=CHCH_2Cl$ was stable to the reaction conditions.

k Addition of Bz_2O_2 at 90° resulted in a 46% yield of product while addition of hydroquinone at 90° resulted in a 0% yield of product.

l HBr, hν, 40–50°.

m Basis of assignment of structure: (1) acceptable C and H analyses; (2) ozonolysis yielded CH_2O and material whose melting point and C and H analyses agreed with those of the product of reaction of $CFCl_2CHBrCH_2Br$n with KOH in ethyl cellosolve.

n Basis of assignment of structure: (1) acceptable C and H analyses; (2) prepared by reaction of $CFCl_2CH=CH_2$ with Br_2 in HOAc; (3) diethylamine derivative gave acceptable C and H analyses.

[o] Basis of assignment of structure: (1) reaction with Zn in ethanol gave 31% of a "bromine-free diene" which had acceptable C and H analyses and which gave CH_2O upon ozonolysis; (2) molecular weight (cryoscopically in benzene) = 409.5 (calc, 415.7); (3) acceptable C, H, and F analyses; (4) ir spectrum contained an intense absorption at 1670 ± 5 cm^{-1} ("characteristic for compounds containing the CFCl=CCl group").

[p] Basis of assignment of structure: unknown to us; original paper unavailable to us.

TABLE 18

Photoreaction of Diazo Compounds with Halides

Halide	Diazo compound	Product	% Yield	Ref.
$CHCl_3$	$N_2CHCOOMe$	$CHCl_2CHClCOOMe$	17	99
$BrCCl_3$	$N_2CHCOOMe$	$BrCCl_2CHClCOOMe +$ $CCl_2=CClCOOMe$	<30	99
CCl_4	CH_2N_2	$C(CH_2Cl)_4$ [a]	60	100, 495
$BrCCl_3$	CH_2N_2	$BrCH_2C(CH_2Cl)_3$	38	100
$CHCl_3$	CH_2N_2	$CH_3C(CH_2Cl)_3$	45	100
CCl_3COOMe	CH_2N_2	$(ClCH_2)_3CCOOMe$	60	100

[a] Not produced in the presence of Ph_2NH. N_2 evolution negligible in the dark.

TABLE 19

Reaction of $BrCCl_3$ and CCl_4 with $CCl_3CH=CH_2$

Reaction conditions	Reported product (% yield)	Ref.
$BrCCl_3$, 100°, Bz_2O_2	$BrCCl_2CHClCH_2CCl_3$ (18) [a]	85
$BrCCl_3$, 100°, Bz_2O_2	$BrCCl_2CHClCH_2Cl$ (14) [b]	85
CCl_4, 145°, Pr^iOH, 10 mole % $(Bu^tO)_2$	$CHCl_2CHClCH_2CCl_3$ (4) [c,d]	524

[a] Basis of assignment of structure: (1) heating with H_2SO_4 yielded "some fumaric acid"; (2) reaction with alcoholic KOH at 0° yielded what was reported to be $CCl_2=CHCCl=CCl_2$; (3) reaction with Zn yielded what was reported to be $CCl_2=CHCH_2CCl_3$.

[b] Basis of assignment of structure: reaction with alcoholic KOH yielded what was reported to be $ClCH_2CCl=CCl_2$, which was "identified as the known $Et_2NCH_2CCl=CCl_2 \cdot HCl$."

[c] Basis of identification of product: VPC retention times on two columns compared to those of an independently prepared sample. [e]

[d] 32% $CCl_2=CHCH_2CCl_3$ was also reported. [f]

[e] Prepared in unspecified yield by the reaction of $CCl_3CH=CH_2$ and $BrCCl_3$ at 145° in ether in the presence of 7 mole % $(Bu^tO)_2$. Basis of assignment of structure: (1) acceptable H analysis; (2) NMR spectrum — $\tau = 4.05$ (1H, J = ~ 3 cps), 5.42 (1H, J = 3, ~3, 6.8 cps), 6.46 (1H, J = 3, 16 cps), 6.76 (1H, J = 6.8, 16 cps).

[f] Basis of assignment of structure: NMR spectrum — $\tau = 3.95$ (1H, J = 6.6 cps), 6.43 (1H, J = 6.6 cps).

TABLE 20

Reaction of Chloroalkanes with CH_3CO_3H

Chloroalkane	Reaction conditions	Reported product (% yield)	Controls	Ref.
CH_3CCl_3	?	$ClCH_2CCl_2CCl_2CH_2Cl\,(9)^a$	None	482
$ClCH_2CCl_3$	?	$Cl_2CHCCl_2CCl_2CHCl_2\,(16)^a$	None	482

[a] Basis of identification of product: "The structures . . . were confirmed by their IR spectra, which were identical to the spectra of known samples, and by mixture melting point tests."

1a. Reaction of Haloalkyl Chlorides with Organotin Hydrides

Replacement of a β-hydrogen by chlorine can increase by a factor of about 5 ($\Delta\Delta G^{\ddagger} = 0.9$ kcal/mole) the rate at which a bromine atom is abstracted by $Bu_3Sn\cdot$ (Table 21) [101]. While this result is consistent with the idea that the transition state for reaction of erythro-$CH_3CHClCHBrCH_3$ with $Bu_3Sn\cdot$ reflects unusual stability of $CH_3CHCl\dot{C}HCH_3$, this is just one possible speculative interpretation. $\Delta\Delta G^{\ddagger}$ is not large and could be brought about by any or several of a number of subtle "factors" "operating" in the four species (two reactants and two transition states) of concern. The choice of a particular interpretation would be arbitrary and unjustified.

2. Reaction of Alkyl Chlorides with Br_2

The products resulting from the bromination of the following compounds have been reported: $CH_3CH_2CH_2CH_2Cl$ (Table 22), $CH_3CH_2CHClCH_3$ (Table 23), $CH_3CH_2CH_2Cl$ (Table 24), and $(CH_3)_2CHCH_2Cl$ (Table 25). In addition to many questionable proofs of structure of the products, in no case has the relative amounts of products been shown to equal the relative rates of abstraction of the corresponding hydrogens. Barring coincidences, necessary conditions for this are that each radical, formed by hydrogen abstraction from the substrate, give the corresponding monobromosubstrate to the same extent, that the monobromosubstrates be formed only

TABLE 21

Effect of β-Substituents on Rates of Reaction of Bromides with
Bu_3SnH at 25° [101]

Halide	Relative rate
$CH_3CHBrCH_2CH_3$	0.21
erythro-$CH_3CHBrCHClCH_3$[a,c]	1.00
meso-$CH_3CHBrCHBrCH_3$[b]	1.6 (per Br)
$PhCHBrCH_2Br$	3.1
$PhCHBrCH_2Cl$	2.4

[a] 90% yield of $CH_3CHClCH_2CH_3$ (identification based solely on VPC retention time).

[b] Product is 2-butene, with only a trace of butane.

[c] "The erythro configuration was assigned to the compound by comparison of its infrared spectrum with those of meso-2,3-dichlorobutane, meso-2,3-dibromobutane, d,1-2,3-dichlorobutane, and d,1-2,3-dibromobutane in the 1030-950 cm^{-1} region. The bromochloride and the two meso isomers have similar absorption bands in this region..."

from their respective radicals, and that each initially produced monobromosubstrate survive the reaction to the same extent. Since these conditions, or any approaching them, have not been met (in many cases because the work had other purposes), we must reserve judgment on whether these results are relevant to the question of bridging. If they are assumed to be so and are taken at face value, they indicate no exceptional stabilization (relative rate constants being an exponential, and sensitive, function of relative energies of transition states) of the transition states leading to β-, γ-, or δ-chloroalkyl radicals.

TABLE 22

Reaction of $CH_3CH_2CH_2CH_2X$ with Br_2

X	Reaction conditions	% Yield of reported monobrominated substrate	Distribution of reported monobrominated substrates (%)				Controls	Ref.
			1,1	1,2	1,3	1,4		
Br	60°, hν, liquid[a]	84	0.9[b]	84.5	14.6	–	f	102
Br	100°, gas	43	–	70[k]	30[k]	–	None	113
Cl	60°, hν, liquid[a]	72	22.8[d]	25.3[e]	51.9[c]	–	None	102
Cl	100°, gas	92	25[d2]	25[k]	50[b]	–	None	113
Cl	146°, hν, gas[g]	?	23.0[i]	21.7[b]	55.3[b]	Trace[h]	None	107
F	62°, hν, gas[n]	?	6.7[b]	5.6[b]	87[k]	0.89[b]	None	121
F	107°, hν, gas[n]	?	6.7[b]	6.2[b]	86[k]	0.65[b]	None	121
F	146°, hν, gas[j]	?	10.0[b]	8.9[b]	81.1[k]	–	None	107
F	160°, hν, gas[n]	?	8.8[b]	6.2[b]	83[k]	1.98[b]	None	121
F	185°, hν, gas[n]	?	8.8[b]	7.4[b]	82[k]	1.52[b]	None	121
CF_3	40°, hν, gas[o]	?	0	10.5[k]	89[k]	0.67[b]	None	121
CF_3	76°, hν, gas[o]	?	0	11.6[k]	88[k]	0.66[b]	None	121
CF_3	155°, hν, gas[o]	?	0	12.7[k]	86[k]	1.42[b]	None	121
CF_3	200°, hν, gas[o]	?	0	13.5[k]	84[k]	1.90[b]	None	121
CF_3	230°, hν, gas[o]	?	0	16.5[k]	81[k]	2.31[b]	None	121

$OCOCF_3$	23°, hν, gas[n]	?	2.07[b]	4.90[b]	92.50[b]	0.44[b]	None	119
$OCOCF_3$	88°, hν, gas[n]	?	2.12[b]	6.57[b]	90.36[b]	0.95[b]	None	119
$OCOCF_3$	153°, hν, gas[n]	?	2.78[b]	9.14[b]	86.15[b]	1.95[b]	None	119
$OCOCF_3$	244°, hν, gas[n]	?	1.97[b]	12.24[b]	82.25[b]	3.53[b]	None	119
$OCOCH_3$	150°, hν, gas[n]	?	15.78[l]	23.04[b]	59.96[b]	1.22[l]	None	119
$OCOCl$	160°, hν, gas[a]	?	21[l]	23[b]	55[b]	1.1[l]	None	122
$OCOF$	150°, hν, gas[n]	?	24[l]	19[b]	57[b]	0	None	122
$COOCH_3$	50°, hν, gas[a]	?	35[l]	20[b]	45[b]	?	None	122
$COOCH_3$	160°, hν, gas[a]	?	27[l]	23[b]	50[b]	1.0[l]	None	122
H	146°, hν, gas[m]	?	0.9[l]	49[l]	49[l]	0.9[l]	None	107,114

[a] Substrate/Br_2 = 5.

[b] Basis of assignment of structure: none.

[c] Basis of assignment of structure: (1) refractive index agreed with a published value; (2) same as footnote e2 except 1,3-dibromobutane.

[d] Basis of assignment of structure: (1) refractive index agreed with a published value; (2) PbO_2 oxidation gave butyraldehyde.

[e] Basis of assignment of structure: (1) refractive index agreed with a published value; (2) ir spectrum of bis(S-alkyl)-isothiourea picrate same as that of material prepared from 1,2-dibromobutane.

f " . . . one run was kept in the dark for 7 days at the end of which time most of the bromine remained unreacted. The irradiated reactions were complete in several hours. The subsequent irradiation of this dark run and analysis by v.p.c. showed that the isomer distribution was identical with that of the other runs."

g Substrate/Br_2/N_2 = 10/1/400.

h " . . . identified . . . by addition of authentic material [to the reaction mixture, followed by VPC analysis]."

i " . . . identified . . . by comparison with the [VP] chromatograms from the chlorination of n-butyl bromide."

j Substrate/Br_2/N_2 = 5/1/200.

k Assignment of structure OK.

l Assignment of structure based on VPC retention time.

m <15% conversion; substrate/Br_2/N_2 = 5/0.5-1/80.

n Substrate/Br_2 = 10.

o Substrate/Br_2 = 20.

TABLE 23

Reaction of $CH_3CH_2CHXCH_3$ with Br_2

X	Reaction conditions	% Yield of reported monobrominated substrate	Distribution of reported monobrominated substrates (%)†					Ref.
			1,2	2,2	erythro(meso)-2,3	threo(dl)-2,3	1,3	
Br	60°, hν, liquid	83	–	16.3[a]	58.3[b]	25.4[b]	–	102
Br	100°, gas	51		37[a]	43[e]	20[f]		113
Br	146°, gas[d]	g		100[e]				108
Cl	60°, hν, liquid	?	–	92.1[c]	4.8[a]	3.1[a]		102
F	146°, hν, gas[d]	?		>93[e]	<7[a] (erythro[a]/threo[a] = 64.3/35.7)			108
H	146°, hν, gas[h]	?	0.9[b]	49[b]	49[b]		0.9[b]	107,114

† Controls: none.

[a] Basis of assignment of structure: none.

[b] Basis of assignment of structure: VPC retention time agreed with that of an authentic sample.

[c] Basis of assignment of structure: PbO_2 oxidation gave $CH_3COCH_2CH_3$.

[d] Substrate/Br_2/N_2 = 10/1/400.

[e] Assignment of structure OK.

[f] Basis of assignment of structure: density agreed with a published value.

[g] Only one major VPC peak.

[h] See footnote m to Table 22.

TABLE 24

Reaction of $CH_3CH_2CH_2X$ with Br_2

X	Reaction conditions	% Yield of reported monobrominated substrate	Distribution of reported monobrominated substrates (%)		Controls	Ref.
			1,1	1,2		
Cl	100°, gas	90	50[a]	50[a]	None	113

[a] Assignment of structure OK.

TABLE 25

Reaction of $CH_3CH(CH_3)CH_2X$ with Br_2

X	Reaction conditions	% Yield of reported monobrominated substrate	Distribution of reported monobrominated substrates (%)		Controls	Ref.
			1,1	1,2		
Cl	$h\nu$, CCl_4[a]	98–99	–	100[b]	None	133

[a] Substrate/Br_2 = 5.
[b] Assignment of structure OK.

TABLE 26

Reaction of 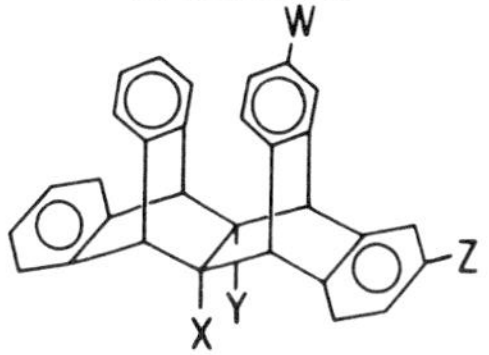 with Br_2 [194]

Reactant				Reaction conditions	Product				% Yield	Relative rate constant	Controls
X	Y	W	Z		X	Y	W	Z			
Br	H	H	H	hν, 12°, CCl₄	Br	Br	H	H	?	2.6	b
H	H	H	H	hν, 12°, CCl₄	H	Br	H	H	?	$(0.5)^a$	b
Br	H	H	H	hν, 72°, CCl₄	Br	Br	H	H	d	1.4	b,c
Cl	H	H	H	hν, 72°, CCl₄	Cl	Br	H	H	d	0.25	None
H	H	H	H	hν, 72°, CCl₄	H	Br	H	H	d	$(0.5)^a$	b,c

[a] The relative rate constant has been divided by 2.

[b] The reactant was "inert to bromine in the dark in the absence of Lewis acids" at an unspecified temperature.

[c] Photobromination of the (X = Y = Z = H, W = NO₂), (X = Y = Z = H, W = Br), and (X = Y = W = H, Z = Cl) compounds to the dibromides yielded none (NMR) of the (W = H, Z = NO₂, X = Y = Br), (W = H, X = Y = Z = Br), or (Z = H, W = Cl, X = Y = Br)[e] compounds, respectively.

d Competitive photobromination of (X = Y = W = Z = H) and (X = Cl, Y = W = Z = H) yielded 94%
[(X = Br, Y = W = Z = H) + (X = Y = Br, W = Z = H) + (X = Cl, Y = Br, W = Z = H)]. Competitive
photobromination of (X = Br, Y = W = Z = H) and (X = Cl, Y = W = Z = H) yielded 98% [(X = Y = Br, W = Z = H) +
(X = Cl, Y = Br, W = Z = H)].

e Conclusion that compound is absent based on assumption that its NMR spectrum would be "similar" to
that of the (X = Y = W = Br, Z = H) compound.

The data in Table 26, for which a preponderance of evidence indicates that the substance of the conditions in the preceding paragraph is met, show a retarding influence of a β-Cl substituent corresponding to $\Delta \Delta G^{\ddagger} = 0.5$ kcal/mole.

3. Reaction of Alkyl Chlorides with Cl_2

The products resulting from the chlorination of the following compounds have been reported: $CH_3CH_2CH_2CH_2Cl$ (Table 27), $CH_3CH_2CH_2Cl$ (Table 28), $CH_3CH_2CH_2CH_2CH_2Cl$ (Table 29), $(CH_3)_2CHCH_2Cl$ (Table 30), $CH_3CH_2CHClCH_3$ (Table 31), $CH_3CH_2CH_2CH_2CH_2CH_2Cl$ (Table 32), $CH_3CH_2CH_2CH_2CH_2CH_2CH_2Cl$ (Table 33), $CH_3CH_2CH_2CH_2CHClCH_3$ (Table 34), and $CH_3CH_2CH_2CHClCH_2CH_3$ (Table 35). The comments in Sec. 5; IC2 apply to these results with regard to β-, γ-, δ-, ε-, ζ-, ϰ-, and θ-chloroalkyl radicals.

Results of the type discussed in the preceding paragraph are capable at most of giving an indication of the stability of a particular transition state relative to another transition state derived from the same substrate. For example, equal extraordinary stabilization of transition states leading to all monochlorosubstrates would be indistinguishable from extraordinary stabilization of none. Data of a type which can allow one to distinguish these possibilities have been obtained for $(CH_3)_3CCl$ (Table 37), $(CH_3)_2CHCH_2Cl$ (Table 37), $CH_3CH_2CH_2CH_2CH_2CH_2Cl$ (Table 38), $CH_3CH_2CH_2CH_2CHClCH_3$ (Table 38), $CH_3CH_2CH_2CHClCH_2CH_3$ (Table 38), and $CH_3CH_2CH_2CH_2CH_2CH_2CH_2CH_2Cl$ (Table 38). If these results, subject to the comments in Sec. 5, IC2, are taken at face value, they permit the conclusion that there is no extraordinary stabilization of the transition states derived from the chlorides in Tables 37 and 38.

4. Reaction of Alkyl Chlorides with $PhICl_2$

It should be noted that the mechanism of chlorination by $PhICl_2$ has not been established, even in outline; particular difficulties arise from the uncertainty regarding the nature of the hydrogen-abstracting radical and the heterogeneity of the reaction mixture. The products resulting from the chlorination of $CH_3CH_2CH_2CH_2Cl$ (Table 39) and $CH_3CH_2CH_2Cl$ (Table 40) have been reported. The comments in Sec. 5; IC2 apply to these results with regard to β-, γ-, and δ-chloroalkyl radicals.

TABLE 27

Reaction of $CH_3CH_2CH_2CH_2X$ with Cl_2

X	Reaction conditions	% Yield of reported monochlorinated substrate	Distribution of reported monochlorinated substrates (%)				Controls	Ref.
			1,1	1,2	1,3	1,4		
Br	35°, hν, gas[f]	?	8.0[g]	trace[g]	66.8[g]	25.2[g]	None	107
Br	60°, hν, liquid[a]	?	5.0[c]	21.8[b,e]	50.3[d]	22.9	None	102
Br	78°, hν, gas[f]	?	7.8[g]	trace[g]	64.9[g]	27.3[g]	None	107
Cl	0°, neat[l]	?	5.3[m]	23[m]	54[m]	17[m]	None	104
Cl	0°, CS$_2$(11.1 M)[l]	?	4.1[m]	20[m]	67[m]	8.3[m]	None	104
Cl	34°, neat[l]	?	6.6[m]	24[m]	51[m]	18[m]	None	104
Cl	34°, CS$_2$(11.1 M)[l]	?	5.5[m]	21[m]	63[m]	10.3[m]	None	104
Cl	35°, hν, gas[h]	?	8.1[i]	25.8[i]	48.6[i]	17.5[i]	None	107
Cl	68°, neat[l]	?	8.0[m]	24[m]	49[m]	20[m]	None	104
Cl	68°, CS$_2$(11.1 M)[l]	?	7.1[m]	23[m]	58[m]	12[m]	None	104
Cl	68°, CS$_2$(7.9 M)[l], hν	?	7.5[m]	23[m]	55[m]	14.4[m]	None	104
Cl	68°, hν, PhH(5.7 M)[l]	?	6.1[m]	24[m]	53[m]	16.9[m]	None	104
Cl	68°, hν, PhH(7.5 M)[l]	?	6.5[m]	24[m]	55[m]	14.6[m]	None	104
Cl	78°, hν, gas[h]	?	9.6[i]	26.1[i]	45.7[i]	18.6[i]	None	107
Cl	80°, liquid	32% HCl	4.7[n]	20.5[n]	52.2[n]	22.6[n]	o	110
Cl	146°, hν, gas[h]	?	9.6[i]	22.5[i]	47.1[i]	20.8[i]	None	107

Radical	Conditions							Ref.
Cl	200°, gas	?	5.4^n	19.0^n	48.0^n	27.6^n	o	110
Cl	202°, PbEt$_4$, gas^r	?	11.5^q	10.5^q	46.5^q	31.5^q	None	112
Cl	312°, PbEt$_4$, gas^r	?	17^q	0	43^q	40^q	None	112
Cl	380°, PbEt$_4$, gas^r	?	19^q	0	47^q	34^q	None	112
Cl	400°, gas	?	19.5^q	0.6^n	51.3^n	28.6^n	o	110
Cl$_2$	80°, liquid	?	1.6^p	11.2^p	47.6^p	39.6^p	None	110
CN	room temp., gas	90	0	75^u	25^u	0	None	116
CN	70°, liquid, hν^{am}	?	2^{ao}	21^{ap}	43^{aq}	34^m	an	513
CN	70°, 50% CCl$_4$, hν^{am}	?	2^{ao}	21^{ap}	43^{aq}	33^m	an	513
F	0°, hν, gas^j	?	9.7^i	20.8^k	52.5^k	17.0^i	None	107
F	35°, hν, gas^j	?	10.7^i	20.8^k	48.6^k	19.9^i	None	107
F	78°, gas^h	?	11.4^i	21.8^k	47.4^k	19.4^i	None	107
F	146°, gas^h	?	12.7^i	21.5^k	45.4^k	20.4^i	None	107
CF$_3$	0°, hν, gas^z	?	0.4^p	15.1^m	63.1^m	21.2^m	None	121
CF$_3$	20°, hν, gas^z	?	0.4^p	15.4^m	63.3^m	20.8^m	None	121
CF$_3$	60°, hν, gas^z	?	0.6^p	17.0^m	61.4^m	21.0^m	None	121
CF$_3$	230°, hν, gas^z	?	0.9^p	18.8^m	58.3^m	21.8^m	None	121
OCOCH$_3$	gas	75	0	0	71^v	29^w	None	116
OCOCH$_3$	0°, hν, MeCN^a	?	0.62^p	21.61^p	51.79^p	25.79^p	None	118
OCOCH$_3$	25°, hν, liquid^{ak}	?	5.5^m	22^m	45.3^m	27.2^m	al	508
OCOCH$_3$	40°, hν, MeCN^a	?	1.33^p	22.80^p	47.51^p	28.42^p	None	118
OCOCH$_3$	40°, hν, gas^x	?	0.88^t	27.27^p	53.28^p	18.58^t	None	119

X	Reaction conditions	% Yield of reported monochlorinated substrate	Distribution of reported monochlorinated substrates (%)				Controls	Ref.
			1,1	1,2	1,3	1,4		
$OCOCH_3$	55°, hν, liquid[ak]	?	5.3[m]	23.3[m]	45.9[m]	25.5[m]	al	508
$OCOCH_3$	100°, hν, gas[x]	?	1.54[t]	27.24[p]	50.70[p]	19.24[t]	None	119
$OCOCH_3$	145°, hν, gas[x]	?	0.83[t]	27.93[p]	50.01[p]	20.19[t]	None	119
$OCOCH_3$	160°, hν, gas[x]	?	1.35[t]	28.12[p]	49.92[p]	20.61[t]	None	119
$OCOCCl_3$	80°, liquid	48	–	5.1[q]	40.4[q]	54.5[q]	None	110
$OCOCF_3$	51°, hν, gas[x]	?	2.79[t]	19.51[p]	56.53[p]	21.18[t]	None	119
$OCOCF_3$	53°, hν, CCl_4 [a]	?	1.96[p]	17.04[p]	47.17[p]	33.84[p]	None	118
$OCOCF_3$	106°, hν, gas[x]	?	3.13[t]	19.93[p]	54.35[p]	22.58[t]	None	119
$OCOCF_3$	184°, hν, gas[x]	?	3.56[t]	20.20[p]	52.90[p]	23.35[t]	None	119
$OCOH$	40°, hν, gas[x]	?	–	19.43[p]	58.71[p]	19.89[t]	None	119
$OCOH$	98°, hν, gas[x]	?	–	20.02[p]	57.22[p]	20.19[t]	None	119
$OCOH$	150°, hν, gas[x]	?	–	18.64[p]	57.74[p]	22.07[t]	None	119
OCH_3	47°, hν, gas[x]	?	44[y]	5.9[p]	37[p]	13[t]	None	120
$COCl$	gas	?	5[q]	65	30[q]	0	None	116,117
$COCl$	20°, hν, liquid[ah]	?	1.5[t]	15–20[t]	50–55[t]	30–35[t]	None	129
$COCl$	20°, hν, PhH[aa]	?	"Very small"	17[q]	71[q]	12[q]	None	125
$COCl$	20°, hν, neat[aa]	?	"Very small"	19[q]	58[q]	23[q]	None	125
$COCl$	20°, hν, neat[ab]	ac	–	19[q]	58[q]	23[q]	None	126
$COCl$	20°, hν, 8.36 M PhH[ab]	ac	–	15[q]	76[q]	10[q]	None	126

COCl	50°, hv, MeCN[a]	?	2.73^p	18.77^p	47.29^p	31.23^p	None	118
COCl	52°, hv, CCl$_4$[a]	?	4.29^p	16.92^p	46.01^p	32.78^p	None	118
COCl	55–60°, hv, gas[x]	?	2.1^t	27^t	51^p	19^t	None	122
COCl	130°, hv, gas[x]	?	2.3^t	26^t	53^p	19^t	None	122
COF	65–70°, hv, gas[x]	?	1.1^t	21^t	57^p	20^t	None	122
COF	100–102°, hv, gas[x]	?	1.4^t	21^t	56^p	22^t	None	122
COF	130°, hv, gas[x]	?	1.6^t	23^t	56^p	20^t	None	122
COF	160°, hv, gas[x]	?	1.4^t	21^t	60^p	17^t	None	122
COOCH$_3$	gas	91	0	71^q	29^q	0	None	116,117
COOCH$_3$	55–60°, hv, gas[x]	?	5.4^t	31^t	45^p	19^t	None	122
COOCH$_3$	100–104°, hv, gas[x]	?	5.5^t	31^t	44^p	20^t	None	122
H	–10°, gas[s]	?	13^t	37^t	37^t	13^t	None	114
H	0°, gas[s]	?	13^t	37^t	37^t	13^t	None	107
H	10°, gas[s]	?	13^t	37^t	37^t	13^t	None	114
H	25°, gas[ai]	?	13	37	37	13	None	136
H	35°, gas[s]	?	14^t	36^t	36^t	14^t	None	107,114
H	78°, gas[s]	?	15^t	35^t	35^t	15^t	None	107,114
H	100°, gas[a]	≤75	16.4^t	33.6^t	33.6^t	33.6^t	aj	135,532
H	146°, gas[s]	?	16^t	34^t	34^t	16^t	None	107,114
H	200°, gas[a]	≤77	17.0^t	33.1^t	33.1^t	17.0^t	aj	135,532
H	210°, gas[ad]	64	22^t	28^t	28^t	22^t	None	128
H	240°, gas[ad]	64	22^t	28^t	28^t	22^t	None	128

X	Reaction conditions	% Yield of reported monochlorinated substrate	Distribution of reported monochlorinated substrates (%)				Controls	Ref.
			1,1	1,2	1,3	1,4		
H	240°, gas[ae]	80	21[t]	29[t]	29[t]	21[t]	None	128
H	240°, gas[af]	80	21[t]	29[t]	29[t]	21[t]	None	128
H	240°, gas[ag]	80	22[t]	27[t]	27[t]	22[t]	None	128
H	240°, gas[a]	78	23[t]	32[t]	32[t]	23[t]	None	128
H	270°, gas[ad]	64	22[t]	28[t]	28[t]	22[t]	None	128
H	270°, gas[a]	86	21[t]	29[t]	29[t]	21[t]	None	128
H	280°, gas[a]	≤83	19[t]	31[t]	31[t]	19[t]	aj	135,532
H	300°, gas[ad]	66	23[t]	27[t]	27[t]	23[t]	None	128
H	300°, gas[a]	90	21[t]	29[t]	29[t]	21[t]	None	128
H	330°, gas[ad]	62	21[t]	29[t]	29[t]	21[t]	None	128
H	330°, gas[a]	92	21[t]	29[t]	29[t]	21[t]	None	128
H	350°, gas[a]	≤ 88	20[t]	30[t]	30[t]	20[t]	aj	135,532
H	360°, gas[ad]	64	19[t]	31[t]	31[t]	19[t]	None	128
H	360°, gas[a]	90	20[t]	30[t]	30[t]	20[t]	None	128
H	390°, gas[ad]	62	19[t]	31[t]	31[t]	19[t]	None	128
H	390°, gas[ae]	76	18[t]	32[t]	32[t]	18[t]	None	128
H	390°, gas[af]	82	20[t]	30[t]	30[t]	20[t]	None	128
H	390°, gas[ag]	84	20[t]	30[t]	30[t]	20[t]	None	128

| H | 390°, gas[a] | 86 | 21[t] 24[t] | 29[t] 26[t] | 21[t] 24[t] | None | 128 |
| H | 450°, gas[a] | ≤85 | | 29[t] 26[t] | 21[t] 24[t] | aj | 135,532 |

[a] Substrate/Cl_2 = 5.

[b] "Approximately 5% of 1,2-dichlorobutane is also produced."

[c] Basis of assignment of structure: see footnote d to Table 22.

[d] Basis of assignment of structure: VPC retention time only slightly different from that of what was reported to be 1-chloro-3-bromobutane (see footnote c to Table 22).

[e] Basis of assignment of structure: as in d, except 1-chloro-2-bromobutane (see footnote e to Table 22).

[f] Substrate/Cl_2/N_2 = 1/1/50.

[g] Basis of assignment of structure: "When dihalides of known structure were available the procedure was to chromatograph the product and then add the individual dihalides one at a time to the reaction product and see which peak has been enlarged . . . However, in many cases synthesis of the dihalides would have been extremely tedious; for instance, no satisfactory synthesis of unsymmetrical 1,2-dihalides is known."

[h] Substrate/Cl_2/N_2 = 10/1/400.

[i] ". . . identified by the individual addition of the authentic dihalide to the reaction products." See footnote g.

[j] Substrate/Cl_2/N_2 = 10/1/100.

k Basis of assignment of structure: "Fluorination of 2-chlorobutane yielded a mixture, the [VP] chromatogram of which had two peaks which coincided with two of the peaks on the chromatogram from . . . [these] results."

l Substrate/Cl$_2$ = 5-10.

m Assignment of structure OK.

n Identified "by comparison with authentic dichlorides."

o The product mixture from a 25° photochlorination was heated to 400°; the proportion of the 1,2-isomer changed only very slightly.

p Basis of assignment of structure: none.

q Basis of assignment of structure not mentioned.

r Substrate/Cl$_2$/N$_2$ = 2/1/3.

s Substrate/Cl$_2$/N$_2$ = 15/1/180-360.

t Assignment of structure based on VPC retention time.

u Basis of assignment of structure: (1) acceptable N analysis; (2) assumed the same trends in density as in what were reported to be the chloropropionitriles (see footnote 1 to Table 28).

v Basis of assignment of structure: comparison of boiling point with a previously published value.

w Basis of assignment of structure: comparison of boiling point and index of refraction with previously published values.

^{x} Substrate/Cl$_2$ = 10/1.

^{y} Calculated from the amount of what was reported to be butyraldehyde (identified on the basis of its VPC retention time).

^{z} Substrate/Cl$_2$ = 6.

^{aa} Substrate/Cl$_2$ = 2.9.

^{ab} Substrate/Cl$_2$ = 10–20.

^{ac} "It was estimated that no more than 5% of higher chlorinated products was present . . ."

^{ad} Substrate/Cl$_2$ = 1.

^{ae} Substrate/Cl$_2$ = 2.

^{af} Substrate/Cl$_2$ = 3.

^{ag} Substrate/Cl$_2$ = 4.

^{ah} Substrate/Cl$_2$ < 3.

^{ai} Variation of the [RH] total/[Cl$_2$] . . . had no effect on the rate constant ratios, provided always that secondary chlorination of the products was avoided."

^{aj} "Nous avons vérifré que les dérivés halogénés primaires ne subissent pas d'isomérisation en dérivés secondaires aux températures atteintes dans nos halogénations et pendant des temps au moins double de la durée des essais."

^{ak} Conversion ≦ 40%.

al Amount of given isomer present varied linearly with percent conversion. Polychlorinated products not detected.

am 25% conversion.

an The percentages reported for each isomer were constant throughout the course of the reaction.

ao Basis of assignment of structure: prepared in 70% yield by reaction of $CH_3CH_2CH_2CH(OH)CN$ with PCl_5.

ap Basis of assignment of structure: Prepared in 70% yield by reaction of (cis + trans)-$CH_3CH_2CH=CHCN$ with HCl.

aq Basis of assignment of structure: prepared in 35% yield by reaction of $CH_3CHClCH_2CH_2Cl$ with NaCN in refluxing dimethoxyethane.

TABLE 28

Reaction of $CH_3CH_2CH_2X$ with Cl_2

X	Reaction conditions	% Yield of reported monochlorinated substrate	Distribution of reported monochlorinated substrates (%)			Controls	Ref.
			1,1	1,2	1,3		
Cl	-78°, γ-rays, liquid[i]	?	9[c]	65[c]	26[c]	None	81
Cl	-78°, hν, liquid[i]	?	9[c]	67[c]	24[c]	None	81
Cl	-78°, liquid	50	10[g]	60[g]	30[g]	None	113
Cl	20°, hν, liquid	?	15.8[c]	53.4[c]	30.8[c]	None	110
Cl	hν, liquid[az]	76-83[ay]	15[g]	59[g]	26[g]	None	507
Cl	25°, γ-rays, liquid[i]	?	9[c]	62[c]	29[c]	None	81
Cl	35°, hν, liquid[i]	?	12[c]	59[c]	29[c]	None	81
Cl	40°, hν, CCl₄	?	11.8[a]	57.6[b]	30.7[b]	None	103
Cl	46°, liquid	?	18.1[c]	49.7[c]	32.2[c]	None	110
Cl	100°, γ-rays, liquid[i]	?	10[c]	63[c]	27[c]	None	81
Cl	100°, liquid	51	20[g]	53[g]	27[g]	None	113
Cl	158°, PbEt₄, gas[h]	?	22[c]	46[c]	32[c]	None	112
Cl	160°, gas[d]	?	21.8[c]	49.7[c]	28.5[c]	None	110
Cl	160°, gas[e]	?	25.1[c]	50.2[c]	24.7[c]	None	110
Cl	260°, PbEt₄, gas[h]	?	27[c]	46[c]	27[c]	None	112
Cl	260°, gas[d]	?	46.7[c]	9.3[c]	44.0[c]	None	110

X	Reaction conditions	% Yield of reported monochlorinated substrate	Distribution of reported monochlorinated substrates (%)			Controls	Ref.
			1,1	1,2	1,3		
Cl	260°, gas[e]	?	38.9[c]	17.8[c]	43.3[c]	None	110
Cl	305°, gas[ba]	66[ay]	25.9[g]	43.7[g]	30.4[g]	None	507
Cl	340°, PbEt$_4$, gas[h]	?	35[c]	22[c]	43[c]	None	112
Cl	342°, gas[ba]	57[ay]	23.2[g]	39.8[g]	37.0[g]	None	507
Cl	360°, gas[d]	?	52.2[c]	4.4[c]	43.4[c]	None	110
Cl	360°, gas[e]	?	48.2[c]	6.6[c]	45.2[c]	None	110
Cl	380°, PbEt$_4$, gas[h]	?	49[c]	0	51[c]	None	112
Cl	400°, gas[d]	?	35.5[c]	4.8[c]	59.7[c]	None	110
Cl	400°, gas[e]	?	36.6[c]	4.7[c]	58.7[c]	None	110
CN	Gas, hν	95	0	51.1[c]	48.9[c]	None	123
CN	Room temp., gas	89	0	69[l]	31[l]	None	116
CN	40°, CCl$_4$	?	2.9[as]	56[as]	41[as]	None	132
CF$_3$	hν	?	–	44[w]	55[am]	None	130
OCOCH$_3$	Gas	85	0	69[n]	31[m]	None	116
OCOCH$_3$	28°, liquid[bb], hν	?	15[g]	46.4[g]	38.6[g]	bc	508
OCOCH$_3$	~80°, CCl$_4$[f]	88(wt.%)	23[g]	42[g]	35[g]	None	111
COCl	Gas, hν	91	6.2[p]	49[q]	44.8[p]	None	123
COCl	Gas	?	0	70[k]	30[j]	None	116,117
COCl	Gas	?	2-4[c]	~55[c]	40-45[c]	None	124

COCl	Liquid	?	$2\text{-}4^c$	50^c	$45\text{-}50^c$	None	124
COCl	$20°$, $h\nu$, PhH^t	?	"Very small"	74^c	26^c	None	125
COCl	$20°$, $h\nu$, neat^t	?	"Very small"	63^c	37^c	None	125
COCl	$20°$, $h\nu$, neat^u	v	3^c	63^c	33^c	None	126
COCl	$20°$, $h\nu$, 8.64 M PhH^u	v	2^c	77^c	21^c	None	126
COCl	$20°$, $h\nu$, liquid^t	?	2^{ax}	$50\text{-}55^{ax}$	$45\text{-}50^{ax}$	None	129
COOCH$_3$	$h\nu$, gas	95	0	57.8^r	42.2^s	None	123
COOCH$_3$	Gas	90	0	69^l	31^l	None	116,117
COOH	Neat	95	5^j	65^k	30^j	None	116,117
COOH	$h\nu$, liquid	92	7.8^o	52.4^o	39.8^j	None	123
COOH	γ-rays^{ap}	79	6^{ar}	61^{ar}	33^{ar}	None	131
COOH	γ-rays^{aq}	75	9^{ar}	56^{ar}	35^{ar}	None	131
H	$25°$, gas^{at}	?	22	56	22	au	136
H	$100°$, gas^{bd}	94	24.4^{aw}	51.2^{aw}	24.4^{aw}	av	135,532
H	$200°$, gas^{bd}	96	24.7^{aw}	50.7^{aw}	24.7^{aw}	av	135,532
H	$280°$, gas^{bd}	96	24.6^{aw}	50.9^{aw}	24.6^{aw}	av	135,532
H	$350°$, gas^{bd}	95	24.7^{aw}	50.6^{aw}	24.7^{aw}	av	135,532
H	$450°$, gas^{bd}	92	29.2^{aw}	41.6^{aw}	29.2^{aw}	av	135,532

[a] Basis of assignment of structure: position in order of VPC elution of products same as that found by Walling and Mayahi [104].

[b] Basis of assignment of structure: (1) see footnote a; (2) VPC retention times of three columns agreed with those of authentic material.

[c] Basis of assignment of structure not mentioned.

[d] Substrate/Cl_2 = 6.

[e] Substrate/Cl_2 = 2.

[f] Substrate/Cl_2/solvent = 2/1/1.

[g] Assignment of structure OK.

[h] Substrate/Cl_2/N_2 = 1/2/3.

[i] ~2 mole % Cl_2.

[j] Basis of assignment of structure: agreement of boiling point with a previously published value (paper cited does not report a boiling point).

[k] Basis of assignment of structure: (1) see j; (2) index of refraction agreed with that of an authentic compound.

[l] Basis of assignment of structure: (1) density agreed with a published value; (2) acceptable N and Cl analyses.

[m] Basis of assignment of structure: index of refraction and boiling point agreed with those of authentic material.

[n] Basis of assignment of structure: none.

[o] Basis of assignment of structure: boiling point agreed with that of authentic material.

[p] Basis of assignment of structure: boiling point agreed with a value published for material whose structure was not established.

[q] Basis of assignment of structure: (1) see footnote p; (2) index of refraction agreed with that of authentic material.

[r] Basis of assignment of structure: comparison of a boiling point (16 mm) with a previously published value (1 atm) (paper cited does not report a boiling point).

[s] Basis of assignment of structure: comparison of a boiling point (16 mm) with that of authentic material (1 atm).

[t] Substrate/Cl_2 = 3.

[u] Substrate/Cl_2 = 10-20.

[v] "It was estimated that no more than 5% of higher chlorinated products was present . . ."

[w] Basis of assignment of structure: (1) acceptable Cl and F analyses; (2) photochlorination yielded what was reported to be $CF_3CH_2CCl_2CH_3$.[x]

[x] Basis of assignment of structure: converted to what was reported to be $CF_3CH_2CF_2CH_3$.[y]

[y] Basis of assignment of structure: obtained from what was reported to be $CCl_3CH_2CF_2CH_3$.[z]

[z] Basis of assignment of structure: (1) obtained from what was reported to be $CH_3CF_2CH_2CHCl_2$;[aa] (2) " . . . shown to have a CCl_3 group by its easy fluorination to $CH_3CF_2CH_2CF_3$ . . . " (see footnote y); (3) Cl analysis acceptable.

aa Basis of assignment of structure: (1) acceptable Cl and F analyses; (2) obtained from what was reported to be $CH_3CF_2CH_2CH_2Cl$[ab]; (3) different from what were reported to be $ClCH_2CF_2CH_2CH_2Cl$[ac] and $CH_3CF_2CHClCH_2Cl$. [ad]

ab Basis of assignment of structure: (1) acceptable Cl and F analyses; (2) different from what were reported to be $CH_3CF_2CHClCH_3$[ae] and $ClCH_2CF_2CH_2CH_3$. [af]

ac Basis of assignment of structure: (1) Cl analysis acceptable; (2) obtained from what was reported to be $ClCH_2CF_2CH_2CH_3$[af]; (3) different from what were reported to be $ClCH_2CF_2CHClCH_3$[ag] and $CHCl_2CF_2CH_2CH_3$. [ah]

ad Basis of assignment of structure: (1) Cl analysis acceptable; (2) obtained from Cl_2 and what was reported to be $CH_3CF_2CH{=}CH_2$. [ai]

ae Basis of assignment of structure: (1) acceptable Cl and F analyses; (2) obtained from what was reported to be $CH_3CCl_2CHClCH_3$[aj]; (3) "The place of the F atoms was ascertained by perchlorination to the known $CCl_3CF_2CCl_2CCl_3$." ''

af Basis of assignment of structure: (1) acceptable Cl analysis; (2) obtained from what was reported to be $ClCH_2CCl_2CH_2CH_3$. [ak]

ag Basis of assignment of structure: (1) acceptable Cl analysis; (2) "The place of the F atoms was demonstrated by perchlorination to the known $CCl_3CF_2CCl_2CCl_3$"; (3) obtained from what were reported to be $ClCH_2CCl_2CHClCH_3$ and $ClCH_2CClFCHClCH_3$.

ah Basis of assignment of structure: (1) See parts 1 and 2 of footnote ag; (2) obtained from what was reported to be $CHCl_2CCl_2CH_2CH_3$. [al]

ai Basis of assignment of structure: obtained from what was reported to be $CH_3CF_2CH_2CH_2Cl$. ab

aj Basis of assignment of structure: obtained from Cl_2 and $CH_3CCl=CHCH_3$.

ak Basis of assignment of structure: obtained from Cl_2 and $CH_2=CClCH_2CH_3$.

al Basis of assignment of structure: obtained from Cl_2 and $CH_3CH_2C\equiv CH$.

am Basis of assignment of structure: (1) acceptable Cl and F analyses; (2) different from what were reported to be $CF_3CH_2CHClCH_3$ w and $CF_3CHClCH_2CH_3$. an

an Basis of assignment of structure: (1) acceptable Cl and F analyses; (2) different from both of the isomers formed in the chlorination of $CF_3CH_2CH_2CH_3$; (3) prepared from HgF_2/HF and what was reported to be $CF_2ClCHClCH_2CH_3$. ao

ao Basis of assignment of structure: prepared from Cl_2 and $CF_2=CHCH_2CH_3$.

ap Substrate/Cl_2 = 1.2.

aq Substrate/Cl_2 = 0.7.

ar "The relative concentrations of the [α, β, and γ] monochlorobutyric acids were determined from the i.r. spectrum, of the reaction mixture." (bands at 8.34, 9.82, and 8.71 μ, respectively)

as Basis of assignment of structure: NMR spectra.

at " . . . variation of the $[RH]_{total}/[Cl_2]$ . . . had no effect on the rate constant ratios, provided always that secondary chlorination of the products was avoided."

au "Decomposition of the products was . . . found to be important in some experiments. In the chlorination of . . . propane above 300° C the experimental values of . . . k_{sec}/k_{pri} were found to be lower than expected and to depend upon the duration of the experiment."

av "Nous avons vérifré que les dérivés halogénés primaires ne subissent pas d'isomérisation en dérivés secondaires aux températures atteintes dans nos halogénations et pendant des temps au moins doubles de la durée des essais."

aw Assignment of structure based on VPC retention time.

ax Boiling points reported.

ay Weight %.

az Substrate/Cl_2 = 1.3-2.0.

ba Substrate/Cl_2 = 1.

bb < 40% conversion.

bc Amount of each isomer present varied linearly with % conversion. Polychlorinated products not detected.

bd Substrate/Cl_2 = 4.5-5.5.

TABLE 29

Reaction of $CH_3CH_2CH_2CH_2CH_2X$ with Cl_2

X	Reaction conditions	% Yield of reported monochlorinated substrate	Distribution of reported monochlorinated substrates (%)					Controls	Ref.
			1,1	1,2	1,3	1,4	1,5		
Cl	70°, gas	?	7[a]	18[a]	31[a]	(31)[a]	13[a]	None	105
OCOCH₃	27°, hν, liquid[m]	?	3.4[h]	12.4[h]	30.7[h]	35.4[h]	18.1[h]	n	508
COCl	0°, hν, MeCN[d]	?	0.58[c]	8.22[c]	28.91[c]	40.31[c]	21.98[c]	None	118
COCl	20°, hν, PhH[e]	?	"Very small"	4.5[a]	20[a]	71[a]	4.7[a]	None	125
COCl	20°, hν, neat[e]	?	"Very small"	8.7[a]	31[a]	44[a]	15.3[a]	None	125
COCl	20°, hν, neat[f]	g	-	9[a]	31[a]	44[a]	16[a]	None	126
COCl	20°, hν, 8.08 M PhH[f]	g	-	5[a]	35[a]	53[a]	7[a]	None	126
COCl	20°, hν, liquid	?	1[a]	5-10[h]	25-30[i]	40-45[h]	20-25[h]	None	129
COCl	52°, hν, CCl₄[d]	?	2.55[c]	10.03[c]	26.88[c]	34.38[c]	25.81[c]	None	118
COCl	52°, hν, MeCN[d]	?	2.70[c]	8.60[c]	27.16[c]	36.60[c]	24.97[c]	None	118
COF	50°, hν, gas[b]	?	1.35[c]	12.56[c]	34.84[c]	38.84[c]	12.89[c]	None	118
COF	105°, hν, gas[b]	?	2.54[c]	13.63[c]	31.98[c]	37.67[c]	14.19[c]	None	118
H	-83°, hν, liquid	1	7.9[a]	26[a]	31[a]	26[a]	7.9[a]	None	506
H	-70°, hν, CS₂	1	1.1[a]	25[a]	47[a]	25[a]	1.1[a]	None	506
H	-40°, hν, liquid[p]	?	12.4[h]	23.6[h]	28.2[h]	23.6[h]	12.4[h]	None	516
H	-20°, hν, liquid[p]	?	13.7[h]	21.7[h]	29.3[h]	21.7[h]	13.7[h]	None	516
H	-20°, hν, CS₂[o]	?	2.8[h]	27.2[h]	40.3[h]	27.2[h]	2.8[h]	None	516
H	0°, hν, liquid[p]	?	12.8[h]	23.8[h]	26.9[h]	23.8[h]	12.8[h]	None	516
H	0°, hν, CCl₄[o]	?	13.4[h]	23.4[h]	26.6[h]	23.4[h]	13.4[h]	None	516

X	Reaction conditions	% Yield of reported monochlorinated substrate	Distribution of reported monochlorinated substrates (%)					Controls	Ref.
			1,1	1,2	1,3	1,4	1,5		
H	0°, hν, PhCH$_3$ [o]	?	5.2[h]	29.9[h]	30.2[h]	29.9[h]	5.2[h]	None	516
H	0°, hν, PhH [o]	?	3.0[h]	29.9[h]	34.4[h]	29.9[h]	3.0[h]	None	516
H	0°, hν, CS$_2$ [q]	?	2.7[h]	28.0[h]	38.8[h]	28.0[h]	2.7[h]	None	516
H	0°, hν, CS$_2$ [o]	?	3.8[h]	26.9[h]	38.7[h]	26.9[h]	3.8[h]	None	516
H	0°, hν, CS$_2$ [r]	?	4.5[h]	27.2[h]	36.6[h]	27.2[h]	4.5[h]	None	516
H	0°, hν, CS$_2$ [s]	?	6.7[h]	26.4[h]	33.9[h]	26.4[h]	6.7[h]	None	516
H	10°, hν, CS$_2$ [o]	?	4.0[h]	27.7[h]	36.7[h]	27.7[h]	4.0[h]	None	516
H	10°, hν, PhH [o]	?	3.0[h]	29.9[h]	34.4[h]	29.9[h]	3.0[h]	None	516
H	20°, hν, liquid [p]	?	13.4[h]	22.8[h]	27.6[h]	22.8	13.4[h]	None	516
H	20°, hν, liquid	?	9.6[a]	26[a]	28[a]	26[a]	9.6[a]	None	506
H	20°, hν, PhH [o]	?	4.4[h]	28.8[h]	33.7[h]	28.8[h]	4.4[h]	None	516
H	20°, hν, CS$_2$	?	3.1[a]	28[a]	38[a]	28[a]	3.1[a]	None	506
H	20°, hν, CS$_2$ [q]	?	4.0[h]	27.2[h]	37.7[h]	27.2[h]	4.0[h]	None	516
H	20°, hν, CS$_2$ [o]	?	4.4[h]	27.9[h]	35.4[h]	27.9[h]	4.4[h]	None	516
H	20°, hν, CS$_2$ [r]	?	5.5[h]	27.2[h]	34.7[h]	27.2[h]	5.5[h]	None	516
H	20°, hν, CS$_2$ [s]	?	8.0[h]	26.4[h]	31.3[h]	26.4[h]	8.0[h]	None	516
H	33°, hν, PhH [o]	?	5.5[h]	27.3[h]	34.5[h]	27.3[h]	5.5[h]	None	516
H	33°, hν, CS$_2$ [o]	?	6.8[h]	26.9[h]	32.6[h]	26.9[h]	6.8[h]	None	516
H	50°, hν, PhH [o]	?	5.8[h]	27.8[h]	32.9[h]	27.8[h]	5.8[h]	None	516
H	100°, gas [d]	86-90	12.9[j]	23.4[j]	27.6[j]	23.4[j]	12.9[j]	k	135, 532
H	200°, gas [d]	86-90	13.0[j]	23.4[j]	27.3[j]	23.4[j]	13.0[j]	k	135, 532

86

H	250°, gas^d	86–90	13.8^j	23.4^j	25.7^j	23.4^j	13.8^j	k	135,532
H	300°, gas^d	86–90	14.1^j	22.7^j	26.5^j	22.7^j	14.1^j	k	532
H	350°, gas^d	86–90	14.0^j	23.7^j	24.7^j	23.7^j	14.0^j	k	135,532
H	450°, gas^d	86–90	18.4^j	20.6^j	22.1^j	20.6^j	18.4^j	k	135,532

^{a} Basis of assignment of structure not mentioned.

^{b} Substrate/Cl_2 = 8 (20 mm).

^{c} Basis of assignment of structure: none.

^{d} Substrate/Cl_2 = 5.

^{e} Substrate/Cl_2 = 2.9.

^{f} Substrate/Cl_2 = 10–20.

^{g} "It was estimated that not more than 5% of higher chlorinated products was present . . ."

^{h} Assignment of structure OK.

^{i} Basis of assignment of structure: p-toluidide gave acceptable C and H analyses.

^{j} Assignment of structure based on VPC retention time.

^{k} "Nous avons vérifié que les dérivés halogénés primaires ne subissent pas d'isomérisation en dérivés secondaires aux températures atteintes dans nos halogénations et pendant des temps au moins doubles de la durée des essais."

l "Di- and higher chlorinated products being almost absent . . ."

m Conversion $\leqq 40\%$.

n Amount of each isomer present varied linearly with % conversion. Polychlorinated products not detected.

o Solvent/substrate/Cl_2 = 16/4/1.

p Substrate/Cl_2 = 4.

q Solvent/substrate/Cl_2 = 40/4/1.

r Solvent/substrate/Cl_2 = 12/4/1.

s Solvent/substrate/Cl_2 = 4/4/1.

TABLE 30

Reaction of $(CH_3)_2CHCH_2Cl$ with Cl_2

X	Reaction conditions	% Yield of reported monochlorinated substrate	Distribution of reported monochlorinated substrates (%)			Controls	Ref.
			1,1	1,2	1,3		
Br	Room temp., hν, CCl₄	?	–	59[c,d]	33[c,e]	None	133
Cl	24°, CCl₄[a]	?	13[b]	39[b]	49[b]	None	106
CN	0°, hν, 8 M in PhH[l]	?	1[n]	83.8[o]	15.2[p]	None	513
CN	30°, hν, liquid[l]	?	3.5[n]	44.4[o]	52.1[p]	m	513
CN	30°, hν, 50% CCl₄[l]	?	4[n]	44[o]	52[p]	m	513
CN	30°, hν, 4 M in PhH[l]	?	2.5[n]	64.7[o]	32.8[p]	None	513
CN	30°, hν, 8 M in PhH[l]	?	1.6[n]	76.5[o]	21.9[p]	None	513
CN	30°, hν, 10 M in PhH[l]	?	1[n]	85.4[o]	13.6[p]	None	513
CN	30°, hν, 8 M in PhCl[l]	?	2[n]	64[o]	34[p]	None	513
CN	30°, hν, 8 M in PhNO₂[l]	?	3[n]	55[o]	42[p]	None	513
CN	30°, hν, 8 M in CS₂[l]	?	1[n]	83[o]	16[p]	None	513
CN	60°, hν, 8 M in PhH[l]	?	1[n]	83.8[o]	15.2[p]	None	513
OCOCH₃	24°, CCl₄[a]	75	0	51[b]	49[b]	None	116
OCOCH₃	25°, hν, liquid[j]	?	15.5[e]	35.7[e]	48.8[e]	k	508
OCOCH₃	85°, hν, liquid[j]	?	17.9[e]	34.9[e]	47.2[e]	k	508
OCOCH₃	95°, hν, liquid[j]	?	18.5[e]	33.4[e]	48.5[e]	k	508

X	Reaction conditions	% Yield of reported monochlorinated substrate	Distribution of reported monochlorinated substrates (%)			Controls	Ref.
			1,1	1,2	1,3		
COCl	24°, CCl$_4$ [a]	?	32[b]	60[b]	8[b]	None	116
COOCH$_3$	24°, CCl$_4$ [a]	?	49[b]	49[b]	2[b]	None	116
COOH	24°, CCl$_4$ [a]	?	26[b]	69[b]	5[b]	None	116
H	−20°, hν, CS$_2$ [q]	?	5.9[e]	82.2[e]	11.9[e]	None	516
H	−15°, hν, liquid [g]	?	22[f]	33[f]	44[f]	None	134
H	−15°, hν, 5.9 M PhCl [g]	?	8[f]	76[f]	16[f]	None	134
H	−10°, hν, liquid [r]	?	24.5[e]	26.4[e]	49.1[e]	None	516
H	−10°, hν, CCl$_4$ [q]	?	24.7[e]	26.0[e]	49.3[e]	None	516
H	−10°, hν, CS$_2$ [q]	?	6.7[e]	79.9[e]	13.4[e]	None	516
H	0°, hν, CCl$_4$ [q]	?	24.3[e]	27.2[e]	48.5[e]	None	516
H	0°, hν, PhH [q]	?	6.2[e]	81.5[e]	12.3[e]	None	516
H	0°, hν, PhCH$_3$ [q]	?	6.4[e]	80.7[e]	12.9[e]	None	516
H	0°, hν, CS$_2$ [s]	?	3.3[e]	90.0[e]	6.7[e]	None	516
H	0°, hν, CS$_2$ [q]	?	7.9[e]	76.2[e]	15.9[e]	None	516
H	0°, hν, CS$_2$ [t]	?	8.0[e]	75.9[e]	16.1[e]	None	516
H	0°, hν, CS$_2$ [u]	?	9.4[e]	71.8[e]	18.8[e]	None	516
H	10°, hν, CS$_2$ [q]	?	8.5[e]	74.4[e]	17.1[e]	None	516
H	24°, CCl$_4$ [a]	?	24[b]	28[b]	48[b]	None	106
H	25°, gas [h]	?	27	46	55	i	136

[a] 2.5% conversion.

[b] Basis of assignment of structure not mentioned.

[c] + 8% $CH_3CBr(CH_3)CH_2Cl$.

[d] Basis of assignment of structure: VPC retention time different and NMR spectrum similar to, but different from, that of $CH_3CBr(CH_3)CH_2Cl$.

[e] Assignment of structure OK.

[f] Assignment of structure based on VPC retention time.

[g] ~0.005 mole of Cl_2/25 ml. of reaction mixture.

[h] ". . . . variation of the $[RH]_{total}/[Cl_2]$. . . had no effect on the rate constant ratios, provided always that secondary chlorination of the products was avoided."

[i] ". . . . in the presence of a tenfold excess of HCl over Cl_2 no change was observed in the rate constant ratios . . . at 0°C and 100°C. We have therefore assumed that the back reaction was negligible $[HCl + R \cdot \rightarrow RH + Cl \cdot]$." "Decomposition of the products was . . . found to be important in some experiments . . . at temperatures above 100°C . . . the experimental values of k_{tert}/k_{pri} . . . were found to be lower than expected and to depend upon the duration of the experiment . . . tert-butyl chloride was found to have almost completely decomposed into isobutene and HCl after 30 min . . . at 194°C."

[j] Conversion ≅ 40%.

[k] Amount of each isomer present varied linearly with % conversion. Polychlorinated products not detected.

^{l} 20% conversion.

^{m} The reported percentages of isomers did not change throughout the reaction. Use of a solvent saturated with HCl gave the same results. Formation of dichloride detected only after 80-85% conversion.

^{n} Basis of assignment of structure: prepared from $(CH_3)_2CHCH(OH)CN$ and PCl_5 in unspecified yield.

^{o} Basis of assignment of structure: (1) prepared in 70% yield from $(CH_3)_2C=CHCN$ and HCl; (2) reaction with refluxing 10% NaOH gave $(CH_3)_2C=CHCN$ in unspecified yield.

^{p} Basis of assignment of structure: none.

^{q} Solvent/substrate/Cl_2 = 16/4/1.

^{r} Substrate/Cl_2 = 4.

^{s} Solvent/substrate/Cl_2 = 40/4/1.

^{t} Solvent/substrate/Cl_2 = 12/4/1.

^{u} Solvent/substrate/Cl_2 = 4/4/1.

TABLE 31

Reaction of $CH_3CH_2CHXCH_3$ with Cl_2

X	Reaction conditions	% Yield of reported monochlorinated substrate	Distribution of reported monochlorinated substrates (%) [†]					Ref.
			1,2	2,2	2,3 erythro(meso)	2,3 threo(dl)	1,3	
Cl	35°, gas[a]	?	4.1[c]	26.3[c]	36.4[c]	14.6[c]	18.6[c]	108
Cl	78°, gas[a]	?	4.3[c]	26.2[c]	35.0[c]	14.8[c]	19.7[c]	108
Cl	200°, Et$_4$Pb, gas[f]	?	0[g]	23[g]	38[g]		39[g]	112
Cl	314°, Et$_4$Pb, gas[f]	?	0[g]	31[g]	22[g]		46[g]	112
Cl	380°, Et$_4$Pb, gas[f]	?	0[g]	40[g]	0		60[g]	112
F	35°, gas[b]	?		35.6[e]	24.7[d]	16.9[d]		108
F	78°, gas[b]	?		33.0[e]	24.2[d]	16.4[d]		108
OCOCH$_3$		70	33.3	0	66		0	116
H			See Table 27					

[†] Controls: none.

[a] Substrate/Cl$_2$/N$_2$ = 11/1/360.

[b] Substrate/Cl$_2$/N$_2$ = 10/1/400.

[c] Basis of assignment of structure: see footnote i of Table 27.

d Basis of assignment of structure: (1) VPC retention times same as those of the products of the reaction of $CH_3CHClCHClCH_3$ with HgO and HF. "It was assumed that the erythro-derivatives were eluted first by analogy with the 2,3-dichlorobutanes."; (2) "by comparing the infrared spectra of the components of the product with those of authentic materials [see (1)]"; (3) assignment of structure, exclusive of stereochemistry, OK.

e Basis of assignment of structure: (1) observation of VP chromatogram of the reaction mixture to which "authentic material" had been added; (2) "by comparing the infrared spectra of the components of the product with those of authentic materials."

f Substrate/Cl_2/N_2 = 2/1/3.

g Basis of assignment of structure not mentioned.

TABLE 32

Reaction of $CH_3CH_2CH_2CH_2CH_2CH_2X$ with Cl_2

X	Reaction conditions	% Yield of reported monochlorinated substrate	Distribution of reported monochlorinated substrates (%)						Controls	Ref.
			1,1	1,2	1,3	1,4	1,5	1,6		
Cl	20°, liquid	?	2.14[c]	11.1[c]	11.6[c]	29.3[c]	28.0[c]	12.7[c]	None	109
Cl	80°, liquid	?	3.0[d]	8.3[d]	19.3[d]	27.3[d]	27.8[d]	14.3[d]	None	110
Cl	260°, gas	?	6.7[d]	2.5[d]	19.9[d]	26.6[d]	30.3[d]	14.0[d]	None	110
Cl	400°, gas	?	9.9[d]	1.4[d]	20.7[d]	27.6[d]	32.8[d]	7.6[d]	None	110
Cl_2		?	0[e]	4.5[e]	10.8[e]	30.6[e]	36.6[e]	17.5[e]	None	110
$OCOCCl_3$	80°, liquid	?	–	–	2.6[b]	31.4[b]	44.4[b]	21.6[b]	None	110
COCl	52°, hν, MeCN[f]	?	1.06[e]	7.69[e]	16.53[e]	20.90[e]	25.92[e]	27.92[e]	None	118
COF	60°, hν, gas[f]	?	1.27[e]	10.83[e]	25.38[e]	25.30[e]	27.49[e]	9.24[e]	None	118
COF	105°, hν, gas[f]	?	1.39[e]	10.24[e]	25.30[e]	25.43[e]	27.61[e]	10.02[e]	None	118
H	20°, liquid[a]	?	11.4[b]	18.7[b]	20[b]	20[b]	18.7[b]	11.4[b]	None	109
H	100°, gas[f]	92–96	11.5[h]	20.7[h]	17.9[h]	17.9[h]	20.7[h]	11.5[h]	g	135, 532
H	200°, gas[f]	92–96	11.8[h]	20.5[h]	17.7[h]	17.7[h]	20.5[h]	11.8[h]	g	135, 532
H	250°, gas[f]	92–96	12.3[h]	20.6[h]	17.2[h]	17.2[h]	20.6[h]	12.3[h]	g	135, 532

X	Reaction conditions	% Yield of reported monochlorinated substrate	Distribution of reported monochlorinated substrates (%)						Controls	Ref.
			1,1	1,2	1,3	1,4	1,5	1,6		
H	300°, gas[f]	92-96	11.6[h]	21.9[h]	16.6[h]	16.6[h]	21.9[h]	11.6[h]	g	135, 532
H	350°, gas[f]	92-96	10.4[h]	23.3[h]	16.4[h]	16.4[h]	23.3[h]	10.4[h]	g	135, 532

[a] $< 5\%$ conversion.

[b] Basis of assignment of structure not mentioned.

[c] Identified "by means of NMR analysis" of samples collected from a VPC.

[d] Identified "by comparison with authentic dichlorides."

[e] Basis of assignment of structure: none.

[f] Substrate/$Cl_2 = 5$.

[g] "Nous avons vérifré que les dérivés halogénes primaires ne subissent pas d'isomérisation en dérivés secondaires aux températures atteintes dans nos halogénations et pendant des temps au moins doubles de la durée des essais."

[h] Assignment of structure based on VPC retention time.

TABLE 33

Reaction of $CH_3CH_2CH_2CH_2CH_2CH_2CH_2CH_2X$ with Cl_2

X	Reaction conditions	Distribution of reported monochlorinated substrates (%)								Controls	Ref.
		1,1	1,2	1,3	1,4	1,5	1,6	1,7	1,8		
Cl	20°, liquid	1.75[c]	5.28[c]	11.5[c]	18.4[c]	17.6[c]	17.7[c]	16.9[c]	11.0[c]	None	109
Cl	80°, liquid	2.8[b]	6.7[b]	12.2[b]	16.0[b]	16.9[b]	15.7[b]	19.6[b]	10.1[b]	None	110
Cl	250°, gas	3.9[b]	2.0[b]	12.6[b]	16.2[b]	17.5[b]	37.3[b]		10.5[b]	None	110
Cl	350°, gas	4.5[b]	0[b]	13.2[b]	21.3[b]	18.3[b]	38.2[b]		4.5[b]	None	110
Cl	450°, gas	4.7[b]	0[b]	12.1[b]	20.7[b]	18.6[b]	36.8[b]		7.1[b]	None	110
H	−20°, liquid[d], hν	14.3[b]	15.4[b]	14.4[b]	13.0[b]	13.0[b]	14.4[b]	15.4[b]	14.3[b]	None	504
H	20°, liquid[a]	14.3[b]	13.4[b]	13.7[b]	13.6[b]	13.6[b]	13.7[b]	13.4[b]	14.3[b]	None	109
H	20°, liquid[d], hν	14.3[b]	15.2[b]	14.2[b]	13.4[b]	13.4[b]	14.2[b]	15.2[b]	14.3[b]	None	504
H	20°, CCl$_4$,[e] hν	13.4[b]	16.0[b]	14.0[b]	13.3[b]	13.3[b]	14.0[b]	16.0[b]	13.4[b]	None	504
H	20°, benzene,[e] hν	9.5[b]	15.5[b]	15.9[b]	13.8[b]	13.8[b]	15.9[b]	15.5[b]	9.5[b]	None	504
H	20°, hν, CS$_2$[e]	7.0[b]	16.0[b]	16.5[b]	14.0[b]	14.0[b]	16.5[b]	16.0[b]	7.0[b]	None	504

[a] < 5% conversion.

[b] Basis of assignment of structure not mentioned.

[c] Identified "by means of NMR analysis" of samples collected from a VPC.

[d] Substrate/Cl_2 = 10.

[e] Substrate/Cl_2 = 5.

TABLE 34

Reaction of 2-Halohexanes with Cl_2

Substrate	Reaction conditions	Distribution of reported monochlorinated substrates (%)						Controls	Ref.
		1,2	1,5	2,2	2,3	2,4	2,5		
$CH_3CH_2CH_2CH_2CHMe-Cl$	20°, liquid	1.76[c]	15.4[c]	3.2[c]	20.2[a]	26.2[b]	33.0[a]	None	109
$CH_3CH_2CH_2CH_2CH(CH_2Cl)-Cl$		5.6[c]	20.0[c]	5.6[c]	12.4[c]	12.5[c]	43.9[c]	None	110
$CH_3CH_2CH_2CH_2CHMe-H$	20°, liquid[d]	11.4[c]	11.4[c]	18.7[c]	20[c]	20[c]	18.7[c]	None	109
$CH_3CH_2CH_2CH_2CHMe-H$	100°, gas	11.5[a]	11.5[a]	20.7[a]	17.9[a]	17.9[a]	20.7[a]	e	135
$CH_3CH_2CH_2CH_2CHMe-H$	200°, gas	11.8[a]	11.8[a]	20.5[a]	17.7[a]	17.7[a]	20.5[a]	e	135
$CH_3CH_2CH_2CH_2CHMe-H$	250°, gas	12.3[a]	12.3[a]	20.6[a]	17.2[a]	17.2[a]	20.6[a]	e	135
$CH_3CH_2CH_2CH_2CHMe-H$	300°, gas	11.6[a]	11.6[a]	21.9[a]	16.6[a]	16.6[a]	21.9[a]	e	135
$CH_3CH_2CH_2CH_2CHMe-H$	350°, gas	10.4[a]	10.4[a]	23.3[a]	16.4[a]	16.4[a]	23.3[a]	e	135

[a] Identified by comparison of VPC retention time with that of an "authentic" compound.

[b] Identified "by 'cross-checking' products from 2- and 3-chlorohexane, since both reactants give the 2,4-isomer . . . "

[c] Basis of assignment of structure not mentioned.

[d] < 5% conversion.

[e] See footnote g to Table 32.

TABLE 35

Reaction of $CH_3CH_2CHXCH_2CH_2CH_3$ with Cl_2

X	Reaction conditions	Distribution of reported monochlorinated substrates (%)						Controls	Ref.
		1,3	1,4	2,3	2,4	3,3	3,4		
Cl	20°, liquid	9.1[c]	15.8[c]	21.8[a]	<29.4[b]	7[c]	>17[a]	None	109
H	20°, liquid[d]	11.4[c]	11.4[c]	18.7[c]	18.7[c]	20[c]	20[c]	None	109
H	100°, gas	11.5[a]	11.5[a]	20.7[a]	20.7[a]	17.9[a]	17.9[a]	e	135
H	200°, gas	11.8[a]	11.8[a]	20.5[a]	20.5[a]	17.7[a]	17.7[a]	e	135
H	250°, gas	12.3[a]	12.3[a]	20.6[a]	20.6[a]	17.2[a]	17.2[a]	e	135
H	300°, gas	11.6[a]	11.6[a]	21.9[a]	21.9[a]	16.6[a]	16.6[a]	e	135
H	350°, gas	10.4[a]	10.4[a]	23.3[a]	23.3[a]	16.4[a]	16.4[a]	e	135

[a] Identified by comparison of VPC retention time with that of an "authentic" compound.

[b] Identified "by 'cross-checking' products from 2- and 3-chlorohexane, since both reactants give the 2,4-isomer . . . "

[c] Basis of assignment of structure not mentioned.

[d] <5% conversion.

[e] See footnote g to Table 32.

TABLE 36

Reaction of $CH_3CH_2CH_2CH_2CH_2CH_2X$ with Cl_2

X	Reaction conditions	% Yield of reported monochlorinated substrates	Distribution of reported monochlorinated substrates (%)							Controls	Ref.
			1,1	1,2	1,3	1,4	1,5	1,6	1,7		
COCl	20°, hν, neat[a]	b	–	5[c]	15[c]	20[c]	25[c]	26[c]	10[c]	None	126
COCl	20°, hν, 7.64 M PhH[a]	b	–	2[c]	13[c]	22[c]	32[c]	29[c]	3.4[c]	None	126
H	15°, liquid[e]	?	9.0[c]	17.2[c]	16.0[c]	15.8[c]	16.0[c]	17.2[c]	9.0[c]	None	127
H	20°, hν, neat[a]	d	7[c]	18[c]	17[c]	16[c]	17[c]	18[c]	7[c]	None	126
H	20°, hν, PhH[a]	d	2.1[c]	17[c]	21[c]	20[c]	21[c]	17[c]	2.1[c]	None	126
H	40°, liquid[e]	?	9.0[c]	17.2[c]	16.0[c]	15.8[c]	16.0[c]	17.2[c]	9.0[c]	None	127
H	80°, liquid[e]	?	12.7[c]	16.4[c]	15.0[c]	13.9[c]	15.0[c]	16.4[c]	12.7[c]	None	127
H	98°, gas[e]	?	10.1[c]	17.8[c]	15.5[c]	13.5[c]	15.5[c]	17.8[c]	10.1[c]	None	127
H	100°, gas	?	10.2[f]	17.8[f]	15.5[f]	13.5[f]	15.5[f]	17.8[f]	10.2[f]	g	135
H	150°, gas[e]	?	10.2[c]	17.6[c]	15.6[c]	13.3[c]	15.6[c]	17.6[c]	10.2[c]	None	127
H	200°, gas[e]	?	10.3[c]	18.0[c]	15.0[c]	13.4[c]	15.0[c]	18.0[c]	10.3[c]	g	127, 135
H	260°, gas[e]	?	11.5[c]	16.7[c]	15.0[c]	13.7[c]	15.0[c]	16.7[c]	11.5[c]	g	127, 135
H	350°, gas	?	8.6[f]	18.5[f]	16.8[f]	12.5[f]	16.8[f]	18.5[f]	8.6[f]	g	135
H	450°, gas	?	7.1[f]	19.1[f]	17.7[f]	12.2[f]	17.7[f]	19.1[f]	7.1[f]	g	135

[a] Substrate/Cl_2 = 10–20.

[b] "It was estimated that no more than 5% of higher chlorinated products was present . . ."

[c] Basis of assignment of structure not mentioned.

[d] "No more than 4% of the total weight of chlorination products consisted of higher chlorinated compounds."

[e] Substrate/Cl_2 = 5.

[f] Assignment of structure based on VPC retention time.

[g] See footnote g to Table 32.

TABLE 37

Reactivity Ratios for Photochlorination[a] of $(CH_3)_3CH$, $(CH_3)_3CCl$, and $(CH_3)_2CHCH_2Cl$ [106]

Reported reaction pair	Reported rate constant ratio
$(CH_3)_2CHCH_2Cl \xrightarrow{k_1} (CH_3)_2CClCH_2Cl$ $(CH_3)_3CH \xrightarrow{k_2} (CH_3)_3CCl$	$k_1/k_2 = 0.75$
$(CH_3)_3CCl \xrightarrow{k_3} (CH_3)_2CClCH_2Cl$ $(CH_3)_3CH \xrightarrow{k_4} (CH_3)_2CHCH_2Cl$	$k_3/k_4 = 0.24$

[a] 2.5% reaction, CCl_4, 24°.

5. Reaction of Alkyl Chlorides with Cl_2O and HOCl

It should be noted that the mechanisms of these chlorinations have not been established. The products resulting from the chlorination of $CH_3CH_2CH_2CH_2Cl$ (Table 41) and $CH_3CH_2CH_2Cl$ (Table 42) have been reported. The comments in Sec. 5;IC2 apply to these results with regard to β-, γ-, δ-chloroalkyl radicals.

TABLE 38

Relative Reactivities of Alkanes and Haloalkanes[a] [109]

Reactants	Reported reactivities (mole/mole)
$CH_3CH_2CH_2CH_2CH_2-Cl/CH_3CH_2CH_2CH_2CH_2CH_2-H$	0.55
$CH_3CH_2CH_2CH_2CHMe-Cl/CH_3CH_2CH_2CH_2CHMe-H$	0.33
$CH_3CH_2CH_2CHEt-Cl/CH_3CH_2CH_2CHEt-H$	0.24
$CH_3CH_2CH_2CH_2CH_2CH_2-Cl/CH_3CH_2CH_2CH_2CH_2CH_2CH_2-H$	0.74

[a] Liquid, 20°.

L. KAPLAN

TABLE 39

Reaction of $CH_3CH_2CH_2CH_2X$ with $PhICl_2$

X	Reaction conditions	% Yield of reported monochlorinated substrate	Distribution of reported monochlorinated substrates (%)				Controls	Ref.
			1,1	1,2	1,3	1,4		
Cl	0°, hν, CCl_4	?	6.2[b]	22.4[b]	50.9[b]	20.9[b]	None	137
Cl	40°, hν, CCl_4[a]	52-57	5.2[b]	24.0[b]	55.9[b]	15.4[b]	None	137
Cl	40°, hν, neat	?	5.0[b]	21.3[b]	61.6[b]	12.2[b]	None	137
Cl	40°, hν, 50% CCl_4	?	Trace[b]	20.2[b]	65.8[b]	14.2[b]	None	137
H	40°, hν, CCl_4	?	3.3[b]	47[c]	47[c]	3.3[b]	None	137

[a] Substrate/$PhICl_2$ = 10.

[b] Basis of assignment of structure: VPC retention time on one column.

[c] Basis of assignment of structure: none.

TABLE 40

Reaction of $CH_3CH_2CH_2X$ with $PhICl_2$

X	Reaction conditions	% Yield of reported monochlorinated substrate	Distribution of reported monochlorinated substrates (%)			Controls	Ref.
			1,1	1,2	1,3		
Cl	40°, hν, CCl$_4$	?	11.8[a]	57.6[a]	30.7[a]	None	137
CN	40°, hν, neat	?	0	57.0[b]	43.0[b]	None	137

[a] Basis of assignment of structure: two of three peaks by VPC retention time on three columns; third by elimination.

[b] Basis of assignment of structure: not mentioned.

TABLE 41

Reaction of $CH_3CH_2CH_2CH_2X$ with Cl_2O and HOCl

X	Reaction conditions	% Yield of reported monochlorinated substrate	Distribution of reported monochlorinated substrates (%)				Controls	Ref.
			1,1	1,2	1,3	1,4		
Cl	Cl_2O, 40°, hν, CCl_4[a]	99	21.6[b]	19.0[b]	50.8[b]	8.6[b]	Table 43	103
Cl	Cl_2O, 40°, hν, CCl_4/H_2O (100/1)	?	17.4[b]	21.0[b]	47.0[b]	14.6[b]	Table 43	103
H	Cl_2O	?	5.8[c]	44[c]	44[c]	5.8[c]	None	103
Cl	HOCl, 40°, hν, CCl_4/H_2O (10/1)	?	7.8[b]	21.7[b]	47.1[b]	23.4[b]	None	103
Cl	HOCl, 40°, hν, CCl_4/H_2O (1/1)	?	7.0[b]	23.2[b]	43.8[b]	26.0[b]	None	103
H	HOCl, 40°, aqueous	?	14[c]	36[c]	36[c]	14[c]	None	103

[a] Substrate/Cl_2O = 10.

[b] Basis of assignment of structure: VPC retention time on one column.

[c] Basis of assignment of structure: not mentioned.

TABLE 42

Reaction of $CH_3CH_2CH_2X$ with Cl_2O and HOCl

X	Reaction conditions	% Yield of reported monochlorinated substrate	Distribution of reported monochlorinated substrates (%)			Controls	Ref.
			1,1	1,2	1,3		
Cl	Cl_2O, 40°, hν, CCl_4	?	42.8[a]	42.4[a]	14.8[a]	Table 44	103
Cl	Cl_2O, 40°, hν, CCl_4/H_2O (100/1)	?	42.0[a]	42.0[a]	16.0[a]	Table 44	103
CN	Cl_2O, 40°, hν, CCl_4	?	20.1[b]	38.9[b]	41.0[c]	None	103
Cl	HOCl, 40°, hν, CCl_4/H_2O (10/1)	?	21.8[a]	46.9[a]	31.3[a]	None	103
Cl	HOCl, 40°, hν, CCl_4/H_2O (1/1)	?	17.2[a]	49.2[a]	33.6[a]	None	103
Cl	HOCl, 40°, hν, C_3H_7Cl/H_2O (1/1)	?	20[a]	50[a]	30[a]	None	103

[a] Basis of assignment of structure: two isomers by VPC retention time on three columns; third by elimination.

[b] Basis of assignment of structure: none.

[c] Basis of assignment of structure: VPC retention time.

TABLE 43

Reaction of 1.8 M $CH_3CH_2CH_2CH_2Cl$ + 0.18 M Cl_2O at 40°

Reaction conditions	Induction period, min	Reaction time, min
Dark, degassed	95	900
Dark, degassed, 0°	120	>3000
Dark, O_2	>5000	–
hν, degassed	3	10
hν, O_2	11	70
0.02 mole % AIBN, degassed	20	80
0.02 mole % AIBN, O_2	60	2000

TABLE 44

Isomer Distribution of $CH_3CH_2CH_2Cl$ at Various Stages in the
Reaction with Cl_2O at 40°

% Active Cl_2 consumed	Reported isomer distribution, %		
	1,1	1,2	1,3
25	60.9	33.5	5.6
32	60.1	33.7	6.2
35	60.1	33.8	6.1
49	59.1	33.9	6.1
50	59.2	34.4	7.0
59	58.2	34.4	6.4
62	56.8	35.2	7.4
89	56.8	35.6	8.0
96	51.8	36.5	7.6
97	50.7	36.9	11.7
98	51.8	37.1	11.1
100	42.8	42.4	14.8

6. Reaction of Alkyl Chlorides with $(CH_3)_3COCl$

The products resulting from the chlorination of $CH_3CH_2CH_2CH_2Cl$ (Table 45) and $CH_3CH_2CH_2Cl$ (Table 46) have been reported. The comments in Sec. 5; IC2 apply to these results with regard to β-, γ-, and δ-chloroalkyl radicals.

7. Reaction of Alkyl Chlorides with Cl_3CSCl

The products resulting from the chlorination of $CH_3CH_2CH_2CH_2Cl$ (Table 47) and $CH_3CH_2CH_2Cl$ (Table 48) have been reported. The comments in Sec. 5; IC2 apply to these results with regard to β-, γ-, and δ-chloroalkyl radicals. Data useful in the sense discussed in paragraph 2 of Sec. 5; IC3, and subject to the provisos and interpretations therein, are in Table 49 [142].

8. Reaction of Alkyl Chlorides with SO_2Cl_2

The products resulting from the chlorination of the following compounds have been reported: $CH_3CH_2CH_2Cl$ (Table 50), $CH_3CH_2CH_2CH_2Cl$ (Table 51), $CH_3CH_2CH_2CH_2CH_2CH_2Cl$ (Table 52), $CH_3CH_2CH_2CH_2CH_2CH_2CH_2CH_2Cl$ (Table 53), $CH_3CH_2CH_2CH_2CHClCH_3$ (Table 54), and $CH_3CH_2CH_2CHClCH_2CH_3$ (Table 55). The comments in Sec. 5; IC2 apply to these results with regard to β-, γ-, δ-, ϵ-, ζ-, η-, and θ-chloroalkyl radicals.

9. Reaction of Haloalkyl Chlorides with Chromous Salts

Much work has been done on the mechanism of the reaction of Cr^{2+} with alkyl halides. However, this work has been concerned mainly with aspects of the mechanism that are not germane to our present purposes. We are concerned with the primal act: Does the encounter of RX with Cr^{2+} lead directly to R·? (Is R· produced, and in that elementary step in which the reactant's C-X bond is broken?) Is this the rate-determining step of the reaction? In order for the possibility to exist that the results of studies of the kinetics of these reactions might be of direct use to us, it is necessary that the transition state of the rate-determining step strongly resemble R·. It is not sufficient for R-X bond breaking to occur in the rate-determining step.

TABLE 45

Reaction of $CH_3CH_2CH_2CH_2X$ with $(CH_3)_3COCl$

X	Reaction conditions	% Yield of reported monochlorinated substrate	Distribution of reported monochlorinated substrates (%)				Controls	Ref.
			1,1	1,2	1,3	1,4		
Cl	40°, hν	?	21.0[a]	19.4[a]	43.9[a]	15.6[a]	None	138
Cl	40°, hν, 4 M CCl₄	?	21.3[a]	19.6[a]	43.6[a]	15.4[a]	None	138
Cl	40°, hν, 4 M PhH	?	21.4[a]	19.0[a]	46.4[a]	13.2[a]	None	138
Cl	40°, hν, 4 M t-C₄H₉Ph	?	23.6[a]	17.3[a]	49.0[a]	10.1[a]	None	138
Cl₂	40°, hν	?	31.3[a]	8.3[a]	38.8[a]	22.5[a]	None	138
H	0°, hν, neat[b]	?	6.0[c]	44[c]	44[c]	6.0[c]	None	140
H	0°, hν, 4 M PhH[b]	?	5.0[c]	45[c]	45[c]	5.0[c]	None	140
H	0°, hν, CCl₄	?	5.2[c]	45[c]	45[c]	5.2[c]	None	141
H	0°, hν, PhCl	?	4.7[c]	45[c]	45[c]	4.7[c]	None	141
H	0°, hν, HOAc	?	6.2[c]	44[c]	44[c]	6.2[c]	None	141
H	0°, hν, MeCN	?	6.0[c]	44[c]	44[c]	6.0[c]	None	141
H	20°, hν, neat[b]	?	7.0[c]	43[c]	43[c]	7.0[c]	None	140
H	20°, hν, 4 M PhH[b]	?	5.9[c]	44[c]	44[c]	5.9[c]	None	140
H	25°, hν, CCl₄	?	6.0[c]	44[c]	44[c]	6.0[c]	None	141
H	25°, hν, PhCl	?	5.8[c]	44[c]	44[c]	5.8[c]	None	141
H	25°, hν, HOAc	?	8.6[c]	41[c]	41[c]	8.6[c]	None	141
H	25°, hν, MeCN	?	7.7[c]	42[c]	42[c]	7.7[c]	None	141

H	32°, hν, gas[b]	?	$8.5^{d,e}$	$41.5^{d,e}$	$41.5^{d,e}$	$8.5^{d,e}$	None	163
H	40°, hν, neat[b]	?	8.0^{c}	42^{c}	42^{c}	8.0^{c}	None	140
H	40°, hν, 8 M CS_2[b]	?	7.4^{c}	42.5^{c}	42.5^{c}	7.4^{c}	None	140
H	40°, hν, 4 M PhH[b]	?	6.9^{c}	43^{c}	43^{c}	6.9^{c}	None	140
H	40°, hν, 4 M t-C_4H_9Ph	?	6.5^{c}	43^{c}	43^{c}	6.5^{c}	None	140
H	40°, hν, CCl_4	?	6.4^{c}	44^{c}	44^{c}	6.4^{c}	None	141
H	40°, hν, PhCl	?	7.2^{c}	43^{c}	43^{c}	7.2^{c}	None	141
H	40°, hν, HOAc	?	10.2^{c}	40^{c}	40^{c}	10.2^{c}	None	141
H	40°, hν, MeCN	?	9.3^{c}	41^{c}	41^{c}	9.3^{c}	None	141
H	41°, hν, gas[b]	?	$8.9^{d,e}$	$41^{d,e}$	$41^{d,e}$	$8.9^{d,e}$	None	163
H	51°, hν, gas[b]	?	$9.1^{d,e}$	$41^{d,e}$	$41^{d,e}$	$9.1^{d,e}$	None	163
H	58°, hν, gas[b]	?	$9.4^{d,e}$	$41^{d,e}$	$41^{d,e}$	$9.4^{d,e}$	None	163
H	70°, hν, PhCl	?	7.2^{c}	43^{c}	43^{c}	7.2^{c}	None	141
H	70°, hν, HOAc	?	12.8^{c}	36^{c}	36^{c}	12.8^{c}	None	141
H	70°, hν, MeCN	?	11.8^{c}	38^{c}	38^{c}	11.8^{c}	None	141

a "Peaks were definitely identified by product isolation (either g.c. or fractional distillation of larger runs) and the determination of physical constants."

b Substrate/$(CH_3)_3$COCl = 10.

c Assignment of structure OK.

d Basis of identification of product: VPC retention time.

e From extrapolation of "reasonably linear" plots of statistically corrected product ratio vs. (light intensity)$^{1/2}$ to light intensity = 0.

TABLE 46

Reaction of $CH_3CH_2CH_2X$ with $(CH_3)_3COCl$

X	Reaction conditions	% Yield of reported monochlorinated substrate	Distribution of reported monochlorinated substrates (%)			Controls	Ref.
			1,1	1,2	1,3		
Cl	40°	?	47.7[a]	41.4[a]	10.9[a]	None	138
CN	40°, neat	?	22.4[a]	43.7[a]	33.9[a]	None	138
$OCOCH_3$	40°	?	29.3[a]	43.0[a]	25.4[a]	None	138
NO_2	40°	?	12.5[a]	33.4[a]	54.0[a]	None	138
COOH	hν	53	20.6[b]	46.4[b]	33.0[b]	None	139

[a] Peaks were definitely identified by product isolation (either by g. c. or fractional distillation of larger runs) and the determination of physical constants."

[b] Basis of assignment of structure not mentioned.

TABLE 47

Reaction of $CH_3CH_2CH_2CH_2X$ with Cl_3CSCl

X	Reaction conditions	% Yield of reported monochlorinated substrate	Distribution of reported monochlorinated substrates (%)				Controls	Ref.
			1,1	1,2	1,3	1,4		
Cl	0°, hv[a]	?	4.7[b]	21.5[b]	63.5[b]	10.3[b]	None	142
H	reflux, hv, neat[c]	13	~4[d]	46[d]	46[d]	~4[d]	None	143
$OCOCH_3$	20°, hv, neat[e]	?	1.9–5.8[f]	22–25[f]	56–62[f]	13.6–14.7[f]	None	515

[a] Substrate/Cl_3CSCl = "large."

[b] Basis of assignment of structure not mentioned.

[c] Substrate/Cl_3CSCl = 3.5.

[d] "The product composition was determined by gc, using samples of authentic materials for comparison."

[e] 20 mole % CCl_3SCl.

[f] Identification of products OK.

TABLE 48

Reaction of $CH_3CH_2CH_2X$ with Cl_3CSCl

X	Reaction conditions	% Yield of reported monochlorinated substrate	Distribution of reported monochlorinated substrates (%)			Controls	Ref.
			1,1	1,2	1,3		
Cl	0°, hν[a]	?	19.3[b]	71.2[b]	9.5[b]	None	142

a Substrate/Cl_3CSCl = "large."

b Basis of assignment of structure not mentioned.

TABLE 49

Reactivities (per H atom) of the Various Positions in 1-Chloroalkanes Relative to the Secondary Hydrogen Atoms in n-Pentane

Substrate	Position			
	1	2	3	4
$CH_3CH_2CH_2CH_2Cl$	0.023	0.108	0.302	0.97
$CH_3CH_2CH_2Cl$	0.024	0.088	0.235	

TABLE 50

Reaction of $CH_3CH_2CH_2X$ with SO_2Cl_2

X	Reaction conditions	% Yield of reported monochlorinated substrate	Distribution of reported monochlorinated substrates (%)			Controls	Ref.
			1,1	1,2	1,3		
Br	Reflux, 0.14 mole % Bz$_2$O$_2$ [j]	72	–	62[k]	38[k]	None	144
Cl	90°, Bz$_2$O$_2$	?	17.9[a]	56.1[a]	23.0[a]	None	110
Cl	Reflux, 1 mole % Bz$_2$O$_2$ [h]	85	–	60[i]	40[i]	None	144
OCOCH$_3$	~80°, Bz$_2$O$_2$ [b]	?	25[d]	46[d]	29[d]	c	111
CCl$_3$	~80°, Bz$_2$O$_2$ [g]	?	8[d]	42[d]	50[e]	c	111
COCl	~80°, Bz$_2$O$_2$, CCl$_4$ [g]	?	3[d]	49[d]	48[f]	c	111
COCl	Reflux, CCl$_4$, Bz$_2$O$_2$	?	15[l]	55[l]	30[l]	None	145
COOH	Reflux, CCl$_4$, Bz$_2$O$_2$	?	10[l]	45[l]	45[l]	None	145

[a] Basis of assignment of structure not mentioned.

[b] Substrate/SO$_2$Cl$_2$ = 5.

[c] "... formation of higher chlorides usually was maintained below 5% of the total chlorination product."

[d] Assignment of structure OK.

e Basis of assignment of structure: acceptable Cl analysis.

f Basis of assignment of structure: none.

g Substrate/SO_2Cl_2 = 10.

h Substrate/SO_2Cl_2 = 3.

i Basis of assignment of structure: boiling point

j Substrate/SO_2Cl_2 = 4.

k Basis of assignment of structure: boiling point and refractive index.

l "The identities of the isomers were confirmed by the preparation of the p-toluides."

TABLE 51

Reaction of $CH_3CH_2CH_2CH_2X$ with SO_2Cl_2

X	Reaction conditions	% Yield of reported monochlorinated substrate	Distribution of reported monochlorinated substrates (%)				Controls	Ref.
			1,1	1,2	1,3	1,4		
Cl	80°, Bz_2O_2	?	6.0[a]	20.6[a]	43.0[a]	30.4[a]	None	110
Cl	~80°, Bz_2O_2[c]	?	7[d]	22[d]	47[d]	24[d]	e	111
Cl	Reflux, 0.1 mole % Bz_2O_2[g]	85	–	25[h]	50[h]	25[h]	None	144
Cl_2	80°, Bz_2O_2[c]	?	2.4[b]	11.5[b]	49.3[b]	36.8[b]	None	110
Cl_2	~80°, Bz_2O_2[c]	?	2[d]	13[d]	48[d]	37[f]	e	111
$COOCH_3$	80°, Bz_2O_2, CCl_4[i]	<96	4.0[k]	31.0[k]	49.0[k]	16.2[k]	j	146

[a] Basis of assignment of structure: "by comparison with authentic dichlorides."

[b] Basis of assignment of structure: none.

[c] Substrate/SO_2Cl_2 = 5.

[d] Assignment of structure OK.

[e] "… formation of higher chlorides usually was maintained below 5% of the total chlorination product."

[f] Basis of assignment of structure: acceptable Cl analysis.

g Substrate/SO_2Cl_2 = 2.5.

h Basis of assignment of structure: boiling point and refractive index.

i Substrate/SO_2Cl_2 = 10.

j " . . . no polychlorination was observed."

k Basis of assignment of structure not mentioned.

TABLE 52

Reaction of $CH_3CH_2CH_2CH_2CH_2CH_2X$ with SO_2Cl_2

X	Reaction conditions	% Yield of reported monochlorinated substrate	Distribution of reported monochlorinated substrates (%)						Controls	Ref.
			1,1	1,2	1,3	1,4	1,5	1,6		
Cl	90°, Bz_2O_2	?	2.8[a]	8.8[a]	18.7[a]	29.0[a]	29.9[a]	10.8[a]	None	110
$OCOCCl_3$	80°, Bz_2O_2	?	–	–	1.9[b]	33.4[b]	37.8[b]	26.9[b]	None	110

[a] Basis of assignment of structure: "by comparison with authentic dichlorides."

[b] Basis of assignment of structure not mentioned.

TABLE 53

Reaction of $CH_3CH_2CH_2CH_2CH_2CH_2CH_2CH_2X$ with SO_2Cl_2

X	Reaction conditions	Distribution of reported monochlorinated substrates (%)								Controls	Ref.
		1,1	1,2	1,3	1,4	1,5	1,6	1,7	1,8		
Cl	90°, Bz_2O_2	2.3[a]	5.6[a]	11.2[a]	15.0[a]	17.0[a]	18.1[a]	20.6[a]	10.1[a]	None	110

[a] Basis of assignment of structure not mentioned.

TABLE 54

Reaction of $CH_3CH_2CH_2CH_2CHXCH_3$ with SO_2Cl_2

X	Reaction conditions	% Yield of reported monochlorinated substrate	Distribution of reported monochlorinated substrates (%)						Controls	Ref.
			2,2	2,3	1,2	2,4	2,5	1,5		
Cl	90°, Bz_2O_2	?	8.9[a]	17.4[a]	0[a]	25.8[a]	26.2[a]	21.7[a]	None	110

[a] Basis of assignment of structure unclear.

TABLE 55

Reaction of $CH_3CH_2CH_2CHXCH_2CH_3$ with SO_2Cl_2

X	Reaction conditions	% Yield of reported monochlorinated substrate	Distribution of reported monochlorinated substrates (%)						Controls	Ref.
			3,3	3,4	2,3	2,4	1,3	1,4		
Cl	80°, hν	?	6.7[a]	23.0[a]	22.7[a]	22.8[a]	7.7[a]	17.1[a]	None	110

[a] Basis of assignment of structure unclear.

In addition to observations which establish that the rate of reaction of RX follows the radical (or cation) order of stability of R and the radical stability (or anionofugacity) of X, there have been two studies directed toward the questions raised above. Kochi and Powers [147] have mentioned two mechanisms which are consistent with the available evidence:

mechanism I

$$RX + Cr^{II}en_2^{2+} \longrightarrow R\cdot + Cr^{III}en_2X^{2+}$$

$$R\cdot + Cr^{II}en_2^{2+} \xrightarrow{\text{fast}} RCr^{III}en_2^{2+}$$

mechanism II

$$RX + Cr^{II}en_2^{2+} \longrightarrow RCr^{IV}en_2X^{2+}$$

$$RCr^{IV}en_2X^{2+} + Cr^{II}en_2^{2+} \xrightarrow{\text{fast}} RCr^{III}en_2^{2+} + Cr^{III}en_2X^{2+}$$

They "postulate" that mechanism I applies, based on the following:

(1) "For a given halide, the rates of reduction of the alkyl group are generally tertiary > secondary > primary, roughly in the order $10^2{:}10{:}1$. For a particular alkyl group, iodides are reduced approximately 3×10^3 faster than chlorides, and bromides have intermediate ($\sim 4 \times 10^2$) reactivity."

(2) "The reactivity of various alkyl halides parallels the inner-sphere reduction of various pentaamminochromium (III) and pentaamminocobalt (III) halides by Cr(II)."

(3) "Analogous mechanisms have been postulated for the reduction of alkyl halides by pentacyanocobaltate (II) complexes and tributyltin hydride. Similar transformations also pertain to the reduction of halides by cobalt(I) complexes."

(4) Mechanism II cannot readily be formulated so as to describe the reduction of diacyl peroxides by $Cr^{II}en_2^{2+}$.

(5) The 1-hexene/methylcyclopentane ratio resulting from the reduction of 6-bromo-1-hexene is proportional to $[Cr^{II}(en)_2^{2+}]$.
We believe that these facts are strongly consistent with, but do not require, mechanism I. (1), including the $3° > 2° > 1°$ order, is not necessarily

inconsistent with mechanism II. (2), (3), and (4) are not directly relevant. (5) is consistent with mechanism II if $(5\text{-hexenyl})Cr^{IV}en_2X^{2+}$ can rearrange to $(\text{cyclopentylcarbinyl})Cr^{IV}en_2X^{2+}$ competitively with reaction with $Cr^{II}en_2^{2+}$. There is no evidence which argues against this possibility.

Barton et al. [148] have "taken considerable pains to demonstrate the existence of a free-radical intermediate in the conversion of progesterone bromohydrin (XIX) to 11 β-hydroxyprogesterone (XX)." They present the

$\underline{\text{XIX}}$ $\underline{\text{XX}}$ $\underline{\text{XXI}}$ $\underline{\text{XXII}}$

following as "definitive evidence for . . . [a] free radical intermediate" in the reduction by $Cr(OAc)_2$:

(1) In the presence of n̲-BuSH, "under conditions where the oxidation of thiol to disulfide . . . is negligible, the reduction produces . . . [BuSSBu] in theoretical yield." " . . . acetic acid, phenol, and hydrochloric acid do not promote the formation of . . . [(XX)]."

(2) "1-Benzyl-1,4-dihydronicotinamide, 1,4-dihydrobenzene, and cyclopentadiene all served as suitable hydrogen donors affording. . . [(XX)] in yields of 67, 67, and 46%, respectively. Triphenyltin hydride [. . . this reagent will convert (XIX) to (XX) at elevated temperature in the absence of $Cr(OAc)_2$.], triphenylsilane, and hypophosphorous acid (yields: 65, 40, and 87%, respectively) are also effective." Singleton and Kochi [151] have pointed out that "these results . . . are not unequivocal evidence of radical trapping since it is possible that an alkylchromium species will also be reduced by these agents."

Barton et al. present the following to "show clearly that reduction of the radical intermediate is competitive with the consumption of a second equivalent of chromous ion to furnish . . . [(XXII) + (XXI)] " in the reduction by $CrCl_2$ in aqueous ethanol: The [(XXI) + (XXII)] / (XX) ratio decreased rapidly with decreasing [$CrCl_2$] in the presence of a fixed concentration of 1,4-dihydrobenzene.

We believe that the results of the two groups, taken together, establish mechanism I based on a preponderance of evidence, but not beyond a reasonable doubt. The burden of proof would rest on a proponent of mechanism II.

Absolute and relative rates of reduction of β-Cl alkyl halides are in Tables 56, 57, and 58. These results indicate an accelerating effect of a β-Cl substituent (Table 59). In view of the problems discussed in Sec. 1;VB, it would be premature, even if the gross features of the mechanism were known, to conclude from these results that β-chloroalkyl radicals are bridged [1;II(a)]. In addition to all of these problems are the following questions: If the intimate details of the transition states of these reactions are sufficiently well-understood so that 1.8-3.2 kcal/mole differences in free energy can be ascribed with confidence to a particular "cause," incipient bridging [1;II(a)], then what is the origin of the 2.2 kcal/mole difference (Table 56) between erythro-$CH_3CH(OTs)CHBrCH_3$ and $CH_3CH_2CHBrCH_3$? the 3.9 kcal/mole difference (Table 57) between $BrCH_2CH_2\overset{+}{N}Me_3$ and $BrCH_2CH_2CH_2CH_3$? (bridging ? - 153) the >2 kcal/mole difference (Table 56) between $BrCH_2CH_2\overset{+}{N}Me_3$ and $BrCH_2CH_2\overset{+}{N}H_3$ (bridging ? - 153). Finally, an alternative "explanation" of the enhanced rates, put forth and labeled "less likely [than bridging]" without elaboration, for the case of a β-bromo substituent [152]("... a one-step loss of a β-bromine atom synchronously with bromine transfer to Cr^{II}.") may apply here. In other words, whatever the mechanism of reaction of $C_nH_{2n+1}X$, the reaction of an alkyl halide with a fugacious β-substituent may not involve a free alkyl radical intermediate.

10. Reaction of Haloalkyl Iodides with PhN=NCPh$_3$

The relative amounts of PhI and PhBr resulting from the reaction of PhN=NCPh$_3$ with RI/BrCCl$_3$ are given in Table 60. These numbers were taken, without justification, to be "the rate of abstraction of iodine from a series of aliphatic iodides relative to the rate of abstraction of bromine from bromotrichloromethane." The "acceleration" "caused by" Cl would correspond to $\Delta\Delta G^{\ddagger} = 0.3$ kcal/mole. In view of the lack of knowledge of the products of the reactions, the lack of adequate control experiments, and the contents of Sec. 1;VB, this small "deviation" provides no cause to consider $ClCH_2CH_2\cdot$ to be bridged [1;II(a)].

TABLE 56

Rates of Reduction of β-Substituted Alkyl Halides by Cr^{II}

Halide	Reaction conditions	Product (% yield)	Rate constant (1 mole^{-1} sec^{-1}) x 10^4	Ref.
meso-$CH_3CHBrCHBrCH_3$	0°, $Cr(ClO_4)_2$, $DMF/H_2O/EtOH$ (85/5/10), 0.90 M $HClO_4$	$CH_3CH=CHCH_3$(?)	800[d]	151, 152
dl-$CH_3CHBrCHBrCH_3$	0°, $Cr(ClO_4)_2$, $DMF/H_2O/EtOH$ (85/5/10), 0.90 M $HClO_4$	$CH_3CH=CHCH_3$(?)	330[d]	151, 152
erythro-$CH_3CHClCHBrCH_3$	0°, $Cr(ClO_4)_2$, $DMF/H_2O/EtOH$ (85/5/10), 0.90 M $HClO_4$	$CH_3CH=CHCH_3$(?)	51 ± 1	151, 152
threo-$CH_3CHClCHBrCH_3$	0°, $Cr(ClO_4)_2$, $DMF/H_2O/EtOH$ (85/5/10), 0.90 M $HClO_4$	$CH_3CH=CHCH_3$(?)	25 ± 0	151, 152
erythro-$CH_3CHOHCHBrCH_3$	0°, $Cr(ClO_4)_2$, $DMF/H_2O/EtOH$ (85/5/10), 0.90 M $HClO_4$	$CH_3CH=CHCH_3$(?)	0.65 ± 0.07[c]	151, 152
threo-$CH_3CHOHCHBrCH_3$	0°, $Cr(ClO_4)_2$, $DMF/H_2O/EtOH$ (85/5/10), 0.90 M $HClO_4$	$CH_3CH=CHCH_3$(?)	0.68 ± 0.15[c]	151, 152
erythro-$CH_3CH(OAc)CHBrCH_3$	0°, $Cr(ClO_4)_2$, $DMF/H_2O/EtOH$ (85/5/10), 0.90 M $HClO_4$	$CH_3CH=CHCH_3$(?)	2.1 ± 0.1	151, 152
threo-$CH_3CH(OAc)CHBrCH_3$	0°, $Cr(ClO_4)_2$, $DMF/H_2O/EtOH$ (85/5/10), 0.90 M $HClO_4$	$CH_3CH=CHCH_3$(?)	2.0 ± 0.2	151, 152

Substrate	Conditions	Product	Rate	Ref.
erythro-$CH_3CH(OTs)CHBrCH_3$	0°, $Cr(ClO_4)_2$, $DMF/H_2O/EtOH$ (85/5/10), 0.90 M $HClO_4$	$CH_3CH=CHCH_3$(?)	7.7	151, 152
threo-$CH_3CH(OTs)CHBrCH_3$	0°, $Cr(ClO_4)_2$, $DMF/H_2O/EtOH$ (85/5/10), 0.90 M $HClO_4$	$CH_3CH=CHCH_3$(?)	6.0	151, 152
$CH_3CH_2CHBrCH_3$	0°, $Cr(ClO_4)_2$, $DMF/H_2O/EtOH$ (85/5/10), 0.90 M $HClO_4$	$CH_3CH_2CH_2CH_3$(?)	0.14 ± 0.01[b]	151, 152
$BrCH_2CH_2Br$	Cr^{II}, 0°–room temp., DMF, 1 M $HClO_4$	C_2H_4[e]	11[d]	153
$BrCH_2CH_2NMe_3^+$	Cr^{II}, 0°–room temp., DMF, 1 M $HClO_4$	C_2H_4[e]	1.6	153
$BrCH_2CH_2NH_3^+$	Cr^{II}, 0°–room temp., DMF, 1 M $HClO_4$	C_2H_4[e]	<0.05	153
$PhCHClCH_2Cl$	29.7°, $CrSO_4$, DMF/H_2O (1/1), 1.0 M $HClO_4$	$PhCH=CH_2$(91)	7700 ± 500	149
$PhCHClCH_3$	29.7°, $CrSO_4$, DMF/H_2O (1/1), 1.0 M $HClO_4$	f	380 ± 130[a]	150
$ClCH_2CH_2Cl$	0°, Cr^{II}en, DMF/H_2O (999/1)	C_2H_4[e]	35[d]	153
$ClCH_2CH_2NMe_3^+$	0°, Cr^{II}en, DMF/H_2O (999/1)	C_2H_4[e]	210	153

[a] "... the typical second-order expressions for equivalent and nonequivalent initial concentrations were plotted ... good linearity was obtained but successive runs were ... [not very reproducible] because of the rapidity of the reaction at the relatively high ... concentrations that were necessary."

b "Followed to 10% completion only, second-order kinetics assumed."

c "Followed to 20% completion only."

d The rate constant has been divided by 2.

e Rate constant defined in terms of appearance of this product.

f PhCH(CH$_3$)CH(CH$_3$)Ph(98).

TABLE 57

Relative Rates of Reduction of β-Substituted Alkyl Bromides by $Cr^{II}en$

Halide	Reaction conditions	Product (% yield)	Relative rate	Ref.
$BrCH_2CH_2Br$	Room temp., DMF/H$_2$O (999/1)	C_2H_4[b]	12.5[a]	153
$BrCH_2CH_2Cl$	Room temp., DMF/H$_2$O (999/1)	C_2H_4[b]	5	153
$BrCH_2CH_2NMe_3^+$	Room temp., DMF/H$_2$O (999/1)	C_2H_4[b]	21	153
$BrCH_2CH_2OTs$	Room temp., DMF/H$_2$O (999/1)	C_2H_4[b]	2	153
$BrCH_2CH_2OCOC_6H_4OCH_3$	Room temp., DMF/H$_2$O (999/1)	C_2H_4[b]	1	153
$BrCH_2CH_2OAc$	Room temp., DMF/H$_2$O (999/1)	C_2H_4[b]	1.0	153
$BrCH_2CH_2CH_2CH_3$	Room temp., DMF/H$_2$O (999/1)	Butane[b]	0.03	153
$BrCH_2CH_3$	Room temp., DMF/H$_2$O (999/1)	C_2H_6[b]	0.04	153

[a] The relative rate has been divided by 2.

[b] Relative rate defined in terms of amount of this product.

TABLE 58

Relative Rates of Reduction of β-Substituted Alkyl Chlorides by Cr^{II}en

Halide	Reaction conditions	Product (% yield)	Relative rate	Ref.
$ClCH_2CH_2Cl$	0°, DMF/H_2O (999/1)	C_2H_4 [b]	4 [a]	153
$ClCH_2CH_2NMe_3^+$	0°, DMF/H_2O (999/1)	C_2H_4 [b]	28	153
$ClCH_2CH_2OTs$	Room temp., DMF/H_2O (999/1)	C_2H_4 [b]	4	153
$ClCH_2CH_2OAc$	Room temp., DMF/H_2O (999/1)	C_2H_4 [b]	1.0	153

[a] The relative rate has been divided by 2.

[b] Relative rate defined in terms of amount of this product.

TABLE 59

Accelerating Effect of β-Chloro Substituents in Reactions of Alkyl Halides with Cr^{II} and $Cr^{II}en$

$(Halide)_1$	faster than $(Halide)_2$	by $\Delta\Delta G^{\ddagger}$(kcal/mole) of[a]
$PhCHClCH_2Cl$	$PhCHClCH_3$	1.8
erythro-$CH_3CHClCHBrCH_3$	$CH_3CH_2CHBrCH_3$	3.2
$BrCH_2CH_2Cl$	$BrCH_2CH_2CH_2CH_3$	3.0

[a] Entries calculated from contents of Tables 56, 57, and 58.

TABLE 60

Reaction of $PhN=NCPh_3$[a] with $RI/BrCCl_3$ at 60° [481]

R	$PhI/PhBr$[c,d]	Controls
$ClCH_2CH_2$	0.52	None
$BrCH_2CH_2$	0.86	e
$ICH_2CH_2CH_2$	0.47[b]	None
HCH_2CH_2	0.33	None
$CH_3CH_2CH_2CH_2$	0.31	None
$PhCH_2CH_2$	0.37	None
$HOOCCH_2CH_2$	0.46	None

[a] $RI/PhN=NCPh_3 \geq 10$.

[b] The ratio has been divided by 2.

[c] A plot of log$(PhI/PhBr)$ vs. σ^* for those RCH_2CH_2I studied showed a large amount of scatter and is not subject to meaningful interpretation.

[d] "The combined yields of iodobenzene and bromobenzene typically averaged 80-90% based on PAT."

[e] "Reaction of . . . [1-bromo-2-iodoethane] with PAT in carbon tetrachloride produces less than 0.5% bromobenzene."

11. Summary

Although there is some evidence (Secs. L5;IC1a,9) which is not inconsistent with β-chloroalkyl radicals being bridged [1;II(a)] and there is
some (Secs. 5;IC2-8) which, if viewed superficially and uncritically, shifts
the burden of proof toward a proponent of bridging [1;II(a)], there is none
for which there is a reasoned, preferred interpretation which indicates
bridging [1;II(a)]. We emphasize separately the thermochemical results
(Sec. 5;IC1) which, taken at face value, indicate that C_2HCl_4· and CCl_3CCl_2·
are not bridged [1;II(a)].

D. Control of Stereochemistry

1. Reaction of Alkyl Chlorides with Br_2

Of those reactions already mentioned, the following, in principle, permit the determination of the degree to which a β-Cl substituent can exert
unusual control over the direction of attack on an alkyl radical:

$$CH_3CH_2CHFCH_3 \longrightarrow CH_3CHBrCHFCH_3 \quad \text{(Table 23)}$$

$$CH_3CH_2CHClCH_3 \longrightarrow CH_3CHBrCHClCH_3 \quad \text{(Table 23)}$$

The reported erythro/threo ratios of products are 1.8 and 1.5, respectively,
corresponding to $\Delta\Delta G^{\ddagger} = 0.4$ and 0.3 kcal/mole, respectively.
In addition to many questionable identifications of products, we are
faced with the problem that in no case has the relative amounts of
diastereomeric products been shown to be equal to their relative rates of
production from their presumed free radical precursors. Barring
coincidences, necessary conditions for this are that each initially produced
diastereomer survive the reaction to the same extent and that the
diastereomeric products be formed only via those routes which produce
them in the same ratio as does the direct reaction of a free β-chloro radical with Br_2. If the results are taken at face value, they indicate that, as
judged by this particular reaction, $CH_3\overset{\cdot}{C}HCHClCH_3$ is not bridged
[1;III(a)].

It has been reported [154] that "The photobromination of (+)-1-chloro-
2-methylbutane, α^{29}_{obsd} + 1.38°, proceeds in a selective manner to yield

(-)-2-bromo-1-chloro-2-methylbutane, α_{obsd}^{29} -1.45°, as the sole dihalide* (~ 97%). It was later reported [235] that product could be obtained with α_{obsd} = - 2.3° and that decreasing $[Br_2]$ led to decreasing activity of the product. It was concluded that a bridged radical was an intermediate. We find evaluation of these results difficult because the optical purity (the classification "selective" [154] is not justified), chemical purity, and % yield of the product are not reported, and because it has not been established that the reaction proceeds via attack on free $CH_3CH_2CH(CH_3)CH_2Cl$ to give free $CH_3CH_2\dot{C}(CH_3)CH_2Cl$ (structure - ?) followed by reaction with an achiral bromine-containing substance to give $CH_3CH_2CBr(CH_3)CH_2Cl$ which is not then asymmetrically destroyed. For these reasons we defer judgment on the meaning of these results with regard to the question of bridging [1; III(a)].

Should information become available in the future which resolves the above uncertainties in a manner consistent with the authors' conclusion, and if the radical is judged to be bridged [1; III(a)], then whether it is bridged [1; I(a)-(c)] would still be an open question. For example, an initially produced $CH_3CH_2\dot{C}(CH_3)CH_2Cl$ radical which is planar about the trisubstituted carbon atom and which has a "normal" C-C-Cl angle, i.e., is not bridged [1; I(c)], would become statistically achiral only on a time scale large compared to the period of rotation about the $\dot{C}$-CH_2Cl bond. In other words, the rotamers of $CH_3CH_2\dot{C}(CH_3)CH_2Cl$ which result directly from $CH_3CH_2CH(CH_3)CH_2Cl$ are chiral, and their racemization is accomplished in part by rotation about the $\dot{C}$-CH_2Cl bond; if such rotation is not much faster than reaction with Br_2, the $CH_3CH_2CBr(CH_3)CH_2Cl$ will be optically active even though the radical is not bridged [1; I(a)-(c)] (Of course, if the radical is not planar about the trisubstituted carbon atom, an additional process becomes necessary for the interconversion of enantiomers.) In order to evaluate the viability of this alternative, it is necessary to know the rates of rotation about the $\dot{C}$-CH_2Cl bond and of reaction of the radical with Br_2. Data which might be helpful in estimating the former are given in Table 61. Based on these data, we estimate the

---

*"The structure of this dihalide was established by dehydrohalogenation with NaOEt in EtOH to $CH_3CH_2C(CH_3)$=CHCl."

potential energy barrier to rotation in two possible models for unbridged $CH_3CH_2\dot{C}(CH_3)CH_2Cl$, $CH_3CH_2CH(CH_3)CH_2Cl$ and $R_2C=CRCH_2Cl$, * to be 3.8 † or ~5.0 ‡ and ~2.3 kcal, respectively. If a preexponential factor of ~10^{12} is used [206, 191, 239], rate constants at 0° in the gas phase of 4 x 10^8 or 4 x 10^7, and 6 x 10^9 sec^{-1}, respectively, are obtained. Unfortunately, there is an even less secure basis for an estimate of the rate constant for reaction of unbridged $CH_3CH_2\dot{C}(CH_3)CH_2Cl$ with Br_2. Available information is collected in Table 62. Skell [235] believes, based on the dependence of the activity of the product on [Br_2], that his data indicate the rate constant for reaction of the species which gives active product to be about

---

* Our estimate is based on the observation (Table 61) that compounds of this type have about the same barrier.

† This is the barrier reported for $(CH_3)_2CH-CH_2Cl$ [241].

‡ Our estimate is based on the assumption that the change in the barrier $CH_3\!\!\mathrel{\mkern-5mu}\overset{\displaystyle f}{}\!\!CH_2X$ as two β-hydrogens are replaced by alkyl groups is about the same for X = Br in a particular case as is the average for X = H, CH_3, Br, Cl, I and F (see Table below).

X	Going from	To	Increases barrier by (kcal)
H	CH_3-CH_3	$(CH_3)_2CH-CH_3$	2 x 0.5
CH_3	$CH_3-CH_2CH_3$	$(CH_3)_2CH-CH_2CH_3$	2 x (0.4-0.6)^a
F	CH_3-CH_2F	$CH_3CH_2-CH_2F$	0.9
Br	CH_3-CH_2Br	$(CH_3)_2CH-CH_2Br$	2 x 0.55
Br	$CH_3CHBr-CH_3$	$CH_3CHBr-CH_2CH_3$	0.5
Cl	CH_3-CH_2Cl	$(CH_3)_2CH-CH_2Cl$	2 x 0.05
Cl	$CH_3CHCl-CH_3$	$CH_3CHCl-CH_2CH_3$	0.9
I	CH_3-CH_2I	$(CH_3)_2CH-CH_2I$	2 x 1.1
I	$CH_3CHI-CH_3$	$CH_3CHI-CH_2CH_3$	1.1

^{a} Estimated from data for $CH_3CH_2CH_2-CH(CH_3)_2$ and $CH_3CH_2-CH(CH_3)CH_2CH_3$.

2 M^{-1} * times that of the conversion of such species into one which gives racemic product.* If the estimated rate constants for reaction of the poor model compounds $CH_3\cdot$, $CF_3\cdot$, $CCl_3\cdot$, $CH_3CH_2\cdot$, and $CHBr_2\dot{C}HBr$ are used, the corresponding ratios of rate constants for the alternative scheme which we are considering are estimated to be 2-300, 0.1-20, 0.1-10, 0.01-1, and 10^{-2} - 10^{-5} M^{-1}, respectively. If the guessed activation energy for reaction of the good model compound, $(CH_3)_3C\cdot$, is used, the ratio is less than 10^{-3} for any reasonable preexponential factor for $(CH_3)_3C\cdot + Br_2$. We cannot emphasize too strongly that all of these esti-mates of rate constants for internal rotation and for reaction with Br_2 are extremely crude. We have of necessity chosen highly approximate models and have estimated their properties by use of highly approximate methods. Our purpose was only to determine whether, based on the very limited data available, the alternative "explanation" presented is a priori unreasonable. It does not appear to be so.

2.   Reaction of Alkyl Chlorides with Cl_2

Of those reactions already mentioned, the following, in principle, per-mit the determination of the degree to which a β-Cl substituent can exert unusual control over the direction of attack on an alkyl radical:

$$CH_3CH_2CHClCH_3 \longrightarrow CH_3CHClCHClCH_3 \quad \text{(Table 31)}$$
$$CH_3CH_2CHFCH_3 \longrightarrow CH_3CHClCHFCH_3 \quad \text{(Table 31)}$$

The reported erythro(meso)/threo(dl) ratios of products are 2.4-2.5 and 1.5, respectively, corresponding to $\Delta\Delta G^{\ddagger} = 0.6$ and 0.2-0.3 kcal/mole, respectively. These results are subject to the reservations expressed in Sec. 5;ID1. If they are taken at face value, they indicate that, as judged by this particular reaction, $CH_3\dot{C}HCHClCH_3$ is not bridged [1;III(a)].

The relative yields of stereoisomers resulting from the chlorination of 2- and 3-chlorohexane have been reported (Table 63) [109]. The reported ratios of the resulting 2,3-isomers correspond to $\Delta\Delta G^{\ddagger} = 0.5$ and

---

*The general aspects of the comments made at similar stages of Sec. 6;ID1 apply here.

TABLE 61

Potential Energy Barriers to Internal Rotation

Compound	Phase	Method	Barrier(s), kcal	Ref.
CH_3-CH_2Br	Gas	Microwave	3.6	199
CH_3-CH_2Br	Gas	Microwave	3.7	200
$(CH_3)_2CH-CH_2Br$	Liquid	Ultrasonic absorption[c]	4.7[a]	244
$CH_3CHBr-CH_3$	Gas	Infrared	4.3	216
$CH_3CHBr-CH_2CH_3$	Liquid	Ultrasonic absorption[c]	4.8[a]	239
$CH_3CHBr-CH_2CH_2CH_3$	Liquid	Ultrasonic absorption[c]	4.8	498
$CH_3CHBr-CH_2CH_2CH_2CH_3$	Liquid	Ultrasonic absorption[c]	4.2	498
$CH_3CH_2CHBr-CH_2CH_3$	Liquid	Ultrasonic absorption[c]	3.6	498
$CH_3CH_2CHBr-CH_2CH_2CH_3$	Liquid	Ultrasonic absorption[c]	3.6	498
$(CH_3)_2CBr-CH_2CH_3$	Liquid	Ultrasonic absorption[c]	5.6	497
CH_3-CH_2Cl	Gas	Microwave	3.6	199
CH_3-CH_2Cl	Gas	Microwave	3.7	237
CH_3-CH_2Cl	Gas	Infrared	3.6	236
CH_3-CH_2Cl	Gas	Infrared	3.8	217
CH_3-CH_2Cl	Gas	Infrared	3.7	216
$(CH_3)_2CH-CH_2Cl$	Liquid	Ultrasonic absorption[c]	3.8[a]	241
$CH_3CHCl-CH_3$	Gas	Thermodynamic	3.5	243
$CH_3CHCl-CH_2CH_3$	Liquid	Ultrasonic absorption[c]	4.4	239

Compound	State	Method	Value	Reference
$CH_3CHCl-CH_2CH_3$	Liquid	Ultrasonic absorption[c]	4.4	498
$CH_3CH_2CHCl-CH_2CH_3$	Liquid	Ultrasonic absorption[c]	3.4	498
$(CH_3)_2CCl-CH_2CH_3$	Liquid	Ultrasonic absorption[c]	3.9	497
CH_3-CH_2I	Gas	Microwave	3.2	242
$(CH_3)_2CH-CH_2I$	Liquid	Ultrasonic absorption[c]	5.4^a	241
$CH_3CHI-CH_3$	Gas	Infrared	4.2	216
$CH_3CHI-CH_2CH_3$	Liquid	Ultrasonic absorption[c]	5.3^a	239
$(CH_3)_2CI-CH_2CH_3$	Liquid	Ultrasonic absorption[c]	8.4	497
CH_3CH_2CN	Gas	Microwave	3.1	407
CH_3CH_2CN	Gas	Microwave	3.3	408
CH_3-CH_3	Gas	Infrared	2.9	227
CH_3-CH_3	Gas	Thermodynamic	2.9	208
$CH_3CH_2-CH_3$	Gas	Thermodynamic	3.4	201
$CH_3CH_2-CH_3$	Gas	Thermodynamic	3.3	202, 203
$CH_3CH_2-CH_3$	Gas	Microwave	3.3-3.6	228, 229
$(CH_3)_2CH-CH_3$	Gas	Microwave	3.9	204
$(CH_3)_2CH-CH_3$	Gas	Microwave	3.9	205
$CH_3CH_2-CH_2CH_3$	Gas	Psychothermodynamic	3.8^a	207
$CH_3CH_2-CH_2CH_3$	Liquid	Ultrasonic absorption[c]	$2.8,^b$ 3.6^a	206
$CH_3CH_2CH_2-CH_2CH_3$	Liquid	Ultrasonic absorption[c]	3.6	206
$CH_3CH_2CH_2-CH(CH_3)_2$	Liquid	Ultrasonic absorption[c]	$3.3,^b$ 4.2^a	209
$CH_3CH_2-CH(CH_3)CH_2CH_3$	Liquid	Ultrasonic absorption[c]	$3.5,^b$ 4.4^a	209
CH_3-CH_2F	Gas	Microwave	3.3	210, 212

Compound	Phase	Method	Barrier(s), kcal	Ref.
CH_3-CH_2F	Gas	Infrared	3.3	211
$CH_3CH_2-CH_2F$	Gas	Microwave	3.6,[b] 4.2[a]	213
$CH_3-CH=CH_2$	Gas	Microwave	2.0	215
$CH_3-CH=CH_2$	Gas	Microwave	2.0	218
$CH_3-CH=CH_2$	Gas	Thermodynamic	2.0	214
$CH_3-CH=CH_2$	Gas	Infrared	2.1	216
$CH_3-CH=CH_2$	Gas	Infrared	2.1	217
$CH_3-C(CH_3)=CH_2$	Gas	Infrared	2.1	219
$CH_3-C(CH_3)=CH_2$	Gas	Thermodynamic	2.4	214
$CH_3-C(CH_3)=CH_2$	Gas	Microwave	2.2	220
$CH_3-CF=CH_2$	Gas	Infrared	2.4	217
$CH_3-CF=CH_2$	Gas	Microwave	2.4	226
$CH_3-CH=CHF$ (trans)	Gas	Microwave	2.2	221
$ClCH_2-CCl=CH_2$	Gas	Dipole moment	2.1	222
$CH_3-CCl=CH_2$	Gas	Microwave	2.7	224
$CH_3-CCl=CH_2$	Gas	Infrared	~2.5	225
$CH_3-CH=CHCl$ (trans)	Gas	Infrared	2.2	223
$CH_3-CH=CHBr$ (trans)	Gas	Infrared	2.1	266
$CH_3-CBr=CH_2$	Gas	Infrared	~2.7	225
$CH_3CH=CHCN$ (trans)	Gas	Microwave	1.9	474

a <u>trans</u>–<u>gauche</u>

b <u>gauche</u>–<u>trans</u>

[c] The "barriers" which result from this method are similar to $\Delta H^{\ddagger}$ and, hence, of different nature than the other entries in this table.

TABLE 62

Estimates of Rate Constants at 0° of R· + Br_2 → RBr + Br·

R	Estimated rate constant, l mole^{-1} sec^{-1}	Method of estimation
CH_3	1×10^{10}	a
CF_3	6×10^8	b
CCl_3	5×10^8	c
CH_3CH_2	8×10^7	d
$CHBr_2CHBr$	2×10^5	e
$(CH_3)_3C$	$E_a \approx 5.7$ kcal/mole	f

a Extrapolation of the data of Kistiakowsky and Van Artsdalen [252] yields $k[CH_3\cdot + HBr \rightarrow CH_4 + Br\cdot]/k[CH_3\cdot + Br_2 \rightarrow CH_3Br + Br\cdot] \approx$ 0.03 at 0°. Fettis and Trotman-Dickenson [248] have reported that $k[HBr + CH_3\cdot \rightarrow Br\cdot + CH_4]/k[CH_3\cdot + I_2 \rightarrow CH_3I + I\cdot] = 10^{-0.30}$ x exp[-1.37/RT]. Combination of these ratios and use of $k[CH_3\cdot + I_2 \rightarrow CH_3I + I\cdot] = 10^{9.9}$ l mole^{-1} sec^{-1} (g) leads to $k[CH_3\cdot + Br_2 \rightarrow CH_3Br + Br\cdot] = 1 \times 10^{10}$ at 0°.

b Reported by Amphlett and Whittle [230].

c Reported by Benson [231]; the result of a psychothermochemical analysis of the results of Sullivan and Davidson [231A].

d Extrapolation of the data of Anderson and Van Artsdalen [251] yields $k[C_2H_5\cdot + HBr \rightarrow C_2H_6 + Br\cdot]/k[C_2H_5\cdot + Br_2 \rightarrow C_2H_5Br + Br\cdot] \approx 0.7$ at 0°. Fettis and Trotman-Dickenson [248] have reported that $k[C_2H_5\cdot + HBr \rightarrow Br\cdot + C_2H_6]/k[C_2H_5\cdot + I_2 \rightarrow C_2H_5I + I\cdot] = 10^{0.23}$ exp[-2.29/RT]. Combination of these ratios and use of $k[C_2H_5\cdot + I_2 \rightarrow C_2H_5I + I\cdot] = 10^{9.5}$ exp[-0.2/RT] l mole^{-1} sec^{-1} (g) leads to $k[C_2H_5\cdot + Br_2 \rightarrow C_2H_5Br + Br\cdot] = 8 \times 10^7$ at 0°.

e Benson et al. [191], employing a psychothermochemical approach, have used the data of Steinmetz and Noyes [245] to estimate $k[CHBr_2\dot{C}HBr \rightarrow Br\cdot + CHBr=CHBr]$ to be $\sim 10^{13}$ exp[-9/RT] sec^{-1}. Steinmetz

and Noyes' data at 30° and 45° lead to an estimated value of
$k[CHBr_2\dot{C}HBr \rightarrow Br\cdot + CHBr=CHBr]/k[CHBr_2\dot{C}HBr + Br_2 \rightarrow$
$CHBr_2CHBr_2 + Br\cdot]$ at 0° of ~ 3 M. When combined with the preceding estimate, this yields $k[CHBr_2\dot{C}HBr + Br_2 \rightarrow CHBr_2CHBr_2 + Br\cdot] \approx$
2×10^5 l mole^{-1} sec^{-1} at 0°.

^{f} Guessed ("tentatively") by Benson and Buss [232].

^{g} Although never determined, but forming one of the cornerstones
of the self-consistency of the experimental determinations of heats of
formation of free radicals and supported indirectly by comparison of its
consequences with independently obtained data, activation energies ≈ 0
for $R\cdot + I_2$ are pervasive in the literature. For example, Flowers and
Benson [246] used $k_{CH_3\cdot} = 10^{9.9}exp[-0.4/RT]$ l mole^{-1} sec^{-1}, Golden
et al. [247] and Fettis and Trotman-Dickenson [248] took $(E_a)_{CH_3\cdot} = 0$,
and Christie [249] reported $k_{CH_3\cdot} \approx 10^{10}$ l mole^{-1} sec^{-1} at 20°; similarly, Hartley and Benson [250] used $k_{C_2H_5\cdot} = 10^{9.50}exp[-0.2/RT]$
l mole^{-1} sec^{-1}.

0.6 kcal/mole, respectively. These results are subject to the reservations
expressed in Sec. 5;IC2. If they are taken at face value, they indicate that,
as judged by this particular reaction, $CH_3CH_2CH_2\dot{C}HCHClCH_3$ and
$CH_3CH_2CH_2CHCl\dot{C}HCH_3$ are not bridged [1; III(a)].

3. Reaction of Alkyl Chlorides with SO_2Cl_2

$$CH_3CH_2CH(CH_3)CH_2Cl \xrightarrow[\text{Bz}_2\text{O}_2]{\text{10 mole \% SO}_2\text{Cl}_2} CH_3CH_2CCl(CH_3)CH_2Cl \quad [155]$$

1060 g, $[\alpha]_D^{23} = +1.65$ 12 ml, $[\alpha]_D^{23} = +0.05$

Indirect, not-too-strong arguments were presented to indicate that the
activity of the product* was due to contamination by $CH_3CH_2CH(CH_3)CHCl_2$.

---

*Identified by comparison of its boiling point at 100 mm and its
density with published values of the boiling point at 1 atm and the density.

Unfortunately, the specific rotation of optically pure product is unknown.

4. Reaction of Haloalkyl Chlorides with Organotin Hydrides

Cis/trans ratios of 2-butenes resulting from the reaction of meso- and dl-$CH_3CHClCHClCH_3$ with $Bu_3SnH(D)$ are given in Table 64. The ratio of ratios resulting from reaction of the two diastereomers corresponds to ~0.2 kcal/mole, an insignificant quantity. However, we cannot draw any conclusions from this fact because of the lack of control experiments (the authors believe "that significant isomerization of the olefins can occur during the reaction"), and the lack of knowledge of the extent to which $CH_3\overset{\bullet}{C}HCHClCH_3$ is a precursor of 2-butene.

5. Reaction of Haloalkyl Chlorides with Chromous Salts

General comments and reservations are in Sec. 5; IC9.

The rates of reaction of erythro- and threo-$CH_3CHClCHBrCH_3$ (Table 56) differ by only 0.4 kcal/mole ($\Delta\Delta G^{\ddagger}$). Cis/trans ratios of 2-butenes resulting from these and other reactions are given in Table 65. The five sets of experiments on $CH_3CHClCHBrCH_3$ and $CH_3CHClCHClCH_3$ yielded ratios of ratios resulting from reaction of each pair of diastereomers corresponding to 0.1, 0.0, 0.0, 0.0, and 0.0 kcal/mole [$\Delta(\Delta\Delta G^{\ddagger})$], respectively, all insignificant quantities. Although these results superficially indicate the absence of a significant "memory effect," they are subject to the reservations expressed in Sec. 5; ID4. The ratio of ratios resulting from $CH_3CHClCHICH_3$ corresponds to 0.4 kcal/mole. Although this selectivity is small energetically, the "memory effect" is real and cannot be ignored. Its significance is obscured, however, by the lack of knowledge of the details of the course of these particular reactions. Possibilities which could obviate the need to consider bridging include (1) one-step formation of 2-butene from $CH_3CHClCHICH_3$; (2) its multistep formation without the intermediacy of $CH_3CHCl\overset{\bullet}{C}HCH_3$; (3) the formation of different distributions of rotamers of $CH_3CHCl\overset{\bullet}{C}HCH_3$ from the erythro- and threo-chloroiodides and their reaction to give 2-butene or a stereochemically stable precursor of 2-butene being competitive with equilibration among themselves. Note that the realization of such possibilities would not imply that $CH_3CHCl\overset{\bullet}{C}HCH_3$ is not bridged [1; III(a)]; it would mean only that

TABLE 63

Reaction of 2- and 3-Chlorohexane with Cl_2, 20°

Reactant	Reported product dichlorohexane	Reported relative yields of stereoisomers		Controls
		(Isomer)$_1$	(Isomer)$_2$	
$CH_3CH_2CH_2CH_2CHClCH_3$	2,3[a]	2.23	1.0	None
	2,4[b]	1.22	1.0	None
	2,5[a]	1.0	1.0	None
$CH_3CH_2CH_2CHClCH_2CH_3$	2,3[a]	2.8	1.0	None

[a] Basis of assignment of structure: VPC retention time compared to that of authentic sample.

[b] Basis of assignment of structure: none.

TABLE 64

Reaction of 2,3-Dichlorobutane with Tributyltin Hydrides (101)

Reagents	Reaction conditions	% Yield of 2-butene	n-Butane		2-Butene trans/cis[b]
			% Yield[b]	% of hydrocarbon product	
dl-dichloride, Bu₃SnH	22°, hν, neat[a]	?	?	99.2(99.4)	3.76(3.00)
dl-dichloride, Bu₃SnH	22°, hν, neat[a]	0.7			3.35[e]
meso-dichloride, Bu₃SnH	22°, hν, neat[a]	?	?	99.0(99.3)	4.88(4.55)
meso-dichloride, Bu₃SnH	22°, hν, neat[c]	?	?(d)	84	2.70
meso-dichloride, Bu₃SnH	22°, hν, neat[a]	0.8			4.82[e]
meso-dichloride, Bu₃SnH	65°, hν, neat[a]	5	39(44)	94.9(95.5)	3.55(2.48)
meso-dichloride, Bu₃SnH	65°, hν, neat[a]	4.8			2.96[e]
meso-dichloride, Bu₃SnH	65°, hν, m-xylene,[a] 1.26 M dichloride	?	26(27)	96.8(96.4)	3.13(2.16)
meso-dichloride, Bu₃SnH	65°, hν, m-xylene,[a] 1.26 M dichloride	3.4			2.67[e]
meso-dichloride, Bu₃SnD	65°, hν, neat[a]	5	42	95.8(93.7,95.2)	2.85(2.70,2.33)
meso-dichloride, Bu₃SnH	70°, hν, m-xylene,[a] 0.38 M dichloride	?	?	95.4(95.5)	2.33(2.57)
meso-dichloride, Bu₃SnH	70°, hν, m-xylene,[a] 0.38 M dichloride	4.5			2.45[e]

[a] $CH_3CHClCHClCH_3/Bu_3SnH = 0.5$.

[b] Value in parenthesis is the result of a duplicate determination. Controls: none.

[c] $CH_3CHClCHClCH_3/Bu_3SnH = 1$.

[d] 86% yield of what was reported to be $CH_3CH_2CHClCH_3$ (identification based on its VPC retention time).

[e] "Results from two or more experiments deviating 4% or less from mean."

TABLE 65

Reaction of 2-Halo-3-Chlorobutane with Chromous Salts

Dihalide	Reaction conditions	2-Butene		Controls	Ref.
		% Yield	trans/cis		
erythro-CH$_3$CHClCHBrCH$_3$	Cr^{II}, 0°, 0.90 M HClO$_4$, DMF/H$_2$O (9/1)	a	2.85	None	152
threo-CH$_3$CHClCHBrCH$_3$	Cr^{II}, 0°, 0.90 M HClO$_4$, DMF/H$_2$O (9/1)	a	2.57	None	152
erythro-CH$_3$CHClCHBrCH$_3$	Cr^{II}en, 25°, DMF/H$_2$O (88/12)	a	2.57	None	152
threo-CH$_3$CHClCHBrCH$_3$	Cr^{II}en, 25°, DMF/H$_2$O (88/12)	a	2.45	None	152
erythro-CH$_3$CHClCHBrCH$_3$	Cr^{II}, 25°, DMSO/H$_2$O (89/11)	a	2.57	None	152
threo-CH$_3$CHClCHBrCH$_3$	Cr^{II}, 25°, DMSO/H$_2$O (89/11)	a	2.03	None	152
meso-CH$_3$CHClCHClCH$_3$	Cr^{II}, 25°, DMSO/H$_2$O (89/11)	a	2.33	None	152
dl-CH$_3$CHClCHClCH$_3$	Cr^{II}, 25°, DMSO/H$_2$O (89/11)	a	2.33	None	152
meso-CH$_3$CHClCHClCH$_3$	Cr^{II}, room temp., DMF/H$_2$O (1/1)	57	2.45	None	149
dl-CH$_3$CHClCHClCH$_3$	Cr^{II}, room temp., DMF/H$_2$O (1/1)	56	2.57	None	149

| erythro-$CH_3CHClCHClCH_3$ | Cr^{II}, room temp., DMF/H_2O (1/1) | >83 | 2.85 | None | 149 |
| threo-$CH_3CHClCHClCH_3$ | Cr^{II}, room temp., DMF/H_2O (1/1) | >82 | 1.44 | None | 149 |

[a] Not reported.

experiments relevant to the question of bridging [1; III(a)] had not been performed.

6. Addition of Cl_2 to Olefins

The addition of Cl_2 to $CF_3CH=CFCF_3$ yielded $CF_3CHClCFClCF_3$ (Table 66) [156], the preference for formation of a particular diastereomer corresponding to 0.2-0.3 kcal/mole. In view of the lack of adequate control experiments, it would not be proper to conclude from this slight specificity that $CF_3CHCl\overset{.}{C}FCF_3$ (or $CF_3CFCl\overset{.}{C}HCF_3$) is weakly bridged [1; III(a)].

Results of a detailed study of the addition of Cl_2 to 2-butene are given in Tables 67 and 68 [157]. Poutsma comments: "The minimum meso/dl ratio obtained from trans-2-butene was 3.2 [$\Delta\Delta G^{\ddagger} = 0.6$ kcal/mole] in an illuminated run while the maximum ratio from the cis-isomer was 1.7 [$\Delta\Delta G^{\ddagger} = 0.3$ kcal/mole] in a similar run . . . the failure to observe identical meso/dl ratios from the two olefins could be the result of (1) a certain fraction of residual stereospecific ionic reaction in all runs imposed on a completely nonstereospecific radical process, or (2) a certain amount of stereoselectivity in the radical reaction. Only 15% of residual ionic reaction could explain the values cited above . . ." In addition, the $\Delta\Delta G^{\ddagger}$ values are not large. The combined picture, including the lack of control experiments relevant to our present purposes, provides no cause to consider $CH_3CHCl\overset{.}{C}HCH_3$ to be bridged [1; III(a)].

7. Reaction of $CH_3CHOHCHClCH_3$ with $(Bu^tO)_2$

See Sec. 6; ID10 and Table 111.

8. Summary

There are no results which indicate even weakly that a β-chloroalkyl radical is bridged [1; III(a)]; there are some which are not inconsistent with the absence of bridging [1; III(a)].

TABLE 66

Reaction of $CF_3CH=CFCF_3$ with Cl_2

| $CF_3CH=CFCF_3$ | Reaction conditions | Product - $CF_3CHClCFClCH_3$ | | |
		% Yield	Reported erythro/threo[a]	Controls
trans	hν, olefin/Cl_2 = 1.03	"Essentially quantitative"	1.44	b
trans/cis = 2.7	hν, olefin/Cl_2 = 1.08	79	1.56	None

[a] Assignment of structure, exclusive of stereochemistry, OK; basis of assignment of stereochemistry: none.

[b] Photoreaction of 26.7 mmoles of trans-$CF_3CH=CFCF_3$ and 16.2 mmoles of Cl_2 yielded 10.3 mmoles of dichloride and 10.4 mmoles of $CF_3CH=CFCF_3$ (> 95% trans).

TABLE 67

Effect of Olefin Concentration on Product Distribution from Chlorination of trans-2-Butene

Mole fraction olefin	Dark CH$_3$CHClCHClCH$_3$ %	Dark meso/dl	Illuminated CH$_3$CHClCHClCH$_3$ %	Illuminated (% Cyclohexyl chloride) [2-butene] / (% CH$_3$CHClCHClCH$_3$) [cyclohexane]	Controls meso/dl [c]
1.00[a,b]	83.5–85.5	6.9–8.9	71.8–75.7		3.2–3.8
0.92[a]	83.9	9.6	71.3		3.9
0.80[d]				0.98	3.85
0.77[d]				1.04	3.95
0.73[a]	79.2		68.1		3.4
0.70[a]		6.6			
0.67[d]				0.81–0.91	3.82–4.07
0.58[d]				0.95	4.04
0.56[a]	77.1	6.3	70.2		4.0
0.50[d]				1.01	3.99
0.39[a]	76.9–78.3	6.1–7.2	69.9		4.0
0.26[a]			72.6		4.0
0.15[a]	82.0–82.6	7.2–7.6	77.6		5.1
0.12[a]			76.6		4.6
0.085[a]	86.6	7.6			
0.079[a]	85.0	7.8	75.5		3.9

148

0.075^a			79.6	5.0
0.070^a	87.7	8.7		
0.050^a	91.2	10.6		
0.045^a			78.2	3.8
0.041^a	91.1	10.8	75.7	3.9
0.030^a	94.3	18.9		

[a] Under N_2, -9°, CCl_2FCF_2Cl solvent.

[b] < 10% conversion.

[c] When reaction run under O_2, meso/dl > 300.

[d] Under N_2, $h\nu$, -9°, cyclohexane solvent.

TABLE 68

Effect of Olefin Concentration on Product Distribution from Chlorination of cis-2-Butene

Mole fraction olefin	Dark		Illuminated			
	$CH_3CHClCHClCH_3$ %	meso/dl	$CH_3CHClCHClCH_3$ %	meso/dl	$\dfrac{(\% \text{ Chlorocyclohexane}) \,[\text{cyclohexane}]}{(\% \ CH_3CHClCHClCH_3) \,[\text{2-butene}]}$	Controls
1.00[a]	82.6–85.1	0.42–0.61	74.8	1.7		b
0.92[a]	79.7	0.84	73.9	1.6		
0.81[c]				1.34	1.03	
0.78[c]				1.20	1.04	
0.68[c]				1.26–1.27	0.98–1.00	
0.58[c]				1.40	0.98	
0.57[a]	79.1	0.74	72.5	1.7		
0.51[c]				1.24	0.93	
0.40[a]	78.8	0.75	74.4	1.4		
0.15[a]	85.2	0.47	77.7	1.2		
0.081[a]	88.0	1.4	82.7	0.85		
0.042[a]	91.2	0.25	88.6	0.63		

[a] Under N_2, –9°, CCl_2FCF_2Cl solvent.

[b] When reaction run under O_2, meso/dl < 0.01.

[c] Under N_2, –9°, cyclohexane solvent.

II. β-CHLOROCYCLOHEXYL RADICALS

A. Energy

1. Reaction of Cyclohexyl Chloride with Cl_2

See Table 69. The comments in Sec. 5;IC2 apply to these results with regard to β-, γ-, and δ-chlorocyclohexyl radicals. Data useful in the sense discussed in paragraph 2 of Sec. 5;IC3, and subject to the provisos and interpretations therein, are in Table 70.

2. Reaction of Br_2 with Cyclohexanes

Reaction of (XXIII) with Br_2 was reported to give in low yield a mixture which was 40% (XXIV)* and 60% (XXV) [†] [280].

3. Reaction of Silver Cyclohexane-1,2-Dicarboxylate with Halogens

See Sec. 6;IIIC3.

B. Control of Stereochemistry

1. Reaction of Cyclohexyl Chloride with Cl_2

See Table 71. In view of the lack of knowledge of details of the mechanism (Sec. 5;ID1) and the lack of adequate control experiments (Sec. 5;ID1), we must withhold judgment on whether these results are relevant to the question of bridging. If they are assumed to be so and are taken at face value, they indicate that the 2-chlorocyclohexyl radical, in contrast to the 2-carboxy-cyclohexyl radical, is strongly bridged [1;III(a)], the largest reported trans/cis ratio of 1,2-dichlorides corresponding to $\Delta\Delta G^{\ddagger} =$ 2.6 kcal/mole.

---

*Basis of identification of product not mentioned.

[†]Basis of identification of product: NMR chemical shift of CH_3 group.

TABLE 69

Reaction of ⬡-X with Cl_2

X	Reaction conditions	% Yield of reported monochlorinated substrate	Distribution of reported monochlorinated substrates (%)				Controls	Ref.
			1,1	1,2	1,3	1,4		
Br	-30°, hν, CS_2	?	6.4[m]	16.6[d,l]	41.7[m]	35.3[m]	None	159
Br	-30°, hν, CCl_4	?	3.1[m]	14.8[c,l]	45.5[m]	36.8[m]	None	159
Br	40°, hν, CCl_4	?	5.9[m]	11.6[b,l]	53.2[m]	29.3[m]	None	159
Cl	-30°, hν, CCl_4	?	3.7[e]	18.8[e]	47.6[f]	30.1[e]	None	159
Cl	-29°, hν, CCl_4[g]	?	1.9[h]	18.6[h]	43.4[h]	36.0[h]	None	160
Cl	-23°, hν, CS_2[g]	?	3.7[h]	12.6[h]	41.6[h]	42.1[h]	None	160
Cl	-13°, hν, CCl_4[g]	?	1.9[h]	20.5[h]	43.1[h]	34.4[h]	None	160
Cl	-10°, hν, CS_2[g]	?	3.9[h]	14.9[h]	42.1[h]	39.0[h]	None	160
Cl	0°, hν, CCl_4[g]	?	2.0[h]	20.7[h]	42.2[h]	35.0[h]	None	160
Cl	3°, hν, CS_2[g]	?	4.0[h]	15.3[h]	41.7[h]	38.8[h]	None	160
Cl	21°, hν, CCl_4[g]	?	2.1[h]	21.3[h]	44.1[h]	32.5[h]	None	160
Cl	23°, hν, CS_2[g]	?	3.9[h]	17.7[h]	42.2[h]	36.1[h]	None	160
Cl	40°, hν, CCl_4	?	5.0[e]	23.9[e]	43.5[f]	27.6[e]	None	159
Cl	42°, hν, CCl_4[g]	?	2.3[h]	22.8[h]	43.6[h]	31.1[h]	None	160
Cl	70°, gas	?	2.7[a]	37[a]	37[a]	24[a]	None	105
Cl	72°, hν, gas[i]	?	2.8[h]	37.8[h]	35.0[h]	24.3[h]	None	160

Cl	99°, hν, gas[i]	?	2.8^h	36.6^h	34.7^h	25.7^h	None	160
Cl	134°, hν, gas[i]	?	2.5^h	35.4^h	34.6^h	27.4^h	None	160
Cl	172°, hν, gas[i]	?	2.5^h	35.2^h	34.7^h	27.5^h	None	160
Cl	196°, hν, gas[i]	?	2.0^h	35.4^h	35.4^h	27.2^h	None	160
Cl	230°, hν, gas[i]	?	2.3^h	40.0^h	32.7^h	25.0^h	None	160
COOH	40°, hν, PhCl[k]	?	4^j	18^j	48^j	30^j	None	161
COOH	80°, hν, PhCl[k]	?	11^j	18^j	45^j	26^j	None	161

[a] Basis of identification not mentioned.

[b] Includes 4.5% 1,2-dichlorocyclohexane.

[c] Includes 0.6% 1,2-dichlorocyclohexane.

[d] Includes 2.6% 1,2-dichlorocyclohexane.

[e] Basis of identification: VPC retention time.

[f] Basis of identification: qualitative appearance of NMR spectrum.

[g] Substrate/Cl_2 = 6.0.

[h] "The isomers were identified by NMR spectroscopy, the spectra agreeing with those published by Russell and co-workers."

[i] Substrate/Cl_2 = 10.

[j] Product identification OK.

[k] ~15% conversion.

[l] Basis of identification of products: trans – comparison of VPC retention time with that of the product of the reaction of cyclohexene with NBS/HCl; cis – none.

[m] Basis of identification of product: none.

TABLE 70

Competitive Chlorination of ⬡–Cl and Cycloheptane (162)

Reaction conditions	[Chlorocyclohexane]		[Cycloheptane]		Reported k_γ/k_C [a,b]	Reported k_δ/k_C [a,b]	Controls
	Initial	Final	Initial	Final			
40°, hν, CCl$_4$	1.00	0.842	0.52	0.358	0.81	1.0	None

[a] "Relative reactivity of a hydrogen in the γ- or δ-position of chlorocyclohexane relative to a cyclohexane hydrogen atom."

[b] "Cycloheptane . . . hydrogen atoms taken as 1.11 . . . times as reactive as a cyclohexane hydrogen atom . . ."

TABLE 71

Reaction of ⬡-x with Cl_2

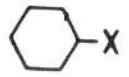

X	Reaction conditions	Reported isomer ratios (trans/cis)			Controls	Ref.
		1,2	1,3	1,4		
Br	-30°, hv, CS_2	>35[e]	1.03[f]	0.78[f]	a	159
Br	-30°, hv, CCl_4	>35[e]	0.78[f]	1.85[f]	a	159
Br	40°, hv, CCl_4	>17[e]	1.23[f]	1.46[f]	a	159
Cl	-30°, hv, CCl_4	41[b]	1.02[c]	1.10[b]	None	159
Cl	-29°, hv, CCl_4[g]	8.7[h]	1.09[h]	0.75[h]	None	160
Cl	-23°, hv, CS_2[g]	4.3[h]	1.04[h]	0.87[h]	None	160
Cl	-13°, hv, CCl_4[g]	11.0[h]	1.11[h]	0.72[h]	None	160
Cl	-10°, hv, CS_2[g]	7.7[h]	1.13[h]	0.91[h]	None	160
Cl	0°, hv, CCl_4[g]	9.2[h]	1.13[h]	0.71[h]	None	160
Cl	3°, hv, CS_2[g]	9.7[h]	1.16[h]	0.91[h]	None	160
Cl	21°, hv, CCl_4[g]	10.4[h]	1.18[h]	0.71[h]	None	160
Cl	23°, hv, CS_2[g]	18[h]	1.17[h]	0.92[h]	None	160

X	Reaction conditions	Reported isomer ratios (trans/cis)			Controls	Ref.
		1,2	1,3	1,4		
Cl	40°, hν, CCl$_4$	11.6[b]	1.25[c]	1.12[b]	None	159
Cl	42°, hν, CCl$_4$[g]	23.4[h]	1.28[h]	0.82[h]	None	160
Cl	72°, hν, gas[i]	18[h]	1.73[h]	0.78[h]	None	160
Cl	99°, hν, gas[i]	10.5[h]	2.01[h]	0.76[h]	None	160
Cl	134°, hν, gas[i]	11.5[h]	1.77[h]	0.76[h]	None	160
Cl	172°, hν, gas[i]	14.3[h]	1.67[h]	0.71[h]	None	160
Cl	196°, hν, gas[i]	16.0[h]	1.32[h]	0.75[h]	None	160
Cl	230°, hν, gas[i]	12.7[h]	1.83[h]	0.71[h]	None	160
COOH	40°, hν, PhCl[k]	1.0[j]	1.0[j]	0.67[j]	None	161
COOH	80°, hν, PhCl[k]	1.3[j]	1.0[j]	0.73[j]	None	161

[a] A significant amount of trans-1,2-dichlorocyclohexane was reported.

[b] Basis of identification of cis- and trans-products: VPC retention time.

[c] Basis of identification of cis- and trans-products: qualitative appearance of NMR spectrum.

[d] Basis of identification of cis- and trans-products not mentioned.

e Basis of identification of products: trans-comparison of VPC retention time with that of the product of the reaction between cyclohexene and NBS/HCl; cis - none.

f Basis of identification of cis- and trans-products: none.

g Substrate/Cl_2 = 6.0.

h "The isomers were identified by NMR spectroscopy, the spectra agreeing with those published by Russell and co-workers."

i Substrate/Cl_2 = 10.

j Identification of products OK.

k ~15% conversion.

III. β-CHLOROCYCLOPENTYL RADICALS

A. Energy

1. Reaction of Cyclopentyl Chloride with Cl_2

See Table 72. The comments in Sec. 5;IC2 apply to these results with regard to β- and γ-chlorocyclopentyl radicals. Data useful in the sense discussed in paragraph 2 of Sec. 5;IC3, and subject to the provisos and interpretations therein, are in Table 73.

2. Reaction of Cyclopentyl Chloride with CCl_3SO_2Cl, SO_2Cl_2, and $(CH_3)_3COCl$

See Table 74. The comments in Sec. 5;IC2 apply to these results with regard to β- and γ-chlorocyclopentyl radicals.

B. Control of Stereochemistry

1. Reaction of Cyclopentyl Chloride with Cl_2

See Table 75. The largest reported trans/cis ratio of 1,2-dichlorides corresponds to $\Delta\Delta G^{\ddagger} > 2.5$ kcal/mole. The comments in Sec. 5;IIB apply to these results with regard to the 2-chlorocyclopentyl radical.

2. Reaction of Cyclopentyl Chloride with CCl_3SO_2Cl, SO_2Cl_2, and $(CH_3)_3COCl$

See Table 76. The comments in Sec. 5;IIIB1 apply to these results.

IV. 3-CHLORO-2-NORBORNYL RADICALS

Control of Stereochemistry

Addition of Cl_2 to Norbornene

Reported products included (XXVI)-(XXXI) (Table 77) [165].

(XXVI)   (XXVII)   (XXVIII)   (XXIX)   (XXX)   (XXXI)

In view of the lack of knowledge of the details of the mechanism (Sec. 5;ID1) and the lack of adequate control experiments (Sec. 5;ID1), we must withhold judgment on whether these results are relevant to the question of bridging. If they are assumed to be so and are taken at face value, they indicate that the 3-chloro-2-norbornyl radical is not bridged [1; III(a)], the largest reported trans/cis ratio of 2,3-dichloronorbornanes corresponding to $\Delta G^{\ddagger} = 0.1$ kcal/mole.

V. 2-CHLOROALLYL RADICALS

Energy

Reaction of Halochloropropenes with Chromous Salts

See Sec. 5;IC9 for general comments. The rates of reduction of 2-halo-3-chloropropenes in Table 78 indicate a slight ($\Delta\Delta G^{\ddagger} = 0.4$ kcal/mole) accelerating effect of a β-chloro substituent. In view of the problems discussed in Sec. 1;VB, it would be premature, even if the gross features of the mechanism were known, to conclude from this result that 2-chloroallyl radicals are even weakly bridged [1;II(a)].

VI. γ-CHLOROALKYL RADICALS

A. Structure

ESR spectra, which have been assigned to γ-chloroalkyl and other radicals on the basis of qualitative consistency with the structures, are summarized in Table 79.

The "normal" α-H coupling constants and small Cl coupling constants attributed to $HO\dot{C}HCH_2CH_2Cl$ and $\cdot CH_2COCH_2Cl$ provide no cause to consider the radicals to be bridged [1; I(a)-(c)]. $\cdot CH_2COCH_2Cl$ is not bridged [1; I(a)] with equivalent geminal hydrogens.

B. Energy

1. Reaction of Alkyl Chlorides with Br_2

See Sec. 5;IC2.

TABLE 72

Reaction of ⬠-x with Cl_2

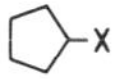

X	Reaction conditions	% Yield of reported monochlorinated substrates	Distribution of reported monochlorinated substrates (%)			Controls	Ref.
			1,1	1,2	1,3		
Br	-30°, hν, CS_2[b]	?	6.0[a]	44.5[a]	49.6[a]	None	162
Br	-30°, hν, CCl_4[b]	?	3.4[a]	43.3[a]	53.3[a]	None	162
Cl	-30°, hν, CCl_4[b]	?	4.2[c]	28.7[c]	67.1[d]	None	162
Cl	40°, hν, CCl_4[b]	?	4.5[c]	31.5[c]	64.0[d]	None	162
Cl	40°, hν, CCl_4[e]	?	4.7[c]	31.8[c]	63.5[d]	None	162
Cl	40°, hν, CCl_4[f]	?	4.5[c]	31.4[c]	64.0[d]	None	162
Cl	40°, hν, CCl_4[g]	?	4.4[c]	31.1[c]	64.4[d]	None	162
Cl	40°, hν, MeCN[b]	?	3.1[c]	29.4[c]	67.5[d]	None	162
Cl	40°, hν, $PhNO_2$[b]	?	4.4[c]	28.1[c]	67.4[d]	None	162
Cl	40°, hν, PhH[b]	?	5.8[c]	28.2[c]	66.0[d]	None	162
Cl	40°, hν, Ph_2O[b]	?	7.9[c]	25.2[c]	66.9[d]	None	162

| Cl | $40°$, $h\nu$, CS_2[b] | ? | 9.4[c] | 29.6[c] | 60.9[d] | None | 162 |
| Cl | $70°$, gas | ? | 2.3[a] | 55[a] | 43[a] | None | 105 |

[a] Basis of identification not mentioned.

[b] 15 "% chlorination."

[c] Basis of identification of product: VPC retention time.

[d] Basis of identification of product: qualitative appearance of NMR spectrum.

[e] 27 "% chlorination."

[f] 47 "% chlorination."

[g] 50 "% chlorination."

TABLE 73

Competitive Chlorination of ⬠-x, ⬡, and ⬠ (162)

X	Competitor	Reaction conditions	[⬠-x] Initial	[⬠-x] Final	Cycloalkane Initial	Cycloalkane Final	Reported k_γ/k_c [a,b]	Controls
Br	Cyclopentane	−30°, hν, CCl$_4$	1.00	0.858	0.21	0.154	0.66	None
Cl	Cyclohexane	40°, hν, CCl$_4$	1.00	0.726	0.16	0.073	0.75	None
Cl	Cyclohexane	40°, hν, PhNO$_2$	1.00	0.861	0.17	0.115	0.75	None

[a] "Relative reactivity of a hydrogen in the γ – . . . position of cyclopentylX relative to a cyclohexane hydrogen atom.

[b] ". . . cyclopentane hydrogen atoms taken as . . . 1.05 times as reactive as a cyclohexane hydrogen atom . . . "

TABLE 74

Chlorination of ⬠-x [162]

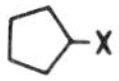

X	Chlorinating agent	Reaction conditions	% Yield of reported monochlorinated substrate	Distribution of reported monochlorinated substrates (%)			Controls
				1,1	1,2	1,3	
Br	$(CH_3)_3COCl$	-30°, hν, CCl_4 [d]	?	8.9[c]	51.0[c]	40.1[c]	None
Cl	$(CH_3)_3COCl$	40°, hν, CCl_4 [d]	?	25.0[a]	33.6[a]	41.5[b]	None
Cl	CCl_3SO_2Cl	80°, hν, CCl_4 [d]	?	14.3[a]	31.3[a]	54.0[b]	None
Cl	SO_2Cl_2	80°, hν, CCl_4 [d]	?	6.4[a]	33.0[a]	60.6[b]	None

[a] Basis of identification of product: VPC retention time.

[b] Basis of identification of product: qualitative appearance of NMR spectrum.

[c] Basis of identification of product not mentioned.

[d] 35 "% chlorination."

TABLE 75

Reaction of $\bigcirc$-x with Cl_2 (162)

X	Reaction conditions	Reported isomer ratios (trans/cis)[h]		Controls
		1,2	1,3	
Br	−30°, hν, CS_2[a]	>89[g]	1.37[g]	None
Br	−30°, hν, CCl_4[a]	>87[g]	1.30[g]	None
Cl	−30°, hν, CCl_4[a]	>57[b]	1.27[c]	None
Cl	40°, hν, CCl_4[a]	>63[b]	1.51[c]	None
Cl	40°, hν, CCl_4[d]	63[b]	1.42[c]	None
Cl	40°, hν, CCl_4[e]	63[b]	1.42[c]	None
Cl	40°, hν, CCl_4[f]	63[b]	1.42[c]	None
Cl	40°, hν, MeCN[a]	22[b]	1.19[c]	None
Cl	40°, hν, $PhNO_2$[a]	22[b]	1.19[c]	None
Cl	40°, hν, PhH[a]	34[b]	1.41[c]	None
Cl	40°, hν, Ph_2O[a]	>50[b]	1.27[c]	None
Cl	40°, hν, CS_2[a]	29[b]	1.41[c]	None

a 15 "% chlorination."

b Basis of identification of product: VPC retention time.

c Basis of identification of product: qualitative appearance of NMR spectrum.

d 27 "% chlorination."

e 47 "% chlorination."

f 50 "% chlorination."

g Basis of identification of product not mentioned.

h % yields not reported.

TABLE 76

Chlorination of ⬠–x [162]

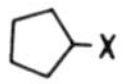

X	Chlorinating agent	Reaction conditions	Reported isomer ratios (trans/cis)[e]		Controls
			1,2	1,3	
Br	$(CH_3)_3COCl$	-30°, hν, CCl_4[a]	>102[d]	3.12[d]	None
Cl	$(CH_3)_3COCl$	40°, hν, CCl_4[a]	> 67[b]	3.44[c]	None
Cl	CCl_3SO_2Cl	80°, hν, CCl_4[a]	14[b]	3.70[c]	None
Cl	SO_2Cl_2	80°, hν, CCl_4[a]	23[b]	3.22[c]	None

[a] 35 "% chlorination."

[b] Basis of identification of product: VPC retention time.

[c] Basis of identification of product: qualitative appearance of NMR spectrum.

[d] Basis of identification of product not mentioned.

[e] % yield not reported.

TABLE 77

Reaction of Norbornene with Cl_2

Reaction conditions	(XXVI)	(XXVII)	(XXVIII)	(XXIX)	(XXX)	(XXXI)	Controls
			"Percentage of total observed products"				
25°, CCl_4, O_2, dark[a]	64.9	25.7	3.6	5.8	0	0	None
25°, CCl_4, $h\nu$, N_2[b]	28.8	5.6	28.5	30.1	5.0	2.1	None
25°, CCl_4, $h\nu$, N_2[c]	40.5	12.7	18.5	22.5	4.1	1.7	None
25°, CCl_4, "radical reaction"[d]	18.8	0	35.4	36.8	6.4	2.7	None
25°, CCl_4, "radical reaction"[e]	16.6	0	33.1	38.8	8.1	3.4	None

[a] Norbornene/Cl_2 = 10–15 ml/0.3–1.0 mmole, 0.022–0.52 mole fraction norbornene.

[b] As above, but 0.077 mole fraction norbornene.

[c] As above, but 0.055 mole fraction norbornene.

[d] (XXVII) and (XXXI) assumed not to be formed via a radical reaction. The corresponding amounts (based on first row of data) of (XXVI), (XXVIII), and (XXIX) were subtracted from the entries in the second row of data.

[e] As in d, but entries obtained from the third row of data.

TABLE 78

Reaction of $ClCH_2CX=CH_2$ with $CrSO_4$

X	Reaction conditions	Product (% yield)	Rate constant, l mole^{-1} min^{-1}	Ref.
Br	29.7°, 1.0 M HClO$_4$, DMF/H$_2$O (1/1)	CH$_3$CH=CH$_2$ (98)	2.6 ± 0.1[a]	149
Cl	29.7°, 1.0 M HClO$_4$, DMF/H$_2$O (1/1)	?	2.2 ± 0.1	149
H	29.7°, 1.0 M HClO$_4$, DMF/H$_2$O (1/1)	?	1.2 ± 0.1	150

[a] "Good [second-order] straight line plots were obtained through 95% completion . . ."

2. Reaction of Alkyl Chlorides with Cl_2

See Sec. 5;IC3. Additional data useful in the sense discussed in paragraph 2 of Sec. 5;IC3, and subject to the provisos and interpretations therein, are given in Tables 80 [168] and 81 [106].

3. Reaction of Alkyl Chlorides with $PhICl_2$

See Sec. 5;IC4.

4. Reaction of Alkyl Chlorides with Cl_2O and $HOCl$

See Sec. 5;IC5.

5. Reaction of Alkyl Chlorides with $(CH_3)_3COCl$

See Sec. 5;IC6. Additional data useful in the sense discussed in paragraph 2 of Sec. 5;IC3, and subject to the provisos and interpretations therein, are given in Table 82 [168].

6. Reaction of Alkyl Chlorides with Cl_3CSCl

See Sec. 5;IC7.

7. Reaction of Alkyl Chlorides with SO_2Cl_2

See Sec. 5;IC8.

8. Reaction of Haloalkyl Chlorides with Chromous Salts

See Sec. 5;IC9 for general comments and reservations. Reported relative rates of reaction of γ-substituted propyl halides with $Cr^{II}en$ are given in Table 83 [169]. These results indicate an accelerating effect of a γ-Cl substituent corresponding to $\Delta\Delta G^{\ddagger} = 1.2$ and 1.8 kcal/mole for $ClCH_2CH_2CH_2I$ and $ClCH_2CH_2CH_2Br$, respectively. In view of the problems discussed in Sec. 2;VB, it would be premature, even if the gross features of the mechanism were known, to conclude from these results that γ-chloroalkyl radicals are bridged [1; II(a)]. In addition to all of these problems is the following question: If the intimate details of the transition states of these reactions are sufficiently well-understood so that 1.2-1.8 kcal/mole differences can be ascribed with confidence to a particular "cause," incipient bridging [1; II(a)], then what is the origin of the $\Delta\Delta G^{\ddagger} = 2.0$ kcal/mole acceleration reported for $BrCH_2CH_2CH_2OTs$?

TABLE 79

ESR Spectra Attributed to γ-Halogen-Containing and Other Radicals

Radical	Method of generation	Coupling Constants (G)			g	Temperature	Ref.
		C_1-H	C_2-H	Cl			
HOĊHCH$_2$Cl	a	16.5 ± 0.4	16.5 ± 0.4	"Lacking"			73
HOĊHCH$_2$OH	a	15.7 ± 0.1	18.6 ± 0.1				73
HOĊHCH$_2$CH$_3$	e	14.9	19.5			22°	78
HOĊHCH$_2$CH$_3$	f	15.3	20.0			29°	76
HOĊH(CH$_2$)$_{10}$CH$_3$	d	17.5 ± 1.0	17.5, 38.6 ± 1.0			-170 to -145°	167
·CH$_2$COCH$_2$Cl	b	19.3, 19.7		0.42	2.0043	-40°	75
·CH$_2$COCH$_3$	c	20.2 ± 0.3			2.004	-100°	79
·CH$_2$COCH$_3$	a	20.3					55
·CH$_2$COCH$_3$	g	19.75, 19.75			2.00441	room	166
·CH$_2$COCH$_3$	h	19.74			2.00446	27°	166
·CH$_2$COCH$_3$	g	19.95, 19.48			2.00443	-49.5°	166
·CH$_2$COCH$_3$	j	19.5 ± 0.4			2.0042		73
·CH$_2$CHO	i	19.0 ± 0.1			2.0045		73

[a] R· generated by mixing in a flow system a solution containing H$_2$SO$_4$ and TiCl$_3$ with a solution containing H$_2$SO$_4$, RH, and H$_2$O$_2$.

[b] Photolysis of a mixture of $ClCH_2CHOHCH_2Cl$ and $(Bu^tO)_2$.

[c] γ-irradiation of acetone at $-196°$.

[d] Deposition of alkali-metal atoms on dodecanol at $77°K$ followed by illumination or slight warming."

[e] $\underline{n}$-C_4H_9OH "oxidized by Ti^{3+} + H_2O_2 system in acidic solution."

[f] Photolysis of $\underline{n}$-C_4H_9OH containing H_2O_2.

[g] Photolysis of acetone containing 1% H_2O_2.

[h] Photolysis of acetone.

[i] As in a, except substrate was $HOCH_2CH_2OH$ or $HOCH_2CH_2Cl$.

[j] As is a, except substrate was $HOCH_2CHOHCH_3$ or $ClCH_2CHOHCH_3$.

TABLE 80

Relative Reactivity of Alkyl Halides toward Cl_2

Reaction site	Substituent	Reported relative reactivity[a,b]
CH_3-	$-CH_2C(CH_3)_3$	(1.00)
CH_3-	$-CH_2CH_2Cl$	0.48

[a] "calculated from the relative yields of the chlorinated products from competitive reaction with binary mixtures of substrates. "

[b] 40°, $h\nu$.

TABLE 81

Reactivity Ratio for Photochlorination[a] of $(CH_3)_3CH$ and $(CH_3)_2CHCH_2Cl$

Reported reaction pair	Reported rate constant ratio
$(CH_3)_2CHCH_2Cl \xrightarrow{6k_1} CH_3CH(CH_2Cl)_2$ $(CH_3)_3CH \xrightarrow{9k_2} (CH_3)_2CHCH_2Cl$	$k_1/k_2 = 0.58$

[a] 2.5% reaction, CCl_4, 24°.

Finally, an alternative "explanation" of the enhanced rates, put forth and labeled "less likely [than bridging]" without elaboration, for the case of a β-bromo substituent [152] (". . . a one-step loss of a β-bromine atom synchronously with bromine transfer to Cr^{II}.") may apply here. In other words, whatever the mechanism of reaction of $C_nH_{2n+1}X$, the reaction of an alkyl halide with a fugacious γ-substituent may not involve a free alkyl radical intermediate.

C.   Control of Stereochemistry

1.   Reaction of Alkyl Chlorides with Cl_2

The relative yields of stereoisomers resulting from the chlorination of 2-chlorohexane have been reported (Table 63) [109]. The reported ratio of

TABLE 82

Relative Reactivity of Alkyl Halides toward $(CH_3)_3COCl$

Reaction site	Substituent	Reported relative reactivity[a,b]
CH_3-	$-CH_2C(CH_3)_3$	(1.00)
CH_3-	$-CH_2CH_2CH_2Cl$	1.19
CH_3-	$-CH_2CH_2Cl$	0.86
$ClCH_2-$	$-CH_2CH_2CH_3$	(1.00)
$ClCH_2-$	$-CH_2CH_3$	0.73
$ClCH_2-$	$-CH_2CH_2CH_2Cl$	0.63
$ClCH_2-$	$-CH_2CH_2Cl$	0.38

[a] "calculated from the relative yields of chlorinated products from competitive reaction with binary mixtures of substrates."

[b] 40°, hν.

the resulting 2,4-isomer corresponds to $\Delta\Delta G^{\ddagger} = 0.1$ kcal/mole. These results are subject to the reservations expressed in Sec. 5;IC2. If they are taken at face value, they indicate that, as judged by this particular reaction, $CH_3CH_2\overset{\cdot}{C}HCH_2CHClCH_3$ is not bridged [1; III(a)].

2. Reaction of Norbornenes with $BrCCl_3$ and CCl_4

Upon the future availability of data for other systems the data in Table 121 may become useful in the present context.

3. Reaction of Dibenzobicyclo[2.2.2]octatrienes with $BrCCl_3$ and Cl_3CSO_2Cl

The data in Table 84 are subject to the comments in the preceding section.

TABLE 83

Relative Rates of Reaction of γ-Substituted Propyl
Halides with $Cr^{II}en^a$ [169]

Y	Reported relative rates	
	$YCH_2CH_2CH_2I$	$YCH_2CH_2CH_2Br$
I	$30^{b,c}$	
Br		$35^{b,d}$
Cl	7^e	20^f
OTs		30^g
CH_3	(1)	(1)

[a] "In aqueous DMF solution at room temperature."

[b] Relative rate has been divided by 2.

[c] 94% conversion to propane and cyclopropane under different reaction conditions.

[d] 98% conversion to propane and cyclopropane under different reaction conditions.

[e] 94-100% conversion to propane and cyclopropane under different reaction conditions.

[f] 98-100% conversion to propane and cyclopropane under different reaction conditions.

[g] 100% conversion to propane and cyclopropane under different reaction conditions.

4. Reaction of Halodibenzobicyclo[2.2.2]octadienes with Bu_3SnH

Reaction of what was reported to be a cis/trans mixture of

 *

with Bu_3SnH in the presence of 3 mole % Bz_2O_2 in reflux-

ing benzene gave 81% (trans/cis > 4) of what was reported to be

---

*See footnote j to Table 84.

* . This result is subject to the comment in the preceding

section.

VII.   γ-CHLOROCYCLOHEXYL RADICALS

A.   Energy

Reaction of Cyclohexyl Chloride with Cl_2

See Sec. 5; IIA1.

B.   Control of Stereochemistry

Reaction of Cyclohexyl Chloride with Cl_2

See Table 71. In view of the lack of knowledge of details of the mechanism (Sec. 5; ID1) and the lack of adequate control experiments (Sec. 5; ID1), we must withhold judgment on whether these results are relevant to the question of bridging. If they are assumed to be so and are taken at face value, they indicate that the 3-chlorocyclohexyl radical is at most weakly bridged [1; III(a)], the largest reported trans/cis ratio of 1,3-dichlorides corresponding to $\Delta\Delta G^{\ddagger} = 0.6$ kcal/mole.

VIII.   γ-CHLOROCYCLOPENTYL RADICALS

A.   Energy

1.   Reaction of Cyclopentyl Chloride with Cl_2

See Sec. 5; IIIA1.

2.   Reaction of Cyclopentyl Chloride with CCl_3SO_2Cl, SO_2Cl_2, and $(CH_3)_3COCl$

See Sec. 5; IIIA2.

B.   Control of Stereochemistry

1.   Reaction of Cyclopentyl Chloride with Cl_2

See Table 75. The largest reported trans/cis ratio of 1,3-dichlorides corresponds to $\Delta\Delta G^{\ddagger} = 0.9$ kcal/mole. The comments in

---

*See footnote i to Table 84.

TABLE 84

X	Addendum	Reaction conditions	Reported product ratio, A/B (Z, Y, total % yield)	Controls	Ref.
H	BrCCl$_3$	Neat, 105°, 1.7 mole % Bz$_2$O$_2$	>11 (CCl$_3$, Br, >92)[b]	a	486
H	BrCCl$_3$	Neat, 50 or 105°, 4–8 mole % Bz$_2$O$_2$	> 4 (CCl$_3$, Br, 80–90)[b]	h	487
H	BrCCl$_3$	1 M in PhCl, 105°, 4–8 mole % Bz$_2$O$_2$	> 4 (CCl$_3$, Br, 80–90)[b]	h	487
H	BrCCl$_3$	0.15 M in PhCl, 105°, 8 mole % Bz$_2$O$_2$, hν	> 3 (CCl$_3$, Br, 64–72)[b]	h	487
H	Cl$_3$CSO$_2$Cl	Neat, 132°, 4 mole % Bz$_2$O$_2$	> 6 (CCl$_3$, Cl, >85)[i]	h	487
H	Cl$_3$CSO$_2$Cl	1 or 2 M in PhCl, 132°, 4–8 mole % Bz$_2$O$_2$	> 3 (CCl$_3$, Cl, 64–77)[i]	h	487
H	Cl$_3$CSO$_2$Cl	0.15 M in PhCl, 132°, 4–8 mole % Bz$_2$O$_2$	> 0.6 (CCl$_3$, Cl, 12–14)[i]	h	487
Cl	BrCCl$_3$	Neat, 105°, 0.7 mole % Bz$_2$O$_2$	2.5 (CCl$_3$, Br, 66)[j]	h	487
Cl	BrCCl$_3$	Neat, 50°, hν	3.0 (CCl$_3$, Br, 80–90)[j]	h	487
Cl	BrCCl$_3$	1 or 2 M in PhCl, 105°, 0.7 mole % Bz$_2$O$_2$	2.5 (CCl$_3$, Br, 80–90)[j]	h	487
Br[k]	Cl$_3$CSO$_2$Cl	Neat, or 0.15, 1, or 2 M in PhCl, 132°, 4 mole % Bz$_2$O$_2$	3.5–3.7 (CCl$_3$, Cl, 80–90)[j]	h	487
Br[k]	Cl$_3$CSO$_2$Cl	2 M in PhCl, 80°, 4 mole % Bz$_2$O$_2$	3.1 (CCl$_3$, Cl, 80–90)[j]	h	487

a None.

b Basis of identification of A: (1) acceptable C and H analyses; (2) NMR spectrum — τ = 5.27 (d, J = 2.0 cps, 1H), 5.58 (d, J = 2.8 cps, 1H), 5.73 (d of d, J = 2.8, 4.4 cps, 1H), 6.72 (d of d, J = 2.0, 4.4 cps, 1H);

(3) refluxing with NaOEt/EtOH/dioxane gave 75% of what was reported to be and 25% of what was reported to be

^{c} Basis of assignment of structure: (1) acceptable C and H analyses; (2) 1613(s), 1622(s) cm^{-1}; (3) NMR spectrum — τ = 4.73 (s, 1H), 5.18 (d, J = 3.4 cps, 1H), 5.42 (d, J = 3.4 cps, 1H); (3) refluxing in benzene with 1 equivalent of Bu$_3$SnH gave 85% of what was reported to be ; (4) reaction with LiCl in DMF at 90-95° gave 96% of what was reported to be ; (5) refluxing in MeOH with 160 mole % AgClO$_4$ gave 91% of what was reported to be

^{d} Basis of assignment of structure: (1) acceptable C and H analyses; (2) 1630 (w) cm^{-1}; (3) NMR spectrum — τ = 4.53 (d, J = 2.2 cps, 1H), 4.85 (d, J = 6.4 cps, 1H); 4) 256 mμ (ϵ 2400), 267 (2550), 272 (3200), 278 (3570); (5) reaction with silica gel or HCl/ether or heating at 165-170° gave 100% of the material described in footnote f; (6) see footnote c(4); (7) refluxing in MeOH gave 56% of the material described in footnote g; (8) see footnote c(5).

^{e} Basis of assignment of structure: (1) acceptable C and H analyses; (2) 1625(s) cm^{-1}; (3) NMR spectrum — τ = 4.76 (s, 1H), 5.67 (d, J = 2.9, 1H), 7.55 (d, 1H).

^{f} Basis of assignment of structure: (1) acceptable C and H analyses; (2) NMR spectrum — τ=4.68 (s, 1H), 5.20 (d, J=3.5 cps, 1H), 5.42 (d, J=3.5 cps, 1H); (3) refluxing in PhCl with 1 equivalent of Bu$_3$SnH gave 100% of the material described in footnote e; (4) see footnote c(5).

g Basis of assignment of structure: (1) acceptable C and H analyses; (2) 1630(s) cm^{-1}; (3) NMR spectrum — $\tau = 4.81$ (s, 1H), 5.40 (d, J = 3.6, 1H), 5.96 (d, J = 3.6 cps, 1H), 6.67 (s, 3H).

h Products "stable under the reaction conditions."

i Basis of identification of product [486]: (1) acceptable C and H analyses; (2) NMR spectrum — $\tau = 5.21$ (d, J = 2.0 cps, 1H), 5.60 (d, J = 2.9 cps, 1H), 5.68 (d of d, J = 2.9, 4.2 cps, 1H), 6.81 (d of d, J = 2.0, 4.2 cps); (3) refluxing in NaOEt/EtOH gave 84% of the material described in footnote f.

j Basis of identification of A and B in mixture: (1) acceptable C and H analyses; (2) portion of NMR spectrum attributed to A — $\tau = 5.07$ (s, 1H), 5.13 (d, J = 1.2 cps, 1H), 6.04 (d, J = 1.2 cps, 1H); (3) portion of NMR spectrum attributed to B — $\tau = 5.07$ (s, 1H), 5.13 (d, J = 1.2 cps, 1H), 6.35 (d, J = 1.2 cps, 1H).

k Basis of assignment of structure: (1) acceptable C and H analyses; (2) NMR spectrum — $\tau = 4.99$ (d, J = 2.2 cps, 1H), 5.08 (d, J = 6.3 cps, 1H); (3) prepared by reaction of what was reported to be the A (X = H, Y = Z = Br) compound[l] with Bu^{t}OK/DMSO at room temperature.

l Basis of assignment of structure: (1) acceptable C and H analyses; (2) NMR spectrum — $\tau = 5.6$–5.9 (m, 4H); (3) prepared by photoreaction of the X = H compound with $BrCl_2CCCl_2Br$ in benzene at 70–80°.

Sec. 5; VIIB apply to these results with regard to the 3-chlorocyclopentyl radical.

IX. γ-CHLOROCYCLOBUTYL RADICALS

Control of Stereochemistry

Reaction of 1,3-Dichlorocyclobutanes with Organotin Hydrides

Results are given in Table 85. The very small stereoselectivity provides no cause to consider the γ-chlorocyclobutyl radicals to be bridged [1; III(a)].

X. δ-CHLOROPROPARGYL RADICALS

Structure

The ESR spectrum which was assigned to $\cdot CH_2C \equiv CCH_2Cl$ on the basis of its qualitative consistency with the structure is summarized, along with those attributed to other propargyl radicals, in Table 86. The α- and β-hydrogen coupling constants are not unusual. The Cl coupling constant may be discussed in terms of the usual "hyperconjugative" or "spin polarization" "effects," obviating the need to consider a bridged structure, or it may be discussed in terms of bridging. On balance, while not inconsistent with a slightly bridged [1; I(c)] structure for $\cdot CH_2C \equiv CCH_2Cl$, these data provide no indication that the molecule is bridged [1; I(a)-(c)]; it is not bridged [1; I(a)] with equivalent geminal hydrogens.

XI. δ-CHLOROALKYL RADICALS

A. Energy

1. Reaction of Alkyl Chlorides with Br_2

    See Sec. 5; IC2.

2. Reaction of Alkyl Chlorides with Cl_2

    See Sec. 5; IC3.

TABLE 85

Reaction of 1,3-Dichlorocyclobutanes with Bu_3SnH

Substrate	Reaction conditions	Product		Controls	Ref.
		Structure (% yield)	cis/trans		
(1,3-dichloro-cyanocyclobutane-d_2)	100–130°	(80)	~2	None	414
(1,3-dichloro-tetramethylcyclobutane)	Reflux, benzene, 0.7 mole % AIBN	(79)	1.7 or 0.6	None	415

TABLE 86

ESR Spectra Attributed to δ-Halogen-Containing and Other Radicals

Radical	Method of generation	Coupling constants (G)			Temperature	Ref.
		C_1-H	C_4-H	Cl		
·$CH_2C{\equiv}CCH_2Cl$	a	17.6	9.7	4.9	~-60°	75
·$CH_2C{\equiv}CCH_3$	g	18.54	12.37		-68°	172
·$CH_2C{\equiv}CCH_2CH_3$	h	18.51	12.34		-108°	172
·$CH_2C{\equiv}CH$	b	19.0			~-40°	56
·$CH_2C{\equiv}CH$	c	17.8			-196°	170
·$CH_2C{\equiv}CH$	d	18.86			~-180°(?)	171
·$CH_2C{\equiv}CH$	e	18.92			-120°	172,174
·$CH_2C{\equiv}CH$	i	16.6			-196°	173
·$CH_2C{\equiv}CPh$	f	16.40			?	172
·$CH_2C{\equiv}CPh$	j	12.3			-196°	173

[a] "They were mostly prepared from the corresponding alkyl bromides by reaction with triethylsilyl radicals, in a few cases hydrogen atoms were abstracted from the parent alkane by t-butoxy radicals."

[b] Photolysis of $BrCH_2C{\equiv}CH$, $(Bu^tO)_2$, and Et_3SiH.

[c] Photolysis of allene, allyl alcohol, propargyl alcohol, or 1,3-dichloropropene.

[d] Electron radiolysis of 5% allene in propane.

e Photolysis of $(Bu^tO)_2$ and allene or methylacetylene.

f Photolysis of $(Bu^tO)_2$ and $PhC \equiv CCH_3$.

g Photolysis of $(Bu^tO)_2$ and 2-butyne or 1,2-butadiene.

h Photolysis of $(Bu^tO)_2$ and 1,2-pentadiene.

i γ-irradiation of propyne.

j γ-irradiation of $PhC \equiv CCH_3$.

3. Reaction of Alkyl Chlorides with $PhICl_2$

   See Sec. 5;IC4.

4. Reaction of Alkyl Chlorides with Cl_2O and $HOCl$

   See Sec. 5;IC5.

5. Reaction of Alkyl Chlorides with $(CH_3)_3COCl$

   See Sec. 5;VIB5.

6. Reaction of Alkyl Chlorides with Cl_3CSCl

   See Sec. 5;IC7.

7. Reaction of Alkyl Chlorides with SO_2Cl_2

   See Sec. 5;IC8.

B. Control of Stereochemistry

Reaction of Alkyl Chlorides with Cl_2

The relative yields of stereoisomers resulting from the chlorination of 2-chlorohexane have been reported (Table 68) [109]. The reported ratio of the resulting 2,5-isomer corresponds to $\Delta\Delta G^\ddagger = 0.0$ kcal/mole. These results are subject to the reservations expressed in Sec. 5;IC2. If they are taken at face value, they indicate that, as judged by this particular reaction, $CH_3\overset{\cdot}{C}HCH_2CH_2CHClCH_3$ is not bridged [1; III(a)].

XII. δ-CHLOROCYCLOHEXYL RADICALS

A. Energy

Reaction of Cyclohexyl Chloride with Cl_2

   See Sec. 5; IIA1.

B. Control of Stereochemistry

Reaction of Cyclohexyl Chloride with Cl_2

See Table 71. In view of the lack of knowledge of details of the mechanism (Sec. 5;ID1) and the lack of adequate control experiments (Sec. 5;ID1), we must withhold judgment on whether these results are relevant to the question of bridging. If they are assumed to be so and are taken at face value, they indicate that the 4-chlorocyclohexyl radical is not bridged [1; III(a)], the largest cis/trans ratio of 1,4-dichlorides corresponding to $\Delta\Delta G^{\ddagger} = 0.3$ kcal/mole.

XIII. MISCELLANEOUS γ- and δ-CHLORO RADICALS

Structure

The ESR spectrum attributed to the o-chlorophenoxymethyl radical on the basis of qualitative consistency with the structure is described in Table 87. The "normal" coupling constants provide no cause to consider the radical to be bridged [1;I(a)-(c)].

The room temperature ESR spectra of (XXXII) and (XXXIII) in CCl_4 have been reported [522]. All of the reported g-factors were "normal," falling between 2.0025 and 2.0028, and would provide no cause to consider the radicals to be bridged [1; I(a)-(c)].

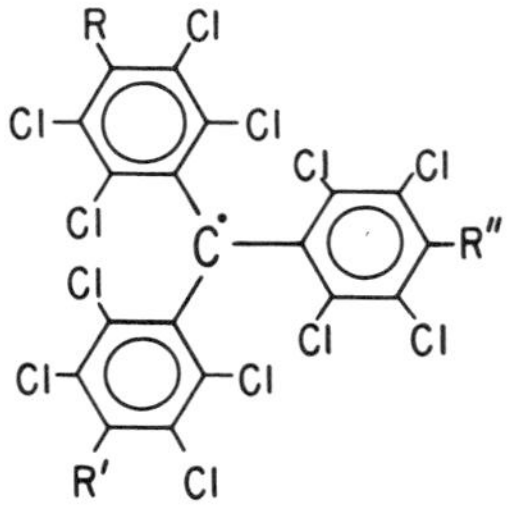

(XXXII)

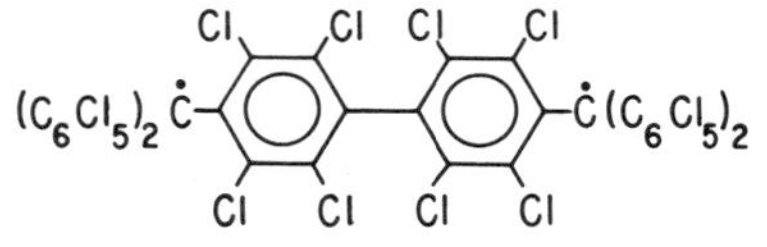

(XXXIII)

$(R = R' = R'' = Cl; R = C_6Cl_5, R' = R'' = Cl; R = R' = C_6Cl_5, R'' = Cl; R = R' = R'' = C_6Cl_5)$

TABLE 87

ESR Spectra Attributed to [a] 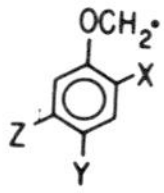 [472]

X	Y	Z	Coupling Constants (G)			
			α-H	$\underline{o}$-H	$\underline{m}$-H	$\underline{p}$-H
Cl	H	H	17.86	0.79	0.43, <0.15	0.43
Br	H	H	17.86	0.74	0.44, 0.12	0.43
Br	Br	OCH_3	17.73	0.86	— , <0.15	—
H	H	H	17.77	0.54	0.27	0.49
H	Br	H	17.77	0.61	0.27	—
H	Cl	H	17.80	0.61	0.28	—
H	F	H	17.61	0.62	0.30	—
H	OCH_3	H	17.36	0.52	0.26	—
H	H	F	17.83	0.59	0.22	0.43
H	H	OCH_3	17.61	0.56	0.25	0.40

[a] Generated by photolysis of 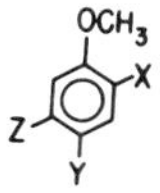 and $(Bu^tO)_2$ at -35°.

XIV. ε-, ζ-, η-, AND θ-CHLOROALKYL RADICALS

Energy

1. Reaction of Alkyl Chlorides with Cl_2

    See Sec. 5;IC3.

2. Reaction of Alkyl Chlorides with SO_2Cl_2

    See Sec. 5;IC8.

Chapter 6

BROMO-SUBSTITUTED RADICALS

I. β-BROMOALKYL RADICALS

A. Structure

ESR spectra attributed to the 3-bromo-2-butyl and the 4-bromo-3-hexyl radicals were obtained by observation of the reaction mixture resulting from the photolysis at 77°K of 2-butene + HBr (or DBr) and of cis-3-hexene + HBr [175]. "The spectra for both isomers [of 2-butene] show identical seven line patterns, but the splittings for the trans compound are 12.6 gauss and those in the cis, 10.8 gauss." "The spectrum of the photolyzed cis-3-hexene-HBr mixture gave a poorly resolved five line spectrum with a hyperfine constant of 8.0 gauss." It was considered unlikely that the spectra were of molecules of the type RĊHCHBrR (R = CH_3, C_2H_5) because (in the 2-butene experiments) "if . . . [such a structure] were correct, one would expect six equally spaced lines with a splitting of about 25 gauss" and (in the 3-hexene experiments) a " radical of . . . [such] structure would also yield a five line spectrum, but the magnitude of coupling would be expected to be about 25 gauss . . ." It was concluded that "in each case an analysis of the hyperfine pattern strongly suggested a bridged [1; I(a)] type intermediate . . ."

Symons [176] "has suggested an interpretation not considered originally, which, if correct, invalidates the conclusions outlined above": "The spectrum reported for 2-butene radicals . . . is in reasonable accord with expectation for the radical $\cdot CH_2CH=CHCH_3$." In addition: "The major drawback of the bridging bromine atom concept is that there is no indication, in the spectra discussed above, of any anisotropy in the g-values. Appreciable anisotropy would be expected for the bromine atom adducts,

187

together with a shift in g_{av} to values somewhat greater than that of the free-spin . . . Furthermore, there is no indication of the expected characteristic hyperfine coupling to bromine." (For comments relevant to the final point see Ref. 195.)

Symons' comments are reinforced by the observations [177, 178] that reaction of H· with 1, 3-butadiene at 77°K results in the formation of material whose ESR spectrum, assigned to · $CH_2CH=CHCH_3$, is very similar to that reported by Abell and Piette [175] in the trans-2-butene + HBr (or DBr) case.

See also Section [6; ID(10)].

B. 1, 2-Br Migrations

See the first two paragraphs of Sec. 5;IB for a general discussion. Observations and data have been tabulated (only those products possibly indicative of a Br shift) according to the following possible means of obtaining β-bromoalkyl radicals: radiolysis or photolysis of alkyl bromides (Table 88), addition of thiols to bromoolefins (Table 89), reaction of alkyl bromides with Cl_2 (Table 90), and reaction of alkyl bromides with $(CH_3)_3COCl$ (Table 91).

C. Energy

1. "Direct" Tests of Bridging

As illustrated by statements made concerning the thermochemistry of the following reactions, β-bromoalkyl radicals have frequently been treated, without justification, as if they were not bridged [1; II(a)]: $(CH_3)_2CBrCH_2$· → Br· + $(CH_3)_2C=CH_2$ (184); $BrCH_2CH_2$· + HBr → $BrCH_2CH_3$ + Br· [185,186, 189]; Br· + $CH_2=CH_2$ → $BrCH_2CH_2$· [185,186,190]; Br· + $CH_2=CHR$ → $BrCH_2\dot{C}HR$ [187]; $BrCH_2CH_3$ → $BrCH_2CH_2$· + H· [188]; Br· + cis-CHBr=CHBr → $CHBr_2\dot{C}HBr$ [191].

The results of a recent study [181] of the radiolysis and photolysis of $(CH_3)_2CHCH_2Br$ to give $(CH_3)_3CBr$, $(CH_3)_2C=CH_2$, HBr, and other products have been treated so as to give "experimental" values of the free energies of formation of $(CH_3)_2CBrCH_2$· and $(CH_3)_2\dot{C}CH_2Br$. Since this is the only work which permits the direct application of definition 1; II(a), it will be discussed in detail in spite of the lack of detailed probing of the mechanism and absence of control experiments. The data were analyzed in terms of a mechanism which, although not unreasonable, is unsubstantiated and has reasonable alternatives. Within that mechanism, assumptions were made which, although not unreasonable, are unsubstantiated. Their analysis is outlined below.

The reaction was considered [181] to be proceeding as indicated in Eqs. (2)-(8):

$$(CH_3)_2CHCH_2Br \xrightarrow{\text{h}\nu \text{ or } \gamma} (CH_3)_2CHCH_2 \cdot + Br \cdot \tag{2}$$

$$(CH_3)_2CHCH_2 \cdot + (CH_3)_2CHCH_2Br \longrightarrow (CH_3)_3CH + (CH_3)_2\dot{C}CH_2Br \tag{3}$$

$$(CH_3)_2CHCH_2 \cdot + HBr \longrightarrow (CH_3)_3CH + Br \cdot \tag{4}$$

$$Br \cdot + (CH_3)_2CHCH_2Br \rightleftharpoons HBr + (CH_3)_2\dot{C}CH_2Br \tag{5}$$

$$Br \cdot + (CH_3)_2C=CH_2 \rightleftharpoons (CH_3)_2\dot{C}CH_2Br \tag{6}$$

$$Br \cdot + (CH_3)_2C=CH_2 \rightleftharpoons (CH_3)_2CBrCH_2 \cdot \tag{7}$$

$$(CH_3)_2CBrCH_2 \cdot + HBr \longrightarrow (CH_3)_3CBr + Br \cdot \tag{8}$$

If it is assumed that equilibria (5), (6), and (7) are maintained throughout most of the reaction, *

$$\frac{[(CH_3)_2\dot{C}CH_2Br]}{[Br \cdot]} = K_6[(CH_3)_2C=CH_2] = \frac{K_5[(CH_3)_2CHCH_2Br]}{[HBr]} \tag{I}$$

and

$$\frac{[(CH_3)_2\dot{C}CH_2Br]}{[(CH_3)_2CBrCH_2 \cdot]} = \frac{K_6}{K_7} \tag{II}$$

---

* $[(CH_3)_2C=CH_2]$ and $[HBr]$ were found to reach the same constant values very early in the reaction in both the photolysis and the radiolysis, the concentrations being the same between and within experiments.

TABLE 88

Radiolysis and Photolysis of Alkyl Bromides

Reactant	Reported product	Reaction conditions	Yield of product	Controls	Ref.
$CH_3CH_2CH_2CH_2Br$	$CH_3CH_2CHBrCH_3$	γ-rays, 25° [g]	G = 10	None	81
$CH_3CH_2CH_2CH_2Br$	$CH_3CH_2CHBrCH_3$	γ-rays, ~25° [i]	G = 0.2 [j]	None	182
$CH_3CH_2CH_2CH_2Br$	$CH_3CH_2CHBrCH_3$	γ-rays, 30°	G = 500,[a] 300 [b]	c	179
$CH_3CH_2CH_2CH_2Br$	$CH_3CH_2CHBrCH_3$	γ-rays, 100° [h]	G = ~80	None	81
$CH_3CH_2CH_2Br$	$CH_3CHBrCH_3$	γ-rays, 30° [d]	G = 2.4	None	179
$CH_3CH_2CH_2Br$	$CH_3CHBrCH_3$	γ-rays, room temp. [e]	G = 2.1	None	81,180
$CH_3CH_2CH_2Br$	$CH_3CHBrCH_3$	γ-rays, 100° [f]	G = 7	None	81
$CH_3CH_2CHBrCH_3$	$CH_3CH_2CH_2CH_2Br$	γ-rays, ~25° [i]	G = 0.9 [j]	None	182
$(CH_3)_2CHCH_2Br$	$(CH_3)_3CBr$	γ-rays, room temp.	G = 40	None	181
$(CH_3)_2CHCH_2Br$	$(CH_3)_3CBr$	hν	Φ=1.6	None	181

[a] Dose = 6.9×10^{19} eV g^{-1} at 3.8×10^{17} eV g^{-1} min^{-1}.

[b] Dose = 1.3×10^{17} eV g^{-1} at 4.5×10^{15} eV g^{-1} min^{-1}.

^{c} Irradiation (dose = 6.8×10^{19} eV g^{-1}) of CH$_3$CH$_2$CH$_2$CH$_2$CH$_2$Br containing 2.5×10^{-3} M I$_2$ at 31° or 83° caused G to drop to less than 0.1; irradiation (dose = 8.8×10^{17} eV g^{-1}) in the presence of 1.1×10^{-3} M DPPH at 30° caused G to drop to less than 0.1.

^{d} Dose = 6.9×10^{19} eV g^{-1}.

^{e} Dose = 0.7–9×10^{20} eV g^{-1} at 3.3×10^{18} eV g^{-1} min^{-1}.

^{f} Dose = 0.7–2×10^{20} eV g^{-1} at 3.3×10^{18} eV g^{-1} min^{-1}.

^{g} Dose = 1.6×10^{20} eV g^{-1} at 3.5×10^{18} eV g^{-1} min^{-1}.

^{h} Dose less than 0.7×10^{20} eV g^{-1}.

^{i} 3.2×10^5 r hr^{-1}.

^{j} Basis of identification of product: VPC retention time.

TABLE 89

Reaction of Thiols with Bromoolefins [92]

Olefin	Thiol	Reaction conditions	Product (% yield)	Controls
$CH_2=CHCH_2Br$	PhSH	$h\nu$, 15°, RSH/halide = 1	$PhSCH_2CHBrCH_3$ (44)	None
$CH_2=CHCH_2Br$	PhSH	$h\nu$, 15°, RSH/olefin = 2.0	$PhSCH_2CHBrCH_3$ (77)	None

TABLE 90

Reaction of Alkyl Bromides with Cl_2 [133]

Reactant	Reaction conditions	Reported product (% yield)	Controls
$(CH_3)_3CBr$	$h\nu$, room temp., CCl_4	$(CH_3)_2CClCH_2Br^a$ (?)	None
$(CH_3)_2CHCH_2Br$	$h\nu$, room temp., CCl_4	$(CH_3)_2CBrCH_2Cl^b$ (?)	None

[a] Basis of identification of product: VPC retention time different and NMR spectrum similar to, but different from, that of $(CH_3)_2CBrCH_2Cl$.

[b] Identification of product OK.

TABLE 91

Reaction of Alkyl Bromides with $(CH_3)_3COCl$

Reactant	Reaction conditions	Reported product (% yield)	Controls	Ref.
$(CH_3)_3CBr$	$h\nu$, $-78°$	$(CH_3)_2CClCH_2Br$[a] (92)	None	183
$(CH_3)_3CBr$	$h\nu$, $-78°$	$(CH_3)_2CClCH_2Br$[c] (<40)	d	184
$CH_3CHBrCH_3$	$h\nu$, $-78°$	$CH_3CHClCH_2Br$[b] (15)	None	183

[a] Basis of identification of product: "The structure . . . followed from the base dehydrohalogenation of crude $(CH_3)_2CClCH_2Br$ which converted it to $(CH_3)_2C{=}CHBr$, uncontaminated by $(CH_3)_2C{=}CHCl$. Addition of BrCl to isobutylene also produced $(CH_3)_2CClCH_2Br$."

[b] "... had IR bands identical with those" of an authentic sample.

[c] Basis of identification of product not mentioned.

[d] $[(CH_3)_2CBrCH_2Cl]/[(CH_3)_2CClCH_2Br]$ linear with $[Bu^tOCl]$.

From (I), *

$$[(CH_3)_2C=CH_2] \, [HBr] = \frac{K_5 \, [(CH_3)_2CHCH_2Br]}{K_6} = C \qquad \text{(III)}$$

$$\frac{d \, [Bu^tBr]}{dt} = k_8 \, [(CH_3)_2CBrCH_2\cdot] \, [HBr] \qquad \text{(IV)}$$

If it is assumed that all termination reactions are diffusion-controlled with the same rate constant, k_t, then the rate of termination is given by [†]

$$R_t = 2k_t \, [(CH_3)_2CBrCH_2\cdot + (CH_3)_2\dot{C}CH_2Br + Br\cdot]^2$$

Let R_t = Rate of initiation, R_i. Then

$$(R_i/2k_t)^{1/2} = [(CH_3)_2CBrCH_2\cdot] + [(CH_3)_2\dot{C}CH_2Br] + [Br\cdot] \qquad \text{(V)}$$

Combine I, II, III, IV, and V:**

$$\left(\frac{d \, [Bu^tBr]}{dt}\right)^{-1} = \frac{(K_6/K_7) + 1}{k_8C}\left(\frac{2k_t}{R_i}\right)^{1/2}[(CH_3)_2C=CH_2]$$

$$+ \frac{1}{K_7k_8C}\left(\frac{2k_t}{R_i}\right)^{1/2} \qquad \text{(VI)}$$

---

* "The observation that the plateau concentrations are the same in photolysis and radiolysis, despite the great difference in rates of initiation, is also in accord with expression III."

[†] (1) The factor of 2 accounts for the fact that the radicals are destroyed pairwise. If by rate of termination is meant rate of destruction of radicals, then every act of termination destroys two radicals. A factor of 2 is also included in R_i for similar reasons (see below).

(2) It may appear as if the cross terms are being counted twice. This is necessary in order to be able to use the same k_t throughout.

** It was found that: (1) $d[Bu^tBr]/dt \propto$ (dose rate)$^{1/2}$ at constant $[Me_2C=CH_2]$ (very few points on graph); (2) $(d[Bu^tBr]/dt)^{-1/2} \propto [Me_2C=CH_2]_{plateau}$ at constant dose rate (scatter; points have large uncertainties).

From the slope of a plot of $(d[Bu^tBr]/dt)^{-1}$ vs. $[(CH_3)_2C=CH_2]_{plateau}$,

$$\frac{(K_6/K_7) + 1}{k_8 C} \left(\frac{2k_t}{R_i} \right)^{1/2} = 1.52 \times 10^8 \ sec \ M^{-2} \qquad (VII)$$

R_i was estimated from G(isobutane) and the dose rate.* k_8 was taken to be equal to $5.4 \times 10^6 \ mole^{-1} \ sec^{-1}$.[†] k_t was calculated from the Smoluchowski equation [192] as $5.3 \times 10^9 \ M^{-1} \ sec^{-1}$ by using the measured self-diffusion coefficient of Bu^iBr [193] and by taking the encounter diameter as 3.5 Å. K_6/K_7 was then calculated from (VII) to be 74.** From (VI) and the slope/intercept ratio of the plot which led to (VII),

$$(K_6/K_7 + 1) K_7 = 323 \pm 60 \ M^{-1}.$$

Then, $K_7 = 4 \ M^{-1}$ and $K_6 = 323 \ M^{-1}$. These values, when combined with a value of $C = 2.25 \times 10^{-5} \ M^2$ determined from a plot (approximately linear with significant scatter) of $[HBr]^{-1}$ vs. $[(CH_3)_2C=CH_2]_{plateau}$ yield $K_5 = (7.9 \pm 2.2) \times 10^{-4}$.

The values of ΔG corresponding to K_5, K_6, and K_7 are compared in Table 92 with values which we have calculated. The comparison provides no support for the idea that β-bromoalkyl radicals are bridged [1; II(a)]. If the "observed" and calculated values of ΔG are taken at face value, it may be concluded that $(CH_3)_2CBrCH_2\cdot$ and $(CH_3)_2\overset{\cdot}{C}CH_2Br$ are not bridged [1; II(a)].

---

*"The net effect of [the processes which occur initially in radiolysis] . . . that is of importance for this study is expected to be the predominant rupture of the C-Br bond, as was found in the radiolysis of n-propyl bromide [180]. Therefore the rate of production of isobutane in radiolysis should be a fairly good measure of the rate of initiation."

[†]Deduced from the rate of reaction of the bromoethyl radical with HBr [189].

**"This estimate is probably only good within a factor of 2 . . ."

TABLE 92

Thermochemistry of β-Bromoethyl Radicals

Reaction	Calculated[a] ΔS, eu	Calculated[a] ΔH, kcal/mole	Calculated[a] ΔG, kcal/mole	"Observed" K	"Observed" ΔG, kcal/mole
$Br\cdot + (CH_3)_2CHCH_2Br$ $\rightleftharpoons HBr + (CH_3)_2\dot{C}CH_2Br$	9.9^b	3.5^b	0.5	$(5.7-10.1)\times10^{-4}$	4.1-4.4
$Br\cdot + (CH_3)_2C=CH_2$ $\rightleftharpoons (CH_3)_2CBrCH_2\cdot$	-23.1^c	-6.4^d	0.5	$(0.8-2.5)$ atm^{-1}	$(-0.5)-1.3$
$Br\cdot + (CH_3)_2C=CH_2$ $\rightleftharpoons (CH_3)_2\dot{C}CH_2Br$	-23.7^e	-12.4^f	-5.3	$(108-157)$ atm^{-1}	$(-3.0)-(-2.8)$

[a] Standard states of ideal gas at 1 atm and 25°. Thermochemical data for Me_3CH and $Me_2C=CH_2$ taken from S.W. Benson, F.R. Cruickshank, D.M. Golden, G. R. Haugen, H.E. O'Neal, A.S. Rodgers, R. Shaw, and R. Walsh, Chem. Revs., 69, 279 (1969); those for $Br\cdot$, HBr, $Me_3C\cdot$, $EtBr$, C_2H_6, $H\cdot$, $Et\cdot$, $Me_2CHCH_2\cdot$, and Me_3CBr taken from S.W. Benson, "Thermochemical Kinetics," Wiley, New York, 1968.

[b] Calculated for $Br\cdot + (CH_3)_2CHCH_3 \rightarrow HBr + (CH_3)_2\dot{C}CH_3$.

[c] Calculated assuming $S(Me_2\dot{C}BrCH_2\cdot) = S(Me_2CHCH_2\cdot) + S(EtBr) - S(C_2H_6)$.

[d] Calculated as $\Delta H(H\cdot + Me_2C=CH_2 \rightarrow Me_2CHCH_2\cdot) + D(Me_3C-H) - D(Me_3C-Br)$.

[e] Calculated assuming $S(Me_2\dot{C}CH_2Br) = S(Me_2\dot{C}CH_3) + S(EtBr) - S(C_2H_6)$.

[f] Calculated as $\Delta H(H\cdot + Me_2C=CH_2 \rightarrow Me_2\dot{C}CH_3) + D(Et-H) - D(Et-Br)$.

2. Reaction of Haloalkyl Bromides with Organotin Hydrides

As is seen from the data in Table 21 replacement of a β-hydrogen by Br and a β-chlorine by bromine in the 3-bromo-2-butyl series results in increases in the rate at which bromine is abstracted by $Bu_3Sn\cdot$ which correspond to $\Delta\Delta G^{\ddagger} = 1.2$ and 0.3 kcal/mole, respectively. We believe the 1.2 kcal/mole figure to be subject to the comments in Sec. 5;IC1a. If a position is taken which rejects our comments, however, and if these rate increases are interpreted in terms of incipient bridging [1; II(a)], then that position must take cognizance of the 0.3 kcal/mole figure and contend that, in the sense that Br is a significantly better bridging group than H, it is insignificantly better than Cl. The 0.2 kcal/mole difference in $\Delta G^{\ddagger}$ between the β-Br and β-Cl compounds in the α-bromophenethyl series would also have to be viewed in this way. Finally, "it was shown that . . . [diethyl] monobromosuccinate was reduced to succinate by Bu_3SnH at 0.9 times the rate at which the latter reacted with [diethyl] dl-2,3-dibromo-succinate." This result provides no cause to consider $EtOOC\overset{\cdot}{C}HCHBrCOOEt$ to be bridged [1; II(a)].

3. Reaction of Alkyl Bromides with Br_2

The products resulting from the bromination of $CH_3CH_2CH_2CH_2Br$ (Table 22) and $CH_3CH_2CHBrCH_3$ (Table 23) have been reported. $CH_3CH_2CH_2CH_2Br$ shows a clear preference for formation of vicinally substituted product, a preference not shown by any other butyl derivative (Cl, F, CF_3, $OCOCF_3$, $OCOCH_3$, OCOCl, OCOF, $COOCH_3$, H). However, there is no indication (Sec. 5; IC2) that the relative amounts of products reflect relative rates of abstraction of the corresponding hydrogens, and in view of the report of Tanner et al. (see below), we need not remain uncertain as to whether these results are relevant to the question of bridging by β-, γ-, or δ-bromo substituents. There is probable cause to believe that they are not.

Tanner et al. [38] have studied the bromination of $CH_3CH_2CH_2CH_2Br$ and have reported the results of control experiments designed to determine whether the distribution of isomers of dibromobutanes depends on the extent of reaction (Table 93) or on the presence of HBr (Table 94). If taken at face value, those results (see also Table 22) corresponding to early stages

TABLE 93

Reaction of $CH_3CH_2CH_2CH_2Br$ with Br_2 [a]

% Reaction	% Yield of reported monobrominated substrate	Distribution of reported monobrominated substrates (%)			
		1,1 [b]	1,2 [c]	1,3 [c]	1,4
2	?	25	26	49	–
4	?	20	29	51	–
9	?	19	28	53	–
18	?	20	32	48	–
29	?	11	47	42	–
38	?	3	59	38	–
59	?	3	65	32	–
84	?	2	79	19	–
95	d	Trace	88	12	–

[a] $h\nu$, 8.1 M $CH_3CH_2CH_2CH_2Br$, 1.325 M Br_2.

[b] Identification of product OK.

[c] ". . . identified by comparison of . . . retention times on three different glpc columns with those of commercially available compounds."

[d] "The value reported represents a material balance for the products determined."

TABLE 94

Photoreaction of $CH_3CH_2CH_2CH_2Br$ with Br_2

Time, min	Reaction conditions	% Yield of reported monobrominated substrate	Distribution of reported monobrominated substrates (%)			
			1,1[a]	1,2[b]	1,3[b]	1,4
30	No HBr added	?	20	23	57	—
30	Added HBr (0.34 M)	?	8	66	26	—
60	No HBr added	?	19	23	58	—
60	Added HBr (0.34 M)	?	3	79	17	—
120	No HBr added	?	8	35	56	—
160	Added HBr (0.34 M)	?	4	83	13	—
180	No HBr added	?	35	40	25	—

[a] Identification of product OK.

[b] " . . . identified by comparison of . . . retention times on three different glpc columns with those of commercially available compounds."

of the reaction and the absence of added HBr, obtained with knowledge and under conditions which begin to approach meeting the conditions of section 5; IC2, indicate no exceptional stabilization of the transition states leading to β-, γ-, or δ-bromoalkyl radicals.

The data in Table 26, for which a preponderance of evidence indicates that the substance of the conditions of Sec. 5; IC2 is met, show an accelerating influence of a β-Br substituent corresponding to $\Delta \Delta G^{\ddagger} = 0.9$ kcal/mole at 12°. In view of the problems discussed in Sec. 1; VB it would be premature to interpret this result in terms of weak incipient bridging [1; II(a)] in the transition state.

4. Reaction of Alkyl Bromides with N-Bromosuccinimide

The bromination of $CH_3CH_2CH_2CH_2Br$ has been studied [38] and the results reported of control experiments designed to determine whether the distribution of isomers of dibromobutane depends on the extent of reaction (Table 95). If taken at face value, those results (see also Table 22) corresponding to early stages of the reaction, obtained with knowledge and under conditions which begin to approach meeting the conditions of Sec. 5; IC2, indicate no exceptional stabilization of the transition states leading to β-, γ-, or δ-bromoalkyl radicals.

5. Reaction of Alkyl Bromides with Cl_2

The products resulting from the chlorination of $CH_3CH_2CH_2CH_2Br$ (Table 27) and $CH_3CH(CH_3)CH_2Br$ (Table 30) have been reported. The comments in Sec. 5; IC2 apply to these results with regard to β-, γ-, and δ-bromoalkyl radicals.

6. Reaction of Alkyl Bromides with SO_2Cl_2

The products resulting from the chlorination of $CH_3CH_2CH_2Br$ are given in Table 50. The comments in Sec. 5; IC2 apply to these results with regard to β- and γ-bromoalkyl radicals.

7. Reaction of Haloalkyl Bromides with Chromous Salts

See Sec. 5; IC9 for general comments. Absolute and relative rates of reduction of β-bromoalkyl halides are in Tables 56 and 57. These results indicate a large accelerating effect of a β-Br substituent (Table 96). In

TABLE 95

Reaction of $CH_3CH_2CH_2CH_2Br$ with N-Bromosuccinimide in CH_3CN

Time, hr	"Total yield of dibromides, %"	Distribution of reported monobrominated substrate (%)			
		1,1[a]	1,2[b]	1,3[b]	1,4
0.5	5.5	13	20	68	–
1	8.0	15	19	66	–
2	8.2	15	20	65	–
3	11.7	15	23	61	–
5	22.6	11	43	46	–
8	23.4	9	49	42	–
12	28.2	8	50	41	–
14	64.7	5	64	31	–
24	75.4	4	68	29	–
26	82.0	Trace	82	18	–
32	82.3	2	82	16	–
48		2	82	16	–
72	95.5	Trace	84	16	–
111	105.0	Trace	85	15	–

[a] Identification of product OK.

[b] " . . . identified by comparison of . . . retention times on three different glpc columns with those of commercially available compounds."

view of the problems discussed in Sec. 1;VB, it would be premature, even if the gross features of the mechanism were known, to conclude from these results that β-bromoalkyl radicals are bridged [1; II(a)]. Before beginning to discuss these results in terms of a bridged [1; II(a)] product β-bromoalkyl radical, we would welcome (1) experimental evidence which provides assurance that the transition state resembles the product radical and that the transition state is not being stabilized by polar or other "factors" peculiar to the transition state and not "present" in the product (In addition to "effects" already mentioned in Sec. 1;VB, it is possible to envision as a

TABLE 96

Accelerating Effect of β-Bromo Substituents in Reactions of
Alkyl Halides with Cr^{II} and Cr^{II}en

(Halide)$_1$	faster than	(Halide)$_2$	by	$\Delta\Delta G^{\ddagger}$ (kcal/mole) of[a]
meso-$CH_3CHBrCHBrCH_3$		$CH_3CH_2CHBrCH_3$		4.7
$BrCH_2CH_2Br$		$CH_3CH_2CH_2CH_2Br$		3.6

[a] Entries calculated from contents of Tables 56 and 57.

"contributing factor" a polar resonance structure for the transition state
wherein a negative charge or lone pair of electrons is placed on the
β-substituent.); (2) a response, in the form of experimental evidence, to
the questions posed in the last paragraph of Sec. 5; IC9; and (3) experimental
proof that the reaction does not proceed via loss of the β-Br substituent
concerted with transfer of Br to Cr^{II}. It should be emphasized, however,
that the results in Table 96 are indeed accommodated, albeit not uniquely,
by the hypothesis of incipient bridging [1; II(a)].

8. Reaction of Haloalkyl Bromides with $Co(CN)_5^{3-}$

Much work has been done on the mechanism of the reaction of $Co(CN)_5^{3-}$
with alkyl halides. However, this work has been concerned mainly with
aspects of the mechanism that are not germane to our present purposes.
We are concerned with the primal act: Does the encounter of RX with
$Co(CN)_5^{3-}$ lead directly to R·? (Is R· produced, and is it produced in that
elementary step in which the reactant's C-X bond is broken?) Is this the
rate-determining step of the reaction? In order for the possibility to exist
that the results of studies of the kinetics of these reactions might be of
direct use to us, it is necessary that the transition state of the rate-determining step strongly resemble R·. It is not sufficient for R-X bond breaking to occur in the rate-determining step.

In addition to observations which establish that the rate of reaction of
RX (R = alkyl) follows the radical (or cation) order of stability of R and the
radical stability (or anionofugacity) of X, the following additional information

has been obtained: (1) Reaction of $(CH_3)_2CHI$ and $K_3Co(CN)_5$ in the presence of excess $CH_2=CHCN$ produced "a small quantity of $(CH_3)_2CHCH=CHCN$ and $(CH_3)_2CHCH_2CH_2CN$" [196]. (2) "Similar adducts were isolated [in unspecified yield] from the reaction in which $CH_2=C(CH_3)COOCH_3$ was an additive." [197]* While these additional results are not completely irrelevant to the questions raised in the preceding paragraph, they do not contribute to their answers. We do not believe that the information now available is sufficient to provide such answers, and we regard those aspects of the mechanism of reaction of $Co(CN)_5^{3-}$ with alkyl halides which are of concern to us in the present context to be unknown.

Second-order rate constants for the reaction of $Co(CN)_5^{3-}$ with β-bromoalkyl halides are given in Table 97 and indicate a large accelerating effect of a β-Br substituent [198]. Even if the conclusion of the preceding paragraph is ignored, the meaning of these results is still uncertain because of problems of the type enumerated in Sec. 6; IC7 and, in general, the content of Sec. 1; VB. It would be premature to consider these results to be relevant to the question of bridging or to conclude from them that β-bromoalkyl radicals are bridged [1; II(a)].

9. Reaction of Haloalkyl Iodides with $PhN=NCPh_3$

See Sec. 5; IC10 and Table 60. The "acceleration" "caused by" Br would correspond to $\Delta\Delta G^{\ddagger} = 0.6$ kcal/mole. In view of the lack of knowledge of the products of the reactions, the lack of adequate control experiments, the contents of Sec. 1; VB, and questions of the type mentioned in Sec. 6; IC7, this small "deviation" provides no cause to consider $BrCH_2CH_2\cdot$ to be bridged [1; II(a)].

10. Summary

Although there is some evidence (Secs. 6; IC2, 7, 8) which is not inconsistent with β-bromoalkyl radicals being bridged [1; II(a)] and there is some (Secs. 6; IC1-6) which, if viewed superficially and uncritically, shifts

---

* "Their structures, $(CH_3)_2CHCH_2CH(CH_3)COOCH_3$ and $(CH_3)_2CHCH=C(CH_3)COOCH_3$ were established by mass spectroscopy. In addition, three adducts with molecular weights approximately twice those of the structures shown . . . were found."

TABLE 97

Reaction of Haloalkyl Bromides with $Co(CN)_5^{3-}$ at 25.0° [a]

Halide	k, M^{-1} sec^{-1}	Relative $\Delta G^{\ddagger}$, kcal/mole
meso- $^-OOCCHBrCHBrCOO^-$	78	2.7
$^-OOCCHBrCH_2COO^-$	0.83	0
$^-OOCCHBrCH_2Br$	11.7	1.6
$^-OOCCH_2CH_2Br$	8×10^{-4}	-4.1

[a] Organic products not identified.

the burden of proof toward a proponent of bridging [1; II(a)], there is none
for which there is a reasoned, preferred interpretation which indicates
bridging [1; II(a)].

D. Control of Stereochemistry

1. Reaction of Alkyl Bromides with Br_2

It has been reported [154, 235] that "liquid phase photobromination of
(+)-1-bromo-2-methylbutane, α_{obsd}^{27} +4.89° at 0°, yielded (-)-1,2-
dibromo-2-methylbutane, α_{obsd}^{27} -2.86°," the product being described as
"of high optical purity" [154] * and "probably a pure antipode" ("attempts
to enhance the activity . . . by partial crystallization or asymmetric destruction have not been successful") [235]. "The observed rotation of the
dibromide was -2.86° in reactions carried out between -40 and +40°. One
run, made at 72-80°, produced 1,2-dibromo-2-methylbutane, α_{obsd}^{27} -2.33°.
A similar effect is observed with variations of bromine concentration.
Low concentrations of bromine yield a slightly levorotatory product; as the

---

* "V.p.c. . . . of the crude reaction mixture indicated the presence
of only two components: unreacted monobromide and 1,2-dibromo-2-
methylbutane."

bromine concentration is increased the rotation of the dibromide levels off at -2.86°." These results were considered to "require explanation in terms of a mechanistic scheme wherein the bromine substituent assists the departure of the tertiary hydrogen atom. This assistance manifests itself in the formation of a bridged radical, which is capable of maintaining its stereochemical configuration until reaction with molecular bromine occurs. The effects of temperature and bromine concentration suggest that the bridged radical can isomerize to an open-chain form which racemizes." Also [233]: "The slow addition of bromine to an irradiated, refluxing solution of (+)-1-bromo-2-methylbutane in $CFCl_3$ produced (-)-1,2-dibromo-2-methylbutane, α_{obsd}^{29} -0.80°. The decreased concentration of bromine effectively increases the lifetime of the intermediate radical and allows partial racemization." We find evaluation of these results difficult because the optical purity, chemical purity, and % yield of the product are not reported and because it has not been established that the reaction proceeds via attack on free $CH_3CH_2CH(CH_3)CH_2Br$ to give free $CH_3CH_2\overset{\cdot}{C}(CH_3)CH_2Br$ (structure - ?) followed by reaction with an achiral bromine-containing substance to give $CH_3CH_2CBr(CH_3)CH_2Br$ which is not asymmetrically destroyed. It may be, as is suggested [154], that "this observation requires a modification of the accepted halogenation mechanism." For these reasons we defer judgment on the meaning of these results with regard to the question of bridging [1; III(a)].

If information becomes available in the future which resolves the above uncertainties in a manner consistent with the authors' liberal use of "require" and if the radical is judged to be bridged [1; III(a)], then whether it is bridged [1; I(a)-(c)] would still be an open question. For example, an initially produced $CH_3CH_2\overset{\cdot}{C}(CH_3)CH_2Br$ radical which is planar about the trisubstituted carbon atom and which has a "normal" C-C-Br angle, i.e., is not bridged [1; I(c)], would become statistically achiral only on a time scale large compared to the period of rotation about the $\overset{\cdot}{C}$-CH_2Br bond. In other words, the rotamers of $CH_3CH_2\overset{\cdot}{C}(CH_3)CH_2Br$ which result directly from $CH_3CH_2CH(CH_3)CH_2Br$ are chiral and their racemization is accomplished in part by rotation about the $\overset{\cdot}{C}$-CH_2Br bond; if such rotation is not much faster than reaction with Br_2, the

$CH_3CH_2CBr(CH_3)CH_2Br$ would be optically active even though the radical
is not bridged [1; I(a)-(c)]. (Of course, if the radical is not planar about
the trisubstituted carbon atom, an additional process becomes necessary
for the interconversion of enantiomers.) In order to evaluate the viability
of this alternative, it is necessary to know the rates of rotation about the
$\overset{\cdot}{C}-CH_2Br$ bond and of reaction of the radical with Br_2. Data which might
be helpful in estimating the former are given in Table 61. Based on these
data, we estimate the potential energy barrier to rotation in two possible
models for unbridged $CH_3CH_2\overset{\cdot}{C}(CH_3)CH_2Br$, $CH_3CH_2CH(CH_3)CH_2Br$ and
$R_2C=CR-CH_2Br$,* to be 4.7† or ~5.0* and ~2.3* kcal, respectively. If
a pre-exponential factor of ~10^{12} is used [191, 206, 239], rate constants
at 0° in the gas phase of ~5 x 10^7 and ~6 x 10^9 sec^{-1}, respectively, are
obtained. Unfortunately, there is an even less secure basis for an estimate of the rate constant for reaction of unbridged $CH_3CH_2\overset{\cdot}{C}(CH_3)CH_2Br$
with Br_2. Available information is collected in Table 62.

Skell [235] believes, based on the dependence of the activity of the
product on $[Br_2]$, that his data indicate that the rate constant for reaction
of the species which gives active product is ~150 M^{-1} ‡ times that of the

---

*See the corresponding footnote in Sec. 5; ID1.

†The barrier reported for $(CH_3)_2CH-CH_2Br$ is 4.7 kcal [244].

‡The ratio of rate constants is obtained from a plot of (amount of
l-enantiomer in product)/(amount of d-enantiomer in product) vs.
$[Br_2]$. Such a plot should be linear if Skell's mechanism or the alternative which we are considering is correct. Although a straight line has
been drawn through the points [235], the best smooth curve is clearly
not linear but shows pronounced upward curvature. In addition, the
best straight line has not been drawn through the points but a (1, 0) intercept has been forced. Actually, the l/d ratio is not known but was deduced by assuming the sample of greatest optical purity to be 100%
optically pure. However, both the rate constant ratio calculated from
the slope of the curve and the degree of deviation of the curve from

conversion of such species into one which gives racemic product.** If the estimated rate constants for reaction of the poor model compounds $CH_3\cdot$, $CF_3\cdot$, $CCl_3\cdot$, $CH_3CH_2\cdot$, and $CHBr_2\dot{C}HBr$ are used, the corresponding ratios of rate constants for the alternative scheme which we are considering are estimated to be 2-300, 0.1-20, 0.1-10, 0.01-1, and 10^{-2}-10^{-5} M^{-1}, respectively. If the guessed activation energy for reaction of the good model compound, $(CH_3)_3C\cdot$, is used, the ratio is less than 10^{-3} M^{-1} for any reasonable pre-exponential factor for the reaction of $(CH_3)_3C\cdot$ with Br_2. We cannot emphasize too strongly that all of these estimates of rate constants for internal rotation and for reaction with bromine are extremely crude. We have of necessity chosen highly approximate models and have estimated their properties by use of highly approximate methods. Our purpose was only to answer the question of whether, based on the very limited data available, the alternative "explanation" presented is a priori unreasonable. Keeping in mind the inappropriateness of our models, our very uncertain estimates of the rate constants corresponding to them, and the application of estimates derived from data obtained for systems in the gas phase to a reaction in solution, we would, if forced, state that the alternative scheme which we have been considering can account for the reported results only with some difficulty.

---

linearity are sensitive functions of the assumed rotation of optically pure product. For example, we have found that if the rotation of optically pure product is assumed to be -3.5°, the points (taking the data at face value) fall reasonably well on a straight line, the best straight line does have a (1,0) intercept, and a ratio of ~100 M^{-1} results. Skell's assumed rotation of optically pure product of -2.86° leads to pronounced upward curvature and a ratio of ~150 M^{-1}, while an assumed rotation of -5.7° leads to pronounced downward curvature and a ratio of ~40 M^{-1}.

** The accuracy of such an analysis depends on an ability to maintain constant bromine concentrations in the range 0.001 - 0.04 M during the course of the reaction. Unfortunately, the details of the method used have not been published.

2. Reaction of Alkyl Bromides with N-Bromosuccinimide

"The photobromination of (+)-1-bromo-2-methylbutane (+4.89°) with NBS was studied" at 25° ($CFCl_3$ solvent), 40° (CH_2Cl_2 solvent), and 76° (CCl_4) and gave "1,2-dibromo-2-methylbutane . . . [as] the only dibromide produced" with α_{obs} of -0.25° (28°C), -0.30° (25°C), and -0.06° (35°C) ("photoinitiation and thermal initiation gave identical results"), respectively [233]. These results were interpreted [233] as were those in the preceding section and are subject to our comments therein.

Results of the bromination of 1,2-diphenylethanes are in Table 98. As defined by the meso/dl ratio of products, the stereoselectivity of the reaction is high. While, superficially, this indicates that PhĊHCHBrPh is moderately bridged [1; III(a)], we cannot draw this conclusion since it has not been shown that the ratio of diastereomeric products reflects the ratio in which they would be formed via the direct reaction of PhCHCHBrPh with Br_2 (one complication would be the possible intermediacy of PhCH=CHPh, formed stepwise or concertedly from $PhCH_2CHBrPh$). If the results are taken at face value, they still do not indicate that PhĊHCHBrPh is bridged [1; I(a)-(c)]; alternative "explanations" of the stereospecificity are discussed at length [234]. With regard to the selectivity in the hydrogen-abstraction step, the k_H/k_D value of 0.5, when compared to the isotope effect of $k_H/k_D = 2.65$ reported for the NBS-bromination of $PhCHDCH_3$, indicates a stereoselectivity corresponding to $\Delta\Delta G^{\ddagger} \cong 1$ kcal. In view of the lack of detailed knowledge of the mechanism of the reaction, it cannot be concluded that PhĊHCHBrPh is weakly bridged [1; III(a)]; if it is assumed to be so, there is still no indication that it is bridged [1; I(a)-(c)], alternative "explanations" of the stereospecificity having been discussed in detail [234].

3. Reaction of Alkyl Bromides with $(CH_3)_3COBr$

It has been reported [154] that "bromination [of (+)-1-bromo-2-methylbutane] with t-butyl hypobromite afforded (-)-1,2-dibromo-2-methylbutane, α_{obsd}^{27} -1.8 to -2.0°" and that "bromination of the (+)-bromide with Bu^tOBr at various concentrations demonstrated a plateau phenomenon similar to that shown by bromine." [235] See the preceding section for general comments and reservations. If the species which

TABLE 98

Reaction of PhCHYCHXPh with N–Bromosuccinimide [234]

X	Y	Reaction conditions	Product (% yield)	meso/dl	Controls
H	H	a	PhCHBrCHBrPh (91%, $\underline{meso}$; $\leq 8\%$, $\underline{dl}$)	>11	c
Br	H	a	PhCHBrCHBrPh (98%, $\underline{meso}$)	>50	c
Br	D (erythro)	a	PhCHBrCHBrPh (79% $\underline{meso}$), $k_H/k_D = 0.5$	> 4	c
Br	D (erythro)	b	PhCHBrCHBrPh (91%, $\underline{meso}$)	>10	

a 1 equivalent of NBS, 0.3 mole % Bz_2O_2, CCl_4, reflux.

b 1.2 equivalents of NBS, 1.6 mole % Bz_2O_2, CCl_4, reflux.

c "An experiment was carried out on exactly the same scale . . . with the inclusion of 0.63 g. . . . of $\underline{dl}$–α,α'–dibromobibenzyl. . . [The product] weighed 1.11 g. (89%). . ., pure $\underline{meso}$–α,α'–dibromobibenzyl. The bulk of the $\underline{dl}$–dibromide was recovered . . ."

transfers bromine is $(CH_3)_3COBr$, we cannot analyze these results quantitatively because of the lack of rate data for the reaction $R\cdot + (CH_3)_3COBr \rightarrow$
$RBr + (CH_3)_3CO\cdot$.

4. Reaction of Alkyl Bromides with Cl_2

It has been reported [154] that reaction of optically active
$CH_3CH_2CH(CH_3)CH_2Br$ with Cl_2 in CS_2 produced inactive
$CH_3CH_2CCl(CH_3)CH_2Br$, a result which provides no cause to consider $CH_3CH_2\overset{\cdot}{C}(CH_3)CH_2Br$ to be bridged [1; III(a)].

5. Reaction of Alkyl Bromides with $(CH_3)_3COCl$

It has been reported [154] that reaction of optically active
$CH_3CH_2CH(CH_3)CH_2Br$ with $(CH_3)_3COCl$ in CS_2 produced inactive
$CH_3CH_2CCl(CH_3)CH_2Br$, a result which provides no cause to consider $CH_3CH_2\overset{\cdot}{C}(CH_3)CH_2Br$ to be bridged [1; III(a)].

6. Reaction of Haloalkyl Bromides with Organotin Hydrides

Cis/trans ratios of 2-butenes resulting from the reaction of meso-
and dl-$CH_3CHBrCHBrCH_3$ with Bu_3SnH and Bu_2SnH_2 are given in Table 99.
The largest $\Delta(\Delta\Delta G^{\ddagger})$ corresponding to the reported ratios of ratios was
2.1 kcal/mole with Bu_3SnH and 1.8 kcal/mole with Bu_2SnH_2. Also, as is
shown in Table 100, there is a clear dependence of $\Delta\Delta G^{\ddagger}$ on $[Bu_3SnH]$. It
is difficult for us to discuss these results with reference to the possibility
of bridging [1; III(a)], basically because the effect of the β-Br substituent
on the details of the mechanism of the reaction is not known. Is the
2-butene produced from a free $CH_3\overset{\cdot}{C}HCHBrCH_3$ radical? Might it be produced directly from $CH_3CHBrCHBrCH_3$? Although the data in Table 100
would seem to belie manifestation of a negative response to the first
question in the form of a positive one to the second, there are alternatives,
particularly at low temperature or high concentrations — something similar to whatever is "responsible" for the results of Altman and Nelson
[238], perhaps? If $CH_3CHBr\overset{\cdot}{C}HCH_3$ is assumed to be the dominant precursor of 2-butene, the results may still be explained if the $CH_3CHBr\overset{\cdot}{C}HCH_3 \rightarrow$
2-butene conversion is at least competitive with equilibration among the
rotamers initially produced in different distributions from meso- and dl-
dibromide. In order to investigate this possibility it is necessary to know

TABLE 99

Reaction of $CH_3CHBrCHBrCH_3$ with Butyltin Hydrides [101]

Reagents	Reaction conditions	n–Butane		2–Butene		Controls
		% yield	% of hydrocarbon product	% yield	trans/cis	
meso–Dibromide, Bu$_3$SnH	25°, neat[a]	<0.8	<0.8	16–53	8.90	g,h
dl–Dibromide, Bu$_3$SnH	25°, neat[a]	0.1–0.2	0.5–0.8	16–21	0.513	h
meso–Dibromide, Bu$_3$SnH	25°, neat[b]	Trace	Trace	30	9.00	h
dl–Dibromide, Bu$_3$SnH	25°, neat[b]	Trace	Trace	40	0.250	h
meso–Dibromide, Bu$_3$SnH	102°, neat[a]	4–8	5–15	77–89	2.54	h
dl–Dibromide, Bu$_3$SnH	102°, neat[a]	7–10	8–14	68–83	0.950	h
meso–Dibromide, Bu$_3$SnH	25°, 1.89 M in ether[a]	<1.0	<1.0	?	5.25	h
dl–Dibromide, Bu$_3$SnH	25°, 1.89 M in ether[a]	<1.0	1.0	?	0.695	h
meso–Dibromide, Bu$_3$SnH	Room temp., 1.9 M in n–heptane,[a] Et$_3$N[e]	?	?	?	5.7	h
dl–Dibromide, Bu$_3$SnH	Room temp., 1.9 M in n–heptane,[a] Et$_3$N[3]	?	?	?	0.33	h

Reagents	Reaction conditions	n-Butane		2-Butene		
		% yield	% of hydrocarbon product	% yield	trans/cis	Controls
<u>meso</u>-Dibromide, Bu_3SnH	-78°, hν, 1.89 M in ether[a]	0.8	1.0-2.7	31	10.8	f
<u>dl</u>-Dibromide, Bu_3SnH	-75°, hν, 1.89 M in ether[a]	?	1.0-3.0	?	0.242	f
<u>meso</u>-Dibromide, Bu_2SnH_2	25°, neat[c]	Trace	Trace	50	7.32	
<u>dl</u>-Dibromide, Bu_2SnH_2	25°, neat[c]	Trace	Trace	58-59	0.337	
<u>meso</u>-Dibromide, Bu_2SnH_2	25°, neat[d]	Trace	Trace	25.5	15.1	
<u>dl</u>-Dibromide, Bu_2SnH_2	25°, neat[d]	Trace	Trace	30	0.181	

[a] $CH_3CHBrCHBrCH_3/Bu_3SnH = 0.5.$

[b] $CH_3CHBrCHBrCH_3/Bu_3SnH = 0.1.$

[c] $CH_3CHBrCHBrCH_3/Bu_3SnH_2 = 1.0.$

[d] $CH_3CHBrCHBrCH_3/Bu_2SnH_2 = 0.20.$

[e] 1 equivalent.

f " . . . the 2,3-dibromobutanes did not react thermally with tri-n-butyltin hydride at -78° . . . "

g " . . . addition of [galvinoxyl] . . . in small amounts [0.35-0.73 mole %] extended the induction period of the reaction."

h "When cis-2-butene was allowed to stand for 20 hr. under ambient conditions with tri-n-butyltin hydride, 3.2% isomerization to trans-isomer occurred. . . . Carrying out the reaction of meso-2,3-dibromobutane with tri-n-butyltin hydride in the presence of cis-2-pentene . . . resulted [in] 11-15% isomerization of the cis-2-pentene to the trans isomer."

the rates of rotation about the $CH_3CHBr-\dot{C}HCH_3$ bond and of loss of bromine to give 2-butene. As before (Sec. 6; ID1), consider $CH_3CHBr-CH_2CH_3$ and $R_2C=CR-CHBrCH_3$ as models for the purpose of estimating the barrier. We take the barriers to be 4.8 [239] and ~2.3 kcal,* respectively. If a pre-exponential factor of ~10^{12} is used [191, 206, 239], rate constants at 25° [corresponds to the largest $\Delta(\Delta\Delta G^{\ddagger})$] in the gas phase of 10^8 and ~7×10^9 sec^{-1}, respectively, are obtained. Estimates of rate constants for $-\overset{|}{\underset{|}{\dot{C}}}-\overset{|}{\underset{|}{C}}-Br \rightarrow \overset{\diagdown}{\diagup}C=C\overset{\diagup}{\diagdown} + Br\cdot$ are given in Table 101. Keeping in mind the contents of the last paragraph of Sec. 6; ID1, we believe that it is indeed possible for ejection of $Br\cdot$ to be competitive with internal rotation, especially if the rate constant for ejection of $Br\cdot$ from $CH_3CHBr\dot{C}HCH_3$ is near the upper limit cited in Table 101. Additional support for this possibility is provided by the reported that, in those cases displaying the greatest "memory effect," the 2-butene/butane ratio of products is greater that 10^2. This indicates that the rate of ejection of $Br\cdot$ is not less than 10^2 that of reaction of the (presumed intermediate) $CH_3CHBr\dot{C}HCH_3$ with Bu_3SnH. The rate constants (25°, cyclohexane) for reaction of methyl, n-hexyl, cyclohexyl, and t-butyl radicals with Bu_3SnH have been reported to be 5.8×10^6, 1.0×10^6, 1.2×10^6, and 7.4×10^5 M^{-1} sec^{-1}, respectively [255]. If a value of ~10^6 M^{-1} sec^{-1} is assumed to apply to the reaction under consideration, then, taking into account the approximate mole fraction of Bu_3SnH (Table 99) and the factor of 10^2 cited above, it is clear that ejection of bromine can easily be competitive with internal rotation. Finally, such a competitive situation has been reported to exist in $CHBr_2\dot{C}HBr$ [245].†

---

*Our estimate is based on the observation (Table 61) that compounds of this type have about the same barrier.

†"At [35.56°] the rate constant for exchange [with Br_2] of either [cis- or trans-1,2-dibromoethylene] ... is ... $(4.82 \pm 0.14) \times 10^{-4}$ (mole/ml)$^{-1/2}$ sec^{-1}. At the same temperature the rate constant for the isomerization of the cis-isomer is ... $(2.28 \pm 0.09) \times 10^{-4}$ (mole-ml)$^{-1/2}$ sec^{-1}, and the rate constant for the isomerization of the trans-isomer is ... $(2.98 \pm 0.15) \times 10^{-4}$ (mole/ml)$^{-1/2}$ sec^{-1}. The results seem to demonstrate conclusively that the $C_2H_2Br_3$ radical is

TABLE 100

Reaction of $CH_3CHBrCHBrCH_3$ with Bu_3SnH [101]

Concentration of reagents	$-\Delta\Delta G^{\ddagger}$ (trans/cis)(kcal/mole) for 2-butene	
	from	
	meso	dl
1.89 M dibromide, 3.78 M Bu_3SnH in ether	0.97	-0.22
Neat, dibromide/Bu_3SnH = 0.5	1.30	-0.40
Neat, dibromide/Bu_3SnH = 0.1	1.27	-0.82

Cis/trans ratios of CHCl=CHCl resulting from the reactions of CHBrClCHBrCl with Bu_3SnH are given in Table 102. Since "both mixtures gave the same proportions of cis- and trans-1,2-dichloroethylene within the experimental error," these results, subject to the relevant comments in Sec. 6; ID6, provide no cause to consider CHBrClCHCl to be bridged [1; III(a)].

Cis/trans ratios of EtOOCCH=CHCOOEt resulting from reactions of EtOOCCHBrCHBrCOOEt with Bu_3SnH are given in Table 103. The specificity is nil. However, because of the greater-than-usual uncertainty (in particular, that arising from the results cited in footnote e to Table 103) surrounding the processes occurring during the reaction, we prefer not to draw any conclusions regarding bridging [1; III(a)] from these results.

7. Reaction of Haloalkyl Bromides with Chromous Salts

General comments and reservations are given in Sec. 5; IC9. The effects of solvent and of ligand on the trans/cis ratios of 2-butene resulting

capable of existing in two isomeric forms and that the free energy bar-rier opposing their interconversion is of a height comparable to the height of the barrier opposing dissociation of either radical into a bromine atom and the corresponding isomer of dibromoethylene."

TABLE 101

Estimated Rate Constants at 25° for $-\overset{|}{\underset{|}{C}}-\overset{|}{C}-Br \rightarrow \overset{\backslash}{\underset{/}{C}}=\overset{/}{\underset{\backslash}{C}} + Br\cdot$

Reactant	Estimated rate constant, sec^{-1}	Method of estimation
$BrCH_2CH_2\cdot$	4×10^4	a
$BrCH_2CH_2\cdot$	2×10^5	e
$BrCHDCHD\cdot$	7×10^7	c
$CH_3CHBr\overset{\cdot}{C}HCH_3$	$<3 \times 10^9$ (54°)	d
$CHBr_2\overset{\cdot}{C}HBr$	1×10^7	f

[a] Schmitz et al. [253] have reported $k[BrCH_2CH_2\cdot \rightarrow Br\cdot + C_2H_4]/k[BrCH_2CH_2\cdot + Br_2 \rightarrow Br\cdot + BrCH_2CH_2Br] = 13 \exp[-6.1/RT]$ M. If $BrCH_2CH_2\cdot$ is assumed to react with Br_2 with the same rate constant as $CH_3CH_2\cdot$ (estimated[b] to be 1×10^8 l mole^{-1} sec^{-1} at 25°), we can estimate $k[BrCH_2CH_2\cdot \rightarrow Br\cdot + C_2H_4]$ to be 4×10^4 sec^{-1} at 25°.

[b] See footnote d to Table 62.

[c] Oldershaw and Cvetanović [254] have reported $k[BrCHDCHD\cdot \rightarrow Br\cdot + CHD=CHD]/k[BrCHDCHD\cdot + Br_2 \rightarrow BrCHDCHDBr + Br\cdot] \approx 0.7$ M at 23°. If $BrCHDCHD\cdot$ is assumed to react with Br_2 with the same rate constant as $CH_3CH_2\cdot$ (estimated[b] to be 1×10^8 l mole^{-1} sec^{-1} at 25°), we can estimate $k[BrCHDCHD\cdot \rightarrow Br\cdot + CHD=CHD]$ to be 7×10^7 sec^{-1} at 25°.

[d] Oldershaw and Cvetanović [254] have reported $k[CH_3CHBr\overset{\cdot}{C}HCH_3 \rightarrow Br\cdot + CH_3CH=CHCH_3]/k[CH_3CHBr\overset{\cdot}{C}HCH_3 + Br_2 \rightarrow CH_3CHBrCHBrCH_3 + Br\cdot] = 14$ M at 54°. If $CH_3CHBr\overset{\cdot}{C}HCH_3$ is assumed to react with Br_2 with a rate constant smaller than that of $CH_3CH_2\cdot$ (estimated[b] to be 2×10^8 l mole^{-1} sec^{-1} at 54°), then we can say that $k[CH_3CHBr\overset{\cdot}{C}HCH_3 \rightarrow Br\cdot + CH_3CH=CHCH_3] < 3 \times 10^9$ sec^{-1} at 54°.

[e] Wong and Armstrong [190] have reported that $k[Br\cdot + C_2H_4 \rightarrow BrCH_2CH_2\cdot] = 8.4 \times 10^6$ M^{-1} sec^{-1}. Benson and O'Neal [256] have estimated (1 atm standard state), by use of group additivities, $\Delta H° = -8.8$ kcal/mole and $\Delta S° = -24.1$ eu for the formation of unbridged [1;II(a)] $BrCH_2CH_2\cdot$. Combination of these estimates (K = 36 M^{-1})

and the reported rate constant gives us $k[BrCH_2CH_2 \cdot \rightarrow Br \cdot + C_2H_4] = 2 \times 10^5$ sec^{-1} at 25°.

[f] See footnote e to Table 62.

TABLE 102

Reaction of $CHClBrCHClBr$ with Bu_3SnH, 0° [101]

| Substrate ($\underline{dl}/\underline{meso}$) | $\dfrac{Bu_3SnH}{substrate}$ | CHCl=CHCl | | Controls |
		Reported % yield	Reported cis/trans	
1.2	0.44	98	2.8	
1.2	0.94	98	2.8	
1.2	2.0	93	2.5	a
2.3	0.50	97	4.0	
2.3	1.0	93	3.0	
2.3	2.0	97	3.0	a

[a] "Repetition of this experiment with 1 mol of $\underline{cis}$-1,2-dichloroethene per mol of dibromodichloride did not change the amount of $\underline{trans}$-1,2-dichloroethene in product . . . "

from reductive elimination from $\underline{dl}$- and $\underline{meso}$-2,3-dibromobutane are indicated by the data in Tables 104 and 105, respectively. The largest $\Delta(\Delta\Delta G^{\ddagger})$ corresponding to the reported ratios of ratios is 3.7 kcal/mole (Table 104, DMSO solvent), a large specificity. Its significance is obscured, however, by the lack of firm knowledge of some of the details of the course of these particular reactions. Possibilities which might obviate the need to consider bridging include (1) one-step formation of 2-butene from $CH_3CHBrCHBrCH_3$; (2) its multistep formation without the intermediacy of $CH_3CHBr\dot{C}HCH_3$; (3) the formation of different distributions of rotamers of $CH_3CHBr\dot{C}HCH_3$ from the $\underline{meso}$- and $\underline{dl}$-dibromides and their reaction to give 2-butene or a stereochemically stable precursor of 2-butene being competitive with equilibration among themselves.

TABLE 103

Reaction of EtOOCCHBrCHBrCOOEt with Bu_3SnH (0°, hν) [101]

Dibromide	Bu_3SnH/substrate	Reported Products, %			
		Fumarate	Maleate[a]	Succinate[e]	Controls
meso	4.8	64	9	28	d
dl	4.8	67	11	22	d
dl[b]	6.0	72	5	23	d
dl[c]	6.0	61	3	36	d
meso	9.8	42	7	51	d
dl	9.2	36	9	55	d

[a] "±4-5%."

[b] "Reaction quenched after 4 min."

[c] "Reaction quenched after 8 min."

[d] ". . . dimethyl [sic-?] maleate was not isomerized under the conditions of the reactions."

[e] "No monobromosuccinate was present in the reaction product mixture; . . the monobromosuccinate was reduced by tri-n-butyltin hydride at 0.9 times the rate at which the latter reacted with dl-2,3-dibromosuccinate.. . . When excess organotin hydride was used, the amount of succinate increased with time after all of the dibromide was consumed. Furthermore, diethyl fumarate was shown to react with the hydride under the dehalogenation conditions to produce succinate."

TABLE 104

Reaction of $CH_3CHBrCHBrCH_3$ with Cr^{II}, 25° [152]

Dibromide	Solvent	2-Butene cis/trans[d]
meso	90/10 (v/v) ethanol/H_2O[a]	4.0
dl	90/10 (v/v) ethanol/H_2O[a]	1.8
meso	90/10 (v/v) DMF/H_2O[a]	4.6
dl	90/10 (v/v) DMF/H_2O[a]	0.75
meso	90/10 (v/v) DMSO/H_2O[a]	32
dl	90/10 (v/v) DMSO/H_2O[a]	0.064
meso	88/12 (v/v) pyridine/H_2O[b]	11
dl	88/12 (v/v) pyridine/H_2O[c]	0.22
meso	70/9/12 (v/v/v) 2,4,6-collidine /DMF/H_2O[b]	10
dl	70/9/12 (v/v/v) 2,4,6-collidine /DMF/H_2O[b]	0.22

[a] Dibromide/Cr^{II} = 0.64.

[b] Dibromide/Cr^{II} = 0.14.

[c] Dibromide/Cr^{II} = 0.20.

[d] ". . . the relative amounts of cis- and trans-butene-2 remained invariant throughout the course of the reaction . . . Cis-trans isomerization did not occur under these experimental conditions."

Although possibilities (1) and (2) represent departures from the proposed [152] mechanism of reaction, there is no evidence which requires that they not be considered. While it would be difficult, but not impossible, to reconcile a dependence of stereospecificity on [Cr^{II}] to these possibilities, the minimal dependence observed (Table 106) obviates the need to do so.

TABLE 105

Reaction of $CH_3CHBrCHBrCH_3$ with Cr^{II}, 25° [152]

Dibromide	Ligand, M	2-Butene trans/cis[e]
meso	Ethylenediamine, 10.8[a]	24
dl	Ethylenediamine, 10.8[b]	0.087
meso	Ethylenediamine, 0.18[d]	3.3
dl	Ethylenediamine, 0.18[d]	0.89
meso	Ethanolamine, 10.8[a]	13
dl	Ethanolamine, 10.8[c]	0.16

[a] Dibromide/Cr^{II} = 0.14, solvent 22/78 (v/v) DMF/H_2O.

[b] Dibromide/Cr^{II} = 0.20, solvent 22/78 (v/v) DMF/H_2O.

[c] Dibromide/Cr^{II} = 0.20, solvent 30/70 (v/v) DMF/H_2O.

[d] Dibromide/Cr^{II} = 0.13, solvent 88/12 (v/v) DMF/H_2O.

[e] ". . . the relative amounts of cis- and trans-butene-2 remained invariant throughout the course of the reaction . . . Cis-trans isomerization did not occur under these experimental conditions."

Consider the third possibility and the rates of three processes relevant to it: (a) ejection of Br· from $CH_3CHBr\dot{C}HCH_3$, (b) internal rotation in CH_3CHBr-$\dot{C}HCH_3$, (c) reaction of $CH_3CHBr\dot{C}HCH_3$ with Cr^{II} to give a β-bromoalkylchromium species. If $(rate)_a \geq (rate)_b$ and $(rate)_c$, the observed stereospecificity can be "explained" without recourse to an argument based on bridging. We have already presented a case for $(rate)_a \geq (rate)_b$ (Sec. 6; ID6). With regard to $(rate)_c$ it has been stated, without elaboration, that process (c) involving Cr^{II} "require[s] almost no activation energy" [152] and "is probably diffusion controlled" [265]; the reaction of $CH_2=CHCH_2CH_2CH_2$· with $Cr^{II}en_2^{2+}$ has been estimated to have a rate constant of 4×10^7 M^{-1} sec^{-1} at 25° in 83 vol. % $DMF-H_2O$ [147]. Since a significant specificity is observed under conditions very

TABLE 106

Effect of Concentration on the Stereospecificity of Elimination from dl-$CH_3CHBrCHBrCH_3$

Cr^{II}/dibromide	Temperature, °C	Solvent	Conduct of reaction	2-Butene cis/trans
37	20	89/11(v/v)DMSO/H_2O	a	15.7
6.2	20	89/11(v/v)DMSO/H_2O	a	14.2
1.8	20	89/11(v/v)DMSO/H_2O	a	12.9
3.6	25	97/3(v/v)DMSO/H_2O	b	12.5
1.1	25	97/3(v/v)DMSO/H_2O	b	10.9
0.11	25	97/3(v/v)DMSO/H_2O	b	10.1
0.036	25	97/3(v/v)DMSO/H_2O	b	9.5
37	73	89/11(v/v)DMSO/H_2O	b	7.3
6.2	73	89/11(v/v)DMSO/H_2O	b	7.3
1.8	73	89/11(v/v)DMSO/H_2O	b	7.1
6.2	73	89/11(v/v)DMSO/H_2O	a	7.2
1.8	73	89/11(v/v)DMSO/H_2O	a	7.0
0.11	73	89/11(v/v)DMSO/H_2O	a	6.6

[a] "Normal addition, dibromide (0.24 M) added to Cr^{II} with rapid stirring."

[b] "Inverse addition, Cr^{II} (3 M) added to dibromide."

close to those for which this rate constant has been reported (Table 105),
we will use $k_c \cong 4 \times 10^7$ M^{-1} sec^{-1} for our present purposes. It is seen
that, at the concentrations of chromous used (0.08 – 0.9 M), $(rate)_a$
(Table 101) may be $\geq (rate)_c$.

Since it is conceivable that k_c for $CH_3CHBr\dot{C}HCH_3$ is greater than
that for $CH_2=CHCH_2CH_2CH_2\cdot$, it is possible that $(rate)_a < (rate)_c$,
and we should consider another possibility which would make it unnecessary
to discuss the results in terms of bridging: $(rate)_c \geq (rate)_b$ and $(rate)_a$.
Kochi and Singleton argue against "a mechanism based solely on rate of
bond rotation vs. reaction with Cr^{II}, because of large differences in
selectivities with β substituents [the Cl, OTs, OAc, and OH compounds
exhibit no stereospecificity]. The lifetimes of open β-substituted alkyl
radicals in a Cr^{II} environment should not differ so widely from Br, Cl,
HO, etc." While we agree that the barriers to internal rotation in the
various β-substituted alkyl radicals should be about the same, we mention
four circumstances, incorporating near-equality of barrier heights, under
which $(rate)_c \geq (rate)_b$ and $(rate)_a$ is still a viable alternative to bridging:

(1) While "the carbon-chromium linkage in [the β-bromoalkylchromium
species] is sufficiently stable to maintain stereochemical integrity until
elimination occurs," this situation does not obtain with other β-substituents
(It is not unreasonable to postulate that Br is the best leaving group).

(2) The overall stereospecificity of the reaction is determined in part
by the stereoselectivities with which Cr^{II} produces the diastereomeric
β-bromoalkylchromium species from each rotamer of the β-bromoalkyl
radical and with which a particular diastereomer produces the isomeric
2-butenes. If either stereoselectivity is small, the overall stereospecificity of the reaction will be small, even if internal rotation is slow. It is not
unreasonable for the β-bromo compound to show the greatest stereoselectivity in both steps, perhaps "because of" a greater tendency to associate with
chromium, resulting in a preference, not shared by the other substituents,
for syn-attack of Cr^{II} and syn-elimination of Cr and Br.

(3) k_c is "normal," resulting in $(rate)_b > (rate)_c$, for all substituents
but Br, for which it is abnormally large for "reasons" mentioned in (2).

(4) Cr^{II} reacts unusually rapidly with the β-bromoalkyl radical to
give, not the β-bromoalkyl-chromium species, but 2-butene directly.

This is not unreasonable because of the anticipated [147] unusually high abstractability of Br.

In conclusion, it should be emphasized that the results cited in this section are consistent with bridged $CH_3CHBr\dot{C}HCH_3$.

8. Addition of HBr to Olefins

The data in Table 107 show a pattern of increasing stereospecificity with decreasing temperature for all olefins. For $CHD=CHC_4H_9$, $CH_3CBr=CHCH_3$, and $CH_3CH=CHCH_3$ the reported stereospecificities are large, corresponding to $\Delta(\Delta\Delta G^{\ddagger}) > 2.1$, > 3.4, and > 3.4 kcal/mole, respectively, at $-78°$. In view of the lack of control experiments which demonstrate that the products are formed only via those routes which produce them in the same ratio as does the direct reaction of the free β-bromoalkyl radical with HBr, we defer judgment on the meaning of these results with regard to the question of bridging. " . . . [a] modification of the chain process [which fails to meet the above conditions] involves complexing of the addendum and olefin prior to addition. If the bromine atom attacks the substrate from the side away from the complexed HBr molecule _trans_ addition would be expected because the newly formed radical can transfer immediately . . ." [34] This suggestion is reasonable and is not weakened by reports of complex formation between HBr and propylene at $-170°$ [270] and between HBr and allyl chloride at $-193°$ [271]. In the latter study it was reported that mixtures of HBr and allyl chloride exhibited an absorption at 283 mμ and that "there was a definite correlation between the intensity of the band [assigned to a complex between HBr and allyl chloride] * . . . and conversion [to hydrobrominated olefin]." The suggestion does, however, lack direct support, and, if Br· adds to the olefins of concern with about the same rate constant as it reported [190] for $Br· + C_2H_4$ (10^7 l mole^{-1} sec^{-1}, independent of temperature), it would require a significant

---

*The absorption did not appear in the spectra of allyl chloride or its hydrobromination products. Pure HBr exhibited a band at 297 mμ. The intensity of the 283-mμ absorption was at a maximum when allyl chloride/HBr = 1.

TABLE 107

Addition of HBr to Olefins

Olefin	Reaction conditions	Reported product hydrobromide		Controls	Ref.
		Reported structure (% yield)	threo/erythro ($\underline{dl}$/meso)		
$CF_3CF=CFCF_3$ (<u>trans</u>)	hν, HBr, 11 hr	$CF_3CHFCFBrCF_3$ (?)[b]	0.47[c]	a	156
$CF_3CF=CFCF_3$ (<u>cis</u>)	hν, HBr	$CF_3CHFCFBrCF_3$ (?)[b]	0.61[c]	None	156
$CHD=CHC_4H_9$(>94% <u>trans</u>)	hν, DBr, −60 to −78°	$BrCHDCHDC_4H_9$ (40)	>19[d]	None	267
$CHD=CHC_4H_9$ (>98% <u>cis</u>)	hν, DBr, −60 to −78°	$BrCHDCHDC_4H_9$ (43)	<0.1[d]	None	267
$CH_3CCl=CHCH_3$(<u>cis</u>)	hν, HBr, −20°, pentane	$CH_3CHBrCHClCH_3$ (88)	2.6[e]	f	268
$CH_3CCl=CHCH_3$(95% <u>trans</u>)	hν, HBr, −20°, pentane	$CH_3CHBrCHClCH_3$ (81)	2.9[e]	None	268
$CH_3CCl=CHCH_3$(<u>cis</u>)	HBr, 25–30°, hexane, 6 mole % Bz$_2$O$_2$	$CH_3CHBrCHClCH_3$ (66)	2.3[e]	None	268
$CH_3CBr=CHCH_3$(<u>trans</u>)	hν, −78°, HBr or DBr[h]	$CH_3CHBrCH(D)BrCH_3$ (?)	>100[g]	l,k	34
$CH_3CBr=CHCH_3$(98% <u>cis</u>)	hν, −78°, HBr[h]	$CH_3CHBrCHBrCH_3$ (?)	0.02–0.1[g]	l,k	34
$CH_3CBr=CHCH_3$(98% <u>cis</u>)	hν, −78°, DBr[h]	$CH_3CHBrCDBrCH_3$ (?)	0.03–0.09[g]	l,k	34

$CH_3CBr=CHCH_3$ (trans)	$-63.5°$, $h\nu$, HBr or DBr^i	$CH_3CHBrCH(D)BrCH_3$ (?)	$>15^g$	j,k	34
$CH_3CBr=CHCH_3$ (98% cis)	$-63.5°$, $h\nu$, DBr^i	$CH_3CHBrCDBrCH_3$ (?)	0.1^g	j,k	34
$CH_3CBr=CHCH_3$ (trans)	$-51.6°$, $h\nu$, HBr^i	$CH_3CHBrCHBrCH_3$ (?)	$16-24^g$	j,k	34
$CH_3CBr=CHCH_3$ (98% cis)	$-51.6°$, $h\nu$, HBr^i	$CH_3CHBrCHBrCH_3$ (?)	$0.18-1.2^g$	j,k	34
$CH_3CBr=CHCH_3$ (trans)	$-51.6°$, $h\nu$, DBr^i	$CH_3CHBrCDBrCH_3$ (?)	$11-16^g$	j,k	34
$CH_3CBr=CHCH_3$ (98% cis)	$-51.6°$, $h\nu$, DBr^i	$CH_3CHBrCDBrCH_3$ (?)	$0.54-0.89^g$	j,k	34
$CH_3CBr=CHCH_3$ (trans)	$-41°$, $h\nu$, HBr^i	$CH_3CHBrCHBrCH_3$ (?)	$11-16^g$	j,k	34
$CH_3CBr=CHCH_3$ (98% cis)	$-41°$, $h\nu$, HBr^i	$CH_3CHBrCHBrCH_3$ (?)	$0.23-0.35^g$	j,k	34
$CH_3CBr=CHCH_3$ (98% cis)	$-41°$, $h\nu$, DBr^i	$CH_3CHBrCDBrCH_3$ (?)	$1.4-2.7^g$	j,k	34
$CH_3CBr=CHCH_3$ (trans)	$-30.6°$, $h\nu$, HBr^i	$CH_3CHBrCHBrCH_3$ (?)	$4.6-6.1^g$	j,k	34
$CH_3CBr=CHCH_3$ (98% cis)	$-30.6°$, $h\nu$, HBr^i	$CH_3CHBrCHBrCH_3$ (?)	$1.3-1.8^g$	j,k	34
$CH_3CBr=CHCH_3$ (trans)	$-30.6°$, $h\nu$, DBr^i	$CH_3CHBrCDBrCH_3$ (?)	$4.9-5.7^g$	j,k	34
$CH_3CBr=CHCH_3$ (98% cis)	$-30.6°$, $h\nu$, DBr^i	$CH_3CHBrCDBrCH_3$ (?)	$1.0-1.6^g$	j,k	34
$CH_3CBr=CHCH_3$ (trans)	$-22.8°$, $h\nu$, HBr^i	$CH_3CHBrCHBrCH_3$ (?)	4.6^g	j,k,m	34
$CH_3CBr=CHCH_3$ (98% cis)	$-22.8°$, $h\nu$, HBr^i	$CH_3CHBrCHBrCH_3$ (?)	1.8^g	j,k,m	34
$CH_3CBr=CHCH_3$ (trans)	$-22.8°$, $h\nu$, DBr^i	$CH_3CHBrCDBrCH_3$ (?)	4.3^g	j,k,m	34
$CH_3CBr=CHCH_3$ (98% cis)	$-22.8°$, $h\nu$, DBr^i	$CH_3CHBrCDBrCH_3$ (?)	2.0^g	j,k,m	34
$CH_3CBr=CHCH_3$ (trans)	$0°$, $h\nu$, HBr^i	$CH_3CHBrCHBrCH_3$ (?)	4.9^g	j,k,m	34
$CH_3CBr=CHCH_3$ (98% cis)	$0°$, $h\nu$, HBr^i	$CH_3CHBrCHBrCH_3$ (?)	$2.4-3.0^g$	j,k,m	34

Olefin	Reaction conditions	Reported product hydrobromide		Controls	Ref.
		Reported structure (% yield)	threo/erythro (dl/meso)		
$CH_3CBr=CHCH_3$ (trans)	$0°$, hv, DBr^i	$CH_3CHBrCDBrCH_3$ (?)	3.8^g	j,k,m	34
$CH_3CBr=CHCH_3$ (98% cis)	$0°$, hv, DBr^i	$CH_3CHBrCDBrCH_3$ (?)	$1.9-2.2^g$	j,k,m	34
$CH_3CBr=CHCH_3$ (trans)	$25°$, hv, HBr^i	$CH_3CHBrCHBrCH_3$ (?)	3.5^g	j,k,m	34
$CH_3CBr=CHCH_3$ (trans)	$25°$, hv, DBr^i	$CH_3CHBrCDBrCH_3$ (?)	3.3^g	j,k,m	34
$CH_3CH=CHCH_3$ (cis)	-60 to $-78°$, hv, DBr	$CH_3CHDCHBrCH_3$ (95)	$>49^n$	o	269
$CH_3CH=CHCH_3$ (trans)	-60 to $-78°$, hv, DBr	$CH_3CHDCHBrCH_3$ (100)	$<0.01^n$	o	269

[a] Photoreaction for 15.5 hr of 7.25 mmole of trans-$CF_3CF=CFCF_3$ and 2.97 mmoles of HBr yielded 4 mmoles of $CF_3CF=CFCF_3$ [(trans/cis) = 4.9].

[b] Assignment of gross structure OK.

[c] Identity of threo vs. erythro appears to have been assumed.

[d] Basis of identification of threo vs. erythro: assumed trans addition.

[e] Conversion of bromochloride to 2-butene by Zn assumed to proceed by cis-elimination; conversion to $CH_3CCl=CHCH_3$ (70% yield) by alcoholic KOH assumed to proceed by trans-elimination. The results are consistent with the two reactions proceeding in opposite senses.

f Unreacted chlorobutene had "isomerized to roughly the equilibrium mixture of cis and trans [~80% trans]."

g Identification of products OK.

h ". . . the reactions were carried out in excess liquid addendum as solvent."

i ". . . the olefin was saturated with addendum prior to irradiation and excess addendum was passed through the liquid reaction mixture during the irradiation."

j "Under these conditions . . . the recovered olefin was partly isomerized."

k ". . . ionic addition gives only 2,2-dibromobutane . . ."

l "When this interval [between mixing and irradiation] was ~ 2 min. only small amounts (e. g. ca 1-3%) of 2,2-dibromobutane were formed . . ." "When the reactions were not irradiated . . . 2,2-dibromobutane was formed exclusively."

m ". . . only . . . 2,3-dibromobutane [no 2,2-dibromobutane] was isolated."

n Reaction of what was reported to be the threo-hydrobromide with EtO$^-$/EtOH yielded $CH_3CH=CHCH_3$ (cis), $CH_3CH=CDCH_3$ (cis), $CH_3CH=CHCH_3$ (trans), and $CH_3CH=CDCH_3$ (trans) in yields of 4.1, 0.08, 0.55, and 27%, respectively. Reaction of what was reported to be the erythro-hydrobromide gave the same products in yields of 0.09, 5.8, 5.5, and 0.06%, respectively. Trans-elimination was assumed.

o "Unreacted olefins recovered from reactions stopped short of completion were found to be unisomerized."

fraction of the olefin to be complexed in order that the predominant reaction be between Br· and complexed olefin.

If the β-bromoalkyl radicals were considered to be bridged [1; III(a)], whether they are bridged [1; I(a)-(c)] would still be an open question. For example, "according to . . . [one possible] interpretation the isomeric radicals [produced by addition of Br· to the olefin] are intercepted . . . [by HBr] more rapidly than they undergo rotation about the C_2C_3 bond" [34]. As in Sec. 6; ID1, estimates of rate constants for internal rotation in the gas phase at -78° of 3 x 10^6 and ~3 x 10^9 sec^{-1} may be obtained. Available rate constants for the reaction R· + HBr → RH + Br· are collected in Table 108. In view of the order of magnitude of [HBr] under the conditions corresponding to high stereospecificity, the above alternative interpretation does not appear to be a priori unreasonable. It is important to keep in mind the inappropriateness of our models, our very uncertain estimates of the rate constants corresponding to them, and our application of estimates derived from data obtained for systems in the gas phase to a reaction in solution.

Relevant to the question of whether one could deduce that these β-bromoalkyl radicals are bridged [1; I(a)-(c)] from an observation that they are bridged [1; III(a)] is the question of whether radicals with β-substituents which would not be "expected" to bridge [1; I(a)-(c)] are bridged [1; III(a)]. The addition of CF_3I, $EtOCONCl_2$, and $(CH_3)_3COCl$ to olefins has been studied; results are given in Table 109. In view of the partial equilibration of cis- and trans-$CF_3CH=CHF$ under the reaction conditions and the lack of control experiments relevant to our present purposes, it would not be proper to conclude from these experiments, not performed under conditions which result in high stereospecificity in the HBr additions (see above), that the proposed radical intermediates are not bridged [1; III(a)].

9.   Addition of Br_2 to Olefins

Results of the addition of Br_2 to several olefins are given in Table 110. In view of the uncertain structures of some of the products, and the lack of adequate control experiments, including those relevant to whether the reaction involves free radicals, we must withhold judgment on whether these results are relevant to the question of bridging.

TABLE 108

Estimates of Rate Constants at -78° of $R\cdot + HBr \rightarrow RH + Br\cdot$

R	Estimated rate constant, $l\ mole^{-1}\ sec^{-1}$	Method of estimation
$CH_3\cdot$	1×10^7 (with DBr)	a
$CH_3\cdot$	1×10^8	b,i
$CH_3CH_2\cdot$	8×10^6	d
$CH_3CH_2\cdot$	4×10^5	h
$(CH_3)_3C\cdot$	$E_a \cong 5.7$ kcal/mole	e
$CCl_3\cdot$	2×10^7	f
$CF_3\cdot$	3×10^8	g

a Based on Arrhenius parameters reported by Gac et al. [272].

b Fettis and Trotman-Dickenson [248] have reported that $k[HBr + CH_3\cdot \rightarrow Br\cdot + CH_4]/k[CH_3\cdot + I_2 \rightarrow CH_3I + I\cdot] = 10^{-0.30}$ exp$[-1.37/RT]$. Use of $k[CH_3\cdot + I_2 \rightarrow CH_3I + I\cdot] = 10^{9.9} l\ mole^{-1}\ sec^{-1}$ c leads to $k[HBr + CH_3\cdot \rightarrow Br\cdot + CH_4] = 1 \times 10^8 l\ mole^{-1}\ sec^{-1}$ at -78°.

c See footnote g to Table 62.

d Fettis and Trotman-Dickenson [248] have reported that $k[C_2H_5\cdot + HBr \rightarrow Br\cdot + C_2H_6]/k[C_2H_5\cdot + I_2 \rightarrow C_2H_5I + I\cdot] = 10^{0.23}$ exp$[-2.29/RT]$. Use of $k[C_2H_5\cdot + I_2 \rightarrow C_2H_5I + I\cdot] = 10^{9.5}$ exp$[-0.2/RT] l\ mole^{-1}\ sec^{-1}$ c leads to $k[HBr + C_2H_5\cdot \rightarrow Br\cdot + C_2H_6] = 8 \times 10^6 l\ mole^{-1}\ sec^{-1}$ at -78°.

e Guessed ("tentatively") by Benson and Buss [232].

f Reported by Benson [231]; the result of a psychothermochemical analysis of the data of Sullivan and Davidson [231A].

g Reported by Amphlett and Whittle [230].

h King et al. [480] have reported $k[Br\cdot + C_2H_6 \rightarrow C_2H_5\cdot + HBr] = 10^{11.6}$ exp$[-14.0/RT]$. Combination with ΔG leads to $k[C_2H_5\cdot + HBr \rightarrow C_2H_6 + Br\cdot] = 4 \times 10^5 l\ mole^{-1}\ sec^{-1}$ at -78°.

i The crude estimates of Batt and Cruickshank [529] lead to 5×10^5.

TABLE 109

Addition of CF_3I, $EtOCONCl_2$, and $(CH_3)_3COCl$ to Olefins

Olefin	Addendum	Reaction conditions	Reported Adduct Structure (% yield)	Reported threo/erythro	Controls	Ref.
$CF_3CH=CHF$ (trans)	CF_3I	$h\nu$	$CF_3CHICHFCF_3$ (78)	0.43^c	a	273
$CF_3CH=CHF$ (cis)	CF_3I	$h\nu$	$CF_3CHICHFCF_3$ (42)	0.43^c	b	273
$CF_3CH=CHCH_3$	CF_3I	$h\nu$	$CF_3CHICHMeCF_3$			
(cis/trans = 0.22)			(68)	1.6^n	o	535
(cis/trans = 0.15)			(74)	1.8^n	o	535
(trans)			(75)	2.1^n	p	535
$CH_3CH=CHC\equiv CH$(cis)	$(CH_3)_3COCl$	$h\nu$, 25°, cyclohexane^d	$CH_3CH(OBu^t)CHClC\equiv CH$ (13.4)^{h,i}	$2.2^{h,i}$	g	274
$CH_3CH=CHC\equiv CH$(cis)	$(CH_3)_3COCl$	$h\nu$, 25°, cyclohexane^e	$CH_3CH(OBu^t)CHClC\equiv CH$ (13.3)^{h,i}	$1.9^{h,i}$	g	274
$CH_3CH=CHC\equiv CH$(cis)	$(CH_3)_3COCl$	$h\nu$, 25°, cyclohexane^f	$CH_3CH(OBu^t)CHClC\equiv CH$ (11.9)^{h,i}	$1.8^{h,i}$	g	274
$CH_3CH=CHC\equiv CH$ (trans)	$(CH_3)_3COCl$	$h\nu$, 25°, cyclohexane^d	$CH_3CH(OBu^t)CHClC\equiv CH$ (10.5)^{h,i}	$1.9^{h,i}$	g	274
$CH_3CH=CHC\equiv CH$ (trans)	$(CH_3)_3COCl$	$h\nu$, 25°, cyclohexane^e	$CH_3CH(OBu^t)CHClC\equiv CH$ (10.4)^{h,i}	$2.0^{h,i}$	g	274

PhCH=CHPh (trans)	EtOCONCl$_2$	5-10°, benzene; then NaHCO$_3$	PhCH(NHCOOEt)CHClPh(65)^k	~1^k	1	525
EtCH=CHEt (trans)	EtOCONCl$_2$	35-40°, benzene; then NaHCO$_3$	EtCH(NHCOOEt)CHClEt (40)^m	0.7 or 1.5^m	1	525

[a] Unreacted olefin was 27% cis.

[b] Unreacted olefin was 69% cis.

[c] Identification of products OK.

[d] Enyne/(CH$_3$)$_3$COCl = 52.4.

[e] Enyne/(CH$_3$)$_3$COCl = 13.1.

[f] Enyne/(CH$_3$)$_3$COCl = 4.4.

[g] ". . . no cis-trans isomerization of recovered enyne was detected . . ."

[h] Basis of identification of threo adduct: (1) Hydrogenation gave what was reported[j] to be CH$_3$CH(OBu^t) CHClCH$_2$CH$_3$ in unspecified yield; (2) Ir and NMR spectra consistent with assigned structure.

[i] Basis of identification of erythro adduct: "the n.m.r. spectrum of the mixture of . . . [threo and erythro] was identical with that of . . . [threo] alone except for the appearance of two doublets for the C≡CH protons separated by 1 Hz . . . ."

[j] Basis of assignment of structure: Reaction of the mixture of chlorohydrins resulting from cis 2-pentene and Cl$_2$/H$_2$O with isobutylene/H$_2$SO$_4$ produced in unspecified yield material with the same retention time as produced from cis-2-pentene and Cl$_2$/Bu^{t}OH.

[k] Basis of identification of [threo + erythro] mixture: (1) acceptable C, H, N, and Cl analyses; (2) absorption at 3450, 1740, 1510, 1240, and 1060 cm^{-1}.

l "an induction period is observed followed by a rapid exothermic reaction." "light causes a rate enhancement and . . . oxygen and ethers of quinones slow it down."

m Basis of identification of [threo + erythro] mixture: (1) acceptable C, H, N, and Cl analyses; (2) absorption at 3300, 1710, 1520, and 780 cm^{-1}.

n Structure, exclusive of stereochemistry, OK. What was reported to be the threo isomer gave 66% $CF_3CH=CMeCF_3$ (trans/cis = 7) upon reaction with KOH at 100°. Similar treatment of what was reported to be the erythro isomer gave 84%, trans/cis = 0.6.

o "cis-trans isomerization was . . . not observed when the trans-olefin was irradiated alone."

p Olefin recovered from the reaction mixture was trans.

TABLE 110

Photoaddition of Br$_2$ to Olefins

Olefin	Reaction conditions	Reported product Structure (% yield)	Reported product threo/erythro (dl/meso)	Controls	Ref.
CF$_3$CH=CFCF$_3$ (trans)[a]		CF$_3$CHBrCFBrCF$_3$ (91)	0.52[b]	c	156
CF$_3$CH=CFCF$_3$ (27% cis)[a]		CF$_3$CHBrCFBrCF$_3$ (84)	0.52[b]	None	156
HOCH$_2$CH=CHCH$_2$OH (≥ 95% cis)	0°, CH$_2$Cl$_2$	HOCH$_2$CHBrCHBrCH$_2$OH (72)	> 2.6[d]	g,h	319
HOCH$_2$CH=CHCH$_2$OH(trans)	0°, CH$_2$Cl$_2$	HOCH$_2$CHBrCHBrCH$_2$OH (77)	< 0.30[d]	g,h	319
AcOCH$_2$CH=CHCH$_2$OAc (trans)	20°, CCl$_4$	AcOCH$_2$CHBrCHBrCH$_2$OAc (76)	< 0.32[e]	g	319
AcOCH$_2$CH=CHCH$_2$OAc (cis)	20°, CCl$_4$	AcOCH$_2$CHBrCHBrCH$_2$OAc (76)	< 0.32[e]	g	319
HOCMe$_2$CH=CHCMe$_2$OH (trans)	CH$_2$Cl$_2$	HOCMe$_2$CHBrCHBrCMe$_2$OH (90)	< 0.1[f]	g	319

[a] Olefins with 10.2 and 1–2 cps F–C–C=C–C–F coupling constants assigned cis– and trans–structures, respectively.

[b] Basis of identification of threo vs. erythro: none. Assignment of gross structure OK.

[c] Unreacted olefin had not isomerized.

[d] "... configuration [meso vs. dl] was established by study of its rate of reaction with KI, for it is known that meso-dibromides react more rapidly with KI than (±)-isomers." Also, the mp of what was reported to be the dl product agreed with that published for material which gave an acceptable Br analysis but whose stereochemistry was not established.

[e] See the first sentence in footnote d.

[f] See the first sentence in footnote d. Also, what was reported to be the meso product gave acceptable C, H, and Br analyses.

[g] ". . . [reactants] are isomerized only to a slight extent under the conditions of photochemical bromination."

[h] The products did not isomerize under the reaction conditions.

10.   Reaction of $CH_3CHOHCHBrCH_3$ with $(Bu^tO)_2$

A mixture of epimeric alcohols was allowed to react at 125° with 5-10 mole % $(Bu^tO)_2$ in the presence of propylene oxide or cyclohexene oxide [488]. The relative amounts of residual alcohols were considered to be indicative of "the relative rates at which the hydrogen abstraction reaction from the two isomers occurs." Results are given in Table 111. Since there is no reason to expect the relative amounts of residual alcohols to reflect inversely the relative rates of hydrogen abstraction and in view of the discussion in Sec. 1; VB, the lack of adequate control experiments, and the results in footnote b to Table 111, we do not consider the very small selectivity to represent a slight amount of bridging [1; III(a)] in $CH_3\overset{\cdot}{C}OHCHXCH_3$.

11. Summary

There is some evidence (6; ID4, 5, 9, 10) which provides no indication of bridging [1; III(a)] and some (6; ID1-3, 6-8) whose relevance is unclear if viewed properly but which indicates bridging [1; III(a)] if viewed superficially and uncritically. The latter, taken at face value, are consistent with bridging [1; I(c)], although there are reasonable alternative interpretations.

II.   β-BROMOCYCLOHEPTYL RADICALS

Control of Stereochemistry

Addition of HBr to Cycloheptenes

Results of the addition of HBr to 1-methyl- and 1-bromocycloheptene are given in Table 112. The reported stereoselectivity is large, corresponding to > 2.0 and 1.3 kcal/mole for the 1-methyl- and 1-bromocycloheptenes, respectively. In view of the uncertain structures of the products and the lack of adequate control experiments, we must withhold judgment on whether these results are relevant to the question of bridging. If they are assumed to be so and are taken at face value, they indicate that these β-bromocycloheptyl radicals are slightly bridged [1; III(a)].

TABLE 111

Peroxide-Induced Eliminative Decomposition of $CH_3CHOHCHXCH_3$ [488]

X	"$k_{\underline{threo}}/k_{\underline{erythro}}$"[a]	"$\Delta\Delta G^{\ddagger}$"[a], kcal/mole	Controls
Cl	1.02	0.0	b
Br	1.44	0.3	b
CH_3S	1.26	0.2	b

[a] See text. The reaction was stopped when "25-30% of the less reactive isomer had reacted."

[b] "When heated for several hours at 125° in the absence of tert-butyl peroxide, the isomeric halohydrins and hydroxy sulfides did not yield 2-butanone as a reaction product. All of the compounds did show some degree of thermal instability but decomposed at significantly slower rates than the tert-butyl peroxide induced reactions yielding 2-butanone."

III. β-BROMOCYCLOHEXYL RADICALS

A. Structure

An ESR spectrum, attributed to the 2-bromocyclohexyl radical, was obtained by observation of the reaction mixture resulting from the photolysis at 77°K of cyclohexene and HBr [175]; the spectrum "showed a basic three line spectrum with a splitting of 26 gauss, each line of which appears to be split further into at least five lines, with a splitting of 4 gauss." It was stated that " . . . an analysis of the hyperfine pattern strongly suggested a bridged type intermediate as the initiating free radical.": "The bridged structure is compatible with a large triplet splitting by the two pseudo-axial beta hydrogens . . . and a second splitting by the pseudo-equatorial beta hydrogens, with further splittings by either alpha or gamma hydrogens. The corresponding non-bridged structure would give a basic five line pattern instead of three." This interpretation is weakened by the observation [277] that electron irradiation of

TABLE 112

Addition of HBr to 1-Substituted Cycloheptenes

Substituent	Reaction conditions	Reported product (cis/trans) (% yield)	Controls	Ref.
CH$_3$	hν, ~70°, heptane	1-Bromo-2-methylcycloheptane[a] (>19)[c] (58-71)	None	275
Br[e]	hν, 0-5°	1,2-dibromocycloheptane (9-11)[f,g] (84)	d	276

[a] Basis of identification of product: (1) Not destroyed by the H$_2$O/acetone conditions which destroy 1-bromo-1-methylcycloheptane; (2) reaction with refluxing pyridine yielded what was reported to be 1-methylcycloheptene.[b]

[b] 3-Methylcycloheptene did not isomerize when treated with refluxing py/py·HBr.

[c] Product identified as cis on the basis of its dehydrobromination to what was reported to be 1-methylcycloheptene via an assumed trans-elimination.

[d] "... dibromides were checked for stability under reaction conditions. In no instance was there evidence for any interconversions between cis and trans isomers."

[e] Basis of assignment of structure: Produced in 12% yield by reaction of 1,2-dibromocycloheptane with Me$_3$N.

[f] Basis of identification of trans-dibromide: in reaction mixture: ir band at 12.10 μ, as in material resulting from the addition of Br$_2$ to cycloheptene.

[g] Basis of identification of cis-dibromide in reaction mixture: (1) ir bands at 9.48 and 12.34 μ; (2) reaction mixture gave acceptable C and H analyses.

cyclohexene at -196° C results in the formation of material whose ESR spectrum, assigned to the cyclohexenyl radical, is very similar to that reported above [175].*

B. Energy

1. Reaction of Cyclohexyl Bromides with Br_2

Photobromination [†] of bromocyclohexane at 60° was reported [102] to yield the following dibromides ("% of total dibromides"): 1,1 and/or cis-1,2 [‡] (3.53), trans-1,2 [**] (94.0), trans-1,3 [||] (0.92), cis-1,3 [||] (0.52), trans-1,4 [||] (0.50), cis-1,4 [||] (0.53). Photobromination of I at 31-40° in CCl_4 yielded the following dibromides (relative % yields): II(93), III(7), IV(<3) [281]. These results are subject to the comments in Sec. 5; IC2.

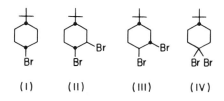

(I)   (II)   (III)   (IV)

---

*Photolysis of cyclohexene at 77° K, followed by warming to -115° C, resulted in the formation of material whose ESR spectrum [278] was "essentially identical" with that resulting from electron irradiation of cyclohexene at -196° C followed by warming to -126° C [277]. Also, the results of Ohnishi and Nitta [277] at -126° C have been reproduced [279].

[†] "The irradiated reactions were complete within a few hours, while . . . dark reaction mixtures still contained large amounts of un-reacted bromine after seven days. The inclusion of HBr in these dark reaction mixtures did not alter the situation."

[‡] Basis of identification of product: none.

[**] Identification of product OK.

[||] Basis of identification of product not mentioned.

If they are taken at face value, a clear preference has been shown for formation of vicinally substituted product. However, there is no indication that the relative amounts of products reflect relative rates of abstraction of the corresponding hydrogens; in view of the report of Tanner et al. [38], we believe that there is doubt as to whether these results are relevant to the question of bridging.

2.   Reaction of Bromocyclohexanes with Chromous Salts

See Sec. 5;IC9 for general comments. Activation parameters for reduction of bromocyclohexanes by Cr^{II} are in Table 113. These results, which indicate a large accelerating effect of a β-Br substituent, are subject to the comments in Sec. 6;IC7.

3.   Reaction of Cyclohexyl Bromide with Cl_2

See Table 69. The comments in Sec. 5;IC2 apply to these results with regard to β-, γ-, and δ-bromocyclohexyl radicals.

C.   Control of Stereochemistry

1.   Reaction of Cyclohexyl Bromides with Br_2

The results cited in Sec. 6;IIIB1 indicate that the 1,2-dibromocyclohexanes are formed with high stereoselectivity. In view of the lack of knowledge of details of the mechanism (Sec. 5;ID1) and the lack of adequate control experiments (Sec. 5;ID1) we must withhold judgment on whether these results are relevant to the question of bridging; in view of the report of Tanner et al. [38], we believe that there is doubt as to whether they are.

2.   Reaction of Cyclohexyl Bromide with Cl_2

See Table 71. These results are subject to the comments in Sec. 6;IIIC1.

3.   Reaction of Silver 1,2-Cyclohexanedicarboxylate with Halogens (The Hunsdiecker Reaction)

Although there have been many reports relevant to the mechanism of the Hunsdiecker reaction, the question of primary concern to us has not yet been answered definitively: What is the immediate precursor of RX? Is

TABLE 113

Reduction of Bromocyclohexanes by Cr^{II} [a]

Cyclohexane	Temperature range, °C	$\Delta H^{\ddagger}$, [b] kcal/mole	$\Delta S^{\ddagger}$, [b] eu
trans-1, 2-Dibromo	-0.6 - 20.0	11.2 ± 0.9	-20.8 ± 2.5
cis-1, 2-Dibromo	10.0 - 29.7	13.2 ± 0.6	-24.2 ± 2.0
Bromo	29.7 - 50.0	14.4 ± 1.3	-27.2 ± 4.5

[a] "In 88% v/v DMF - water and 1.0 M perchloric acid."

[b] "Errors are maximum errors computed by the method of R. C. Petersen, J. H. Markgraf, and S. D. Ross, J. Am. Chem. Soc., 83, 3819 (1961)."

it free R·? Since the reaction mixture is heterogeneous and contains silver and halogen, it might be anticipated that it would be difficult to answer with confidence the latter question in the affirmative. Possibly relevant evidence is discussed below.

Many examples have been reported of the loss of configuration about the carboxyl-bearing carbon during the course of the conversion RCOOAg → RX [282-295] (However, see also Refs. 296 and 306); Sukman [306] has reported the formation (12% yield) of racemic 2,4,6-tribromo-5-methoxy-2'-nitrobiphenyl in the reaction of optically active silver 4,6-dibromo-5-methoxy-2'-nitrodiphenyl-2-carboxylate with Br_2. In all of these studies adequate control experiments relating to the configurational stability of reactants, products, and proposed intermediates other than R· were not reported.

The production of small amounts of RCl as a by-product of the reaction of RCOOAg with Br_2/CCl_4 has been reported [297, 298] and demonstrated [299, 300]. Its origin is unknown.

In order to help distinguish between two of the possible precursors of RX, a free carbonium ion and a free radical, it would be useful to know the fate of those silver salts whose R group would be "expected" to

behave differently as a free radical than as a carbonium ion; some frequently cited examples follow.

The reaction of silver cyclobutanecarboxylate with Br_2 yielded 44-53%
cyclobutyl bromide and 2-10%, 1,2,4-tribromobutane, with 39-40% unac-
counted for [301, 305]. The poor material balance and the presence of
material whose precursor could have been allylcarbinyl or cyclopropylcarbinyl bromide makes it difficult to cite these observations as evidence
against a free cyclobutyl cation intermediate. Reaction of silver β, β-
dimethylbutyrate with Br_2 was reported [302] to give an 83% yield of
neopentyl bromide; both the yield and the structure were assumed. Smith
and Hull [303] found that 62% neopentyl bromide was produced in the same
reaction (t-amyl bromide was absent, but was not reported to be stable to
the reaction conditions); unfortunately, the remaining 38% was not
identified.

Reports have appeared to the effect that substituted cyclohexyl [282]
and norbornyl [283, 285, 304] radicals show the same stereoselectivity as
the Hunsdiecker intermediate in the formation of RX by reaction with some,
but not all, halogen-transfer agents. These observations do not help much
in distinguishing between the various possible free, encumbered, and
adsorbed precursors of RX in the Hunsdiecker reaction.

Sufficient evidence to support the proposition that the immediate precursor of RX is free R· does not exist. It is with this in mind that we discuss the generation of "free" "radicals" by use of the Hunsdiecker reaction.

Results relevant to the stereospecificity of the title reaction are given
in Table 114. The mechanism of this reaction may be unusually complex
(see below). All of the uncertainties, including those which could be
resolved by additional control experiments, make any conclusion from these
results premature. If the observed high stereoselectivity of formation of
trans-dihalide is naively taken at face value and the 2-bromocyclohexyl
radical considered to be bridged [1; III(a)], any attempt to go further and
consider it to be bridged [1; I(a)-(c)] would have to overcome the contents
of Table 115, among other things.

Applequist and Werner [307] have studied the reaction in detail with
the results given in Table 116. The transformations correspond to the

TABLE 114

Reaction of Silver 1,2-Cyclohexanedicarboxylates with Halogens, CCl_4

Substrate	Halogen	Temperature	1,2-diBr trans	cis	1-Br,2-Cl trans	cis	1,2-diCl trans	cis	Controls	Ref.
cis[a]	Br_2	0–25°	43	–					d,e	284
cis	Br_2	0°	14.3	0.1					None	289
cis[a]	Br_2	Reflux	47	–					b,d,e	284
cis	Br_2	77°	27.6	1.3					None	289
trans[a]	Br_2	0–25°	48	–						284
trans	Br_2	0°	12.7	0.1					None	289
trans[a]	Br_2	Reflux	37	–					c	284
cis	$Br_2/Cl_2(1/1)$	0°	21.6	–	5.6	–	Trace	–	f	289
trans	$Br_2/Cl_2(1/1)$	0°	19.6	–	1.6	–	Trace	–	f	289

[a] ". . . [the silver salts] were checked by conversion back to the free acids. The configurations were unchanged."

[b] Refluxing in CCl_4 with $Br_2/AgBr$ led to 60% recovery of all cis material.

[c] Refluxing in CCl_4 with $Br_2/AgBr$ led to 94% recovery of all trans material.

d Cis substrate was allowed to react in the presence of cis-dibromide. The product showed a melting point depression of ~30° and had an "IR spectrum completely identical with that of a synthetic mixture made up of authentic cis- and trans-dibromides. The data could not distinguish partial isomerization from complete retention of configuration, but there could be no doubt that a very large portion of the cis-survived the reaction."

e The cis-diacid was recovered quantitatively with no change in melting point after refluxing in CCl_4 with Br_2/AgBr.

f Cyclohexene yielded 6.5% trans-1,2-dichloro, 46.1% trans-1-bromo-2-chloro-, and 3.3% trans-1,2-dibromocyclohexane.

TABLE 115

Reaction of Silver Methyl Cyclohexane-1,2-Dicarboxylates with Br_2 (0°, CCl_4) [289]

| Substrate | Product 1-Bromo-2-carbomethoxycyclohexane (% yield) | | Controls |
	trans[a]	cis[b]	
cis	18.5	< 5	c
trans	22.1	< 5	c

[a] Assignment of structure ultimately rests on that of what was reported to be trans-5-bromocyclohexene-4-carboxylic acid which is based on agreement of melting point with that published for material characterized solely by acceptable C and H analyses.

[b] Assignment of structure ultimately rests on that of what was reported to be cis-2-bromocyclohexanecarboxylic acid which is based on agreement of melting point with that published for authentic material.

[c] The product of the reaction of cis-1-bromo-2-carbomethoxycyclohexane with equimolar amounts of silver methyl cis-cyclohexane-1,2-dicarboxylate and NaBr in CCl_4 at 0° had an ir spectrum "identical to that of" starting material. "The characteristic IR band of trans at 14.50 μ was not present."

conversion

They discuss the reaction in terms of the scheme included in Table 116:
"In order to accommodate the data . . . and the controls . . . it is necessary to have a mechanistic scheme which includes an inversion pathway
and either a retention pathway or a pathway leading to racemic product.
In . . . [Table 116] is presented a scheme of possibilities . . . assuming
. . . that the bishypobromite IV is an intermediate in all pathways. Decomposition of IV by reactions 9, 13, and 14 . . . appears to be the most
reasonable mechanism for formation of inverted product III. Reactions
9-12, with steric control of entering bromine in steps 10 and 12, represent
the classical Hunsdiecker mechanism . . . without specific neighboring
group involvements, and could be expected to lead to dextrorotatory
dibromide. It is felt that this pathway makes little if any contribution,
since reaction 10 should be favored over 13 by increasing bromine concentration, whereas a threefold increase in initial bromine concentration
(resulting in a larger ratio as the reaction proceeds) did not decrease the
optical activity of the product . . . Furthermore, it will be noted that
whereas under some conditions the activity of the product became almost
nil, it always remained negative. The evidence then suggests most
strongly that there is a racemate producing mechanism in competition with
9, 13, and 14. The most likely racemate-producing mechanisms are those
which involve a cyclohexene intermediate . . . and those which involve a
bridged-bromine radical, V, as an intermediate or as a transition state for
racemization of . . . VI. A strong argument against the participation of V
is that its formation from VI in competition with reaction 14 would be
favored by low bromine concentration, so that the failure of a threefold increase in bromine concentration to change the optical activity of product
. . . argues against reaction 15 in the same way that it argues against
10 . . . Loss of a bromine atom from VI . . . is ruled out . . . again by
the insensitivity of the optical yield to bromine concentration. Reaction 17

TABLE 116

Reaction of Optically Active[a] Silver _trans_-1,2-Cyclohexanedicarboxylate
with Br_2, CCl_4 [307]

| Reaction conditions | Dibromide | | Controls[f,g] |
	Specific rotation, deg.	% yield	
31.3°, 1 equiv Br_2	-1.8 to -4.9 (2 runs)	26-29	c,e
0°, 1 equiv Br_2	-8.4 to -27.0 (6 runs)	14-19	c,d
0°, 1 equiv Br_2, hν	-19.8 to -21.4 (2 runs)	39-41	b,c
0°, 3 equiv Br_2	-24.5 to -25.3 (2 runs)	51-58	c,d

[a] Specific rotation +20.5 to +21.8°.

[b] 32% of the starting diacid was recovered, melting point unchanged.

[c] Treatment of racemic _trans_-dibromide with optically active silver salt resulted in the recovery of 94% racemic _trans_-dibromide.

[d] Treatment of optically active _trans_-dibromide with 1800 mole % AgOAc and 2700 mole % Br_2 resulted in the recovery of 56% _trans_-dibromide which showed 8.1% racemization.

[e] Same as d except 42% racemization. If AgOAc omitted, 82% recovery and no racemization.

^{f} Treatment of 1.0 g of cis-dibromide with 800 mole % Br_2 in CCl_4 resulted in the recovery of 0.92 g of material whose infrared spectrum had all of the bands of the cis-isomer and none of the trans ($<5\%$).

^{g} 38% diacid recovered from one run with "undiminished optical rotation."

might form cyclohexene in a path whose competition with the inversion mechanism (reaction 13) would be independent of bromine concentration." Since the highly erratic optical yields and the fact that only two runs were made with three equivalents of Br_2 make statistically untenable a claim that optical yield is independent of $[Br_2]$, we believe that the data presented do not justify attempts to force any proposed mechanism to conform to a requirement that the optical activity of the product be independent of $[Br_2]$. We present a scheme which accounts for all of the data presented. Its point of departure from that of Applequist and Werner is the inclusion of the 2-(AgOOC)cyclohexyl radical as an intermediate; it eliminates the need to propose steps 13 and 17 but involves a sense of stereoselectivity for the conversion 2-(AgOOC)cyclohexyl → silver 2-bromocyclohexanecarboxylate which, although reasonable, lacks close precedent.

A cyclohexene component of the scheme has been omitted only to enhance the clarity of presentation. If a sufficient fraction of the product is derived from the 2-(AgOOC)cyclohexyl radical and if the COOAg group can direct (through some type of coordination?) the incoming bromine-containing molecule so as to provide sufficient cis stereoselectivity, this scheme is consistent with the results. *

4. Addition of HBr to Cyclohexenes

Results of the addition of HBr to 1-chloro-4-$\underline{t}$-butyl-, 1-chloro-, 1-bromo-, 1-methyl-, 1-chloro-5-$\underline{t}$- butylcyclohexene, and cholesteryl acetate are given in Table 117. Results of the addition of $BrCCl_3$, perfluoroalkyl iodides, $EtOCONCl_2$, EtOCONHCl, and $BrCH(CN)_2$ to cyclohexenes are given in Table 118. In view of the lack of adequate proofs of structure in some cases and, in all cases, of control experiments which demonstrate that the products are formed in the same ratio as from the direct reaction of the free 2-bromocyclohexyl radical with HBr, we defer judgment on the meaning of these results with regard to the question of bridging. If the results are taken at face value, 2-bromocyclohexyl radicals can be moderately (e.g., stereoselectivity of addition to 1-chlorocyclohexene corresponds to $\Delta\Delta G^{\ddagger} = 3.1$ kcal/mole) bridged [1; III(a)] while 2-trichloromethyl-, perfluoroalkyl-, urethano-, or dicyanomethyl-cyclohexyl radicals are not bridged [1; III(a)]. Within this framework of superficiality, one can speculate about this difference in "behavior" in terms of bridging [1; I(a)-(c)] only by Br or in terms of a combination of conformational factors in the substituted cyclohexyl radical and its transition state leading to product, Br being different in that it combines

---

*The results of Applequist and Werner have been confirmed in part by Riemann [289] who has reported that silver $\underline{trans}$-1,2-cyclo-hexanedicarboxylate ($[\alpha]_D^{20} = +17.7°$) reacted with Br_2/CCl_4 at 0° to give 13% $\underline{trans}$-1,2-dibromocyclohexane ($[\alpha]_D^{20} = -18.1°$).

being much "smaller" than the other groups (less preference for the equatorial position) [317] with HBr being a very good hydrogen transfer agent (Table 108).*

5. Addition of Br_2 to Cyclohexenes

Photoaddition of Br_2 to cyclohexene in hexane yielded 92% trans-1,2-dibromocyclohexane [319]. In view of the lack of adequate control experiments, including those relevant to whether the reaction proceeds via a free radical mechanism, we must withhold judgment on whether these results are relevant to the question of bridging.

IV.   β-BROMOCYCLOPENTYL RADICALS

A. Structure

ESR spectra, attributed to the β-bromocyclopentyl radical, were obtained by observation of the reaction mixture resulting from the photolysis at 77°K of cyclopentene and HBr (or DBr) [175]; "a basic five line spectrum with a splitting of 18.8 gauss, each line of which is split further into a triplet with a splitting of 9.4 gauss" was obtained. "In each case an analysis of the hyperfine pattern strongly suggested a bridged type intermediate as the initiating free radical." "The unbridged structure . . . also could be expected to give a five line spectrum due to splitting by the possible equivalent alpha and beta hydrogens, but these would be split further into quintets by the gamma hydrogens if any fine structure were to be observed at all."

Symons [176] "has suggested an interpretation not considered originally, which, if correct, invalidates the conclusions obtained above": "The spectrum reported for the radical derived from cyclopentene . . . is in good accord with expectation for the cyclopentenyl radical . . ." In addition: "The major drawback of the bridging bromine atom concept is that there is

---

*Perfluoroalkyl iodides and $BrCCl_3$ react rapidly, but more slowly than HBr, with radicals [318]: $k(CF_3I + CH_3\cdot) = 10^{10.8}exp(-7.5/RT)$ and $k(BrCCl_3 + CH_3\cdot) = 10^{10.2}exp(-7.1/RT)$ l mole^{-1} sec^{-1}.

TABLE 117

X	Y	Z	Reaction conditions	Reported products		Controls	Ref.
				% Yield	cis/trans (X, Br)		
Cl	Bu^t	H	hν, $-78°$, pentane^a	76	21^d	None	40
Cl	Bu^t	H	hν, -60 to $-70°$, pentane^b	c	32^d	None	40
Cl	H	H	hν, 0 or $-20°$, pentane	84–88	330^g	e	308
Br	H	H	hν, 0 or $-20°$, pentane	76–81	200^g	f	308
Cl	H	Bu^t	Room temp., 5 mole % Bz$_2$O$_2$, pentane	84	> 5^h	None	309
Cholesteryl acetate			20°, O$_2$, CCl$_4$	?		m	310

^{a} HBr/olefin = 1.9, [HBr] = 1.7.

^{b} HBr/olefin = 2.4.

^{c} "virtually complete reaction."

^{d} Basis of identification of cis product OK. Assignment of a trans structure to the other product rests on (1) comparison of rate constant for dehydrohalogenation with that for cis–1, 2-dichlorocyclohexane, (2) assumption that "an axial hydrogen [NMR] absorption will occur at higher field than that of an equatorial hydrogen on a similarly substituted carbon," (3) acceptable H analysis, and (4) acceptable molecular weight determination. If the assignment of a trans structure is incorrect, the cis/trans ratio is greater than that cited in the table.

[e] In ether as solvent or in the presence of $FeCl_3$ or $FeCl_3/Ph_2NH$, 1-bromo-1-chlorocyclohexane is produced in 79, 82, and 79% yields, respectively.

[f] In ether as solvent or in the presence of $FeCl_3$ or $FeCl_3/Ph_2NH$, 1,1-dibromocyclohexane is produced in 84, 84, and 76% yields, respectively.

[g] Identification of products OK.

[h] Basis of identification of all-cis product: (1) same as product obtained in 27% yield from reaction of PBr_3 and what was reported to be trans-2-chloro-trans-4-t-butylcyclohexanol[i]; (2) acceptable C and H analyses; (3) ir absorptions at 680 and 736 cm^{-1} (attributed to axial Br and equatorial Cl stretch, respectively); (4) reaction with Bu^{t}OK/Bu^{t}OH regenerated starting olefin in 90% yield.

[i] Basis of assignment of structure: (1) different from what was reported to be cis-2-chloro-cis-4-t-butylcyclohexanol[j]; (2) acceptable C, H, and Cl analyses; (3) reaction with Bu^{t}OK/Bu^{t}OH gave a 77% yield of what was reported to be trans-4-t-butyl-cyclohexene epoxide[k]; (4) reaction with KOH/Pr^{i}OH gave a 51% yield of what was reported to be trans-2-hydroxy-cis-4-t-butylcyclohexanol.[l]

[j] Basis of assignment of structure: (1) acceptable C, H, and Cl analyses; (2) reaction with KOH/MeOH gave 56% 4-t-butylcyclohexanone.

[k] Basis of assignment of structure: boiling point and index of refraction agreed with those reported for authentic material.

[1] Basis of assignment of structure: (1) acceptable C and H analyses; (2) same as product of saponification of the hydroxyacetate mixture formed by reaction of 4-$\underline{t}$-butylcyclohexene and peracetic acid.

[m] Reaction in $CHCl_3$ under N_2 gave a "good yield" of

TABLE 118

Addition of Radicals to Cyclohexenes

Substrate	Addendum	Reaction conditions	Reported product			Ref.
			Structure (% yield)	trans/cis	Controls	
Cyclohexene	$BrCCl_3$	$8-15°$, γ-rays	CCl_3 / Br (64)	$>5^i$	None	315
Cyclohexene	$BrCCl_3$	Room temp., γ-rays	CCl_3 / Br (50)	$\sim1^a$	None	311
Cyclohexene	$BrCCl_3$	$h\nu$, 4 hr	CCl_3 / Br (60)	1.2^a	c	312
Cyclohexene	CF_3I	$20-40°$, $h\nu$	CF_3 / I (~50)	1.4^a	None	313,314
Cyclohexene	CF_3I	$80°$, 4 mole % AIBN, 15 hr	CF_3 / I (90)	1.95^a	e	313,314
Cyclohexene	CF_3CF_2I	$130°$, 10 mole % $(Bu^tO)_2$	CF_2CF_3 / I (<40)	1.14^a	None	313,314
Cyclohexene	$(CF_3)_2CFI$	$h\nu$, $2°$	$CF(CF_3)_2$ / I (9)	2.6^a	d,g	313,314
Cyclohexene	$(CF_3)_2CFI$	$h\nu$, $25-30°$	$CF(CF_3)_2$ / I (46)	1.5^a	g	313,314

Substrate	Addendum	Reaction conditions	Reported product			
			Structure (% yield)	trans/cis	Controls	Ref.
Cyclohexene	$(CF_3)_2CFI$	50°, 6 mole % AIBN	$CF(CF_3)_2$, I (80)	2.5[a]	f, g	313, 314
Cyclohexene	$(CF_3)_2CFI$	68°, 6–12 mole % AIBN	$CF(CF_3)_2$, I (70–80)	2.5–2.7[a]	f, g	313, 314
Cyclohexene	$CF_3(CF_2)_2I$	–20 to 2°, hν	$(CF_2)_2CF_3$, I (72)	0.82–0.92[a]	h	313, 314
Cyclohexene	$CF_3(CF_2)_2I$	25–35°, hν	$(CF_2)_2CF_3$, I (77)	0.87[a]	h	313, 314
Cyclohexene	$CF_3(CF_2)_2I$	50–71°, 3–7 mole % AIBN	$(CF_2)_2CF_3$, I (87–90)	1.01[a]	f, h	313, 314
Cyclohexene	$CF_3(CF_2)_3I$	47–60°, hν	$(CF_2)_3CF_3$, I (69)	1.0[a]	None	313, 314
Cyclohexene	$CF_3(CF_2)_3I$	75°, 5 mole % AIBN	$(CF_2)_3CF_3$, I (90)	1.07[a]	None	313, 314
Cyclohexene	$CF_3(CF_2)_6I$	hν, 25°	$(CF_2)_6CF_3$, I (75)	1.10[a]	None	313, 314
Cyclohexene	$CF_3(CF_2)_6I$	75°, 5 mole % AIBN	$(CF_2)_6CF_3$, I (88)	1.1[a]	None	313, 314
Cyclohexene	$BrCH(CN)_2$	≤25°, CH_2Cl_2	$CH(CN)_2$, Br (?)	1.0[a]	s	316

Olefin	Reagent	Conditions	Product	(%)			Ref.
Cyclohexene	$BrCH(CN)_2$	$h\nu$, ≤35 to ≤42°, benzene or CH_2Cl_2[j]	[$CH(CN)_2$, Br]	(?)	1.1^a	s	316
Cyclohexene	$BrCH(CN)_2$	$h\nu$, ≤36°, CH_2Cl_2[k]	[$CH(CN)_2$, Br]	(32)	1.1^a	s	316
Cyclohexene	$BrCH(CN)_2$	≤161°, $\underline{o}$-$C_6H_4Cl_2$[m]	[$CH(CN)_2$, Br]	(?)	1.2^a	None	316
Cyclohexene	$EtOCONCl_2$	35–40°, benzene; then $NaHCO_3$	[$NHCO_2Et$, Cl]	(43)	6.2^u	v	525
Cyclohexene	$EtOCONCl_2$	40°; then $KI/Na_2S_2O_3$	[$NHCO_2Et$, Cl]	(?)	8^a	None	526
Cyclohexene	$EtOCONHCl$	$h\nu$, 0–20°, benzene	[$NHCO_2Et$, Cl]	(>80)	0.14^a	None	526
1-Methyl-cyclohexene	$BrCH(CN)_2$	$h\nu$, ≤37°, CH_2Cl_2[n]	[$CH(CN)_2$, CH_3, Br]	(72)	2.1^p	t	316
1-Ethylcyclo-hexene	$BrCH(CN)_2$	$h\nu$, ≤38°, CH_2Cl_2[o]	[$CH(CN)_2$, C_2H_5, Br]	(25)	1.5^q	None	316
[octahydronaphthalene]	$BrCCl_3$	$h\nu$	[CCl_3, Br]	(50)	$>10^{b,a}$	None	312

Substrate	Addendum	Reaction conditions	Reported product		Controls	Ref.
			Structure (% yield)	trans/cis		
(structure)	$BrCH(CN)_2$	$h\nu$, CH_2Cl_2	(structure) $CH(CN)_2$ Br (90)	$>9^r$	None	316

[a] Identification of products OK.

[b] Our estimate, assuming a 5% limit of detectability of the cis-isomers.

[c] A 28/72 cis/trans mixture in $BrCCl_3$ "did not change composition when irradiated under the addition reaction conditions for 22 hr."

[d] Trans/cis cited corresponds to 9% conversion. The ratios corresponding to 14 and 19% conversion are 2.2 and 2.1, respectively.

[e] Reaction for 20 hr gave trans/cis = 1.81.

[f] The trans/cis ratio was an insensitive function of % conversion at 65°.

[g] Equilibrium constant (cis → trans, by extrapolation) = 0.43, 0.53, 0.65, and 0.73 at 2°, 25°, 50°, and 68°, respectively, in benzene.

[h] Equilibrium constant (cis → trans, by extrapolation) = 0.12, 0.17, and 0.29 at 2°, 25°, and 68°, respectively, in benzene.

[i] Basis of assignment of structure: (1) acceptable C, H, and halogen analyses; (2) correlated with product of addition of HBr to 1-cyclohexenecarboxylic acid.

[j-o]

Alkene/$BrCH(CN)_2$ = 1.2–1.9, 39, 1.8, 3.0, 1.8–29, and 1.3, respectively.

[p] Assignment of structure, exclusive of stereochemistry, OK. One isomer [assigned Br and $CH(CN)_2$ trans] gave a 53% yield of 1-methyl-7,7-dicyanonorcarane upon reaction with Et_3N/CH_2Cl_2 at 0°; the other "remained unchanged" under those conditions.

[q] As in p, except substitute "73" for "53" and "ethyl" for "methyl."

[r] Assignment of structure, exclusive of stereochemistry, OK. All other isomers other than the proposed

ruled out. Latter deemed inconsistent with formation of the cyclopropane in 90% yield

upon reaction with Et_3N/CH_2Cl_2 at 0°.

[s] Φ_{30}^0 = 470. , "added to cyclohexene prior to start of the reaction, failed to take part in the

reaction sequence as could be shown by quantitative gas chromatography of the reaction mixture." ". . . the trans/cis ratio did not change even upon longer standing of the reaction mixture."

[t] "Irradiation of in the presence of $BrCH(CN)_2$ failed to produce even a trace of ."

[u] Basis of identification of trans product: absorption at 3400, 1730, and 1510 cm^{-1}. Basis of identification of cis product: none.

[v] "an induction period is observed followed by a rapid exothermic reaction." " . . . light causes a rate enhancement and . . . oxygen and ethers of quinones slow it down."

no indication, in the spectra discussed above, of any anisotropy in the g-values. Appreciable anisotropy would be expected for the bromine atom adducts, together with a shift in g_{av} to values somewhat greater than that of the free-spin . . . Furthermore, there is no indication of the expected characteristic hyperfine coupling to bromine." (For comments relevant to the final point see Ref. 195).

Symons' comments are reinforced by the observations that γ-irradiation at room temperature [479] and electron-irradiation at -143° [279, 499] and -196° [499] of cyclopentene resulted in the formation of material whose ESR spectrum, assigned to the cyclopentenyl radical, is very similar to that reported by Abell and Piette [175] in the cyclopentene + HBr (or DBr) case.

B. Energy

1. Reaction of Cyclopentyl Bromide with Br$_2$

The photobromination of cyclopentyl bromide resulted in a 90% yield of 1,2-dibromocyclopentane [102]. No control experiments were reported. This result is subject to the comments in Sec. 6;IIIB1.

2. Reaction of Cyclopentyl Bromide with Cl$_2$ and (CH$_3$)$_3$COCl

See Sec. 5;IIIA1 (Tables 72 and 73) and 5;IIIA2 (Table 74) and the discussions therein to which the results relating to β- and γ-bromocyclopentyl radicals are subject.

C. Control of Stereochemistry

1. Reaction of Cyclopentyl Bromide with Br$_2$

The photobromination of cyclopentyl bromide resulted in a trans/cis ratio of 1,2-dibromocyclopentanes of > 30 [102]. No control experiments were reported. This result is subject to the comments in Sec. 6;IIIB1.

2. Reaction of Cyclopentyl Bromide with Cl$_2$

See Table 75. The reported trans/cis ratio of 1-bromo-2-chlorocyclopentanes corresponds to $\Delta\Delta G^{\ddagger} > 2.2$ kcal/mole. The comments in Sec. 5;IIB apply to these results with regard to the 2-bromocyclopentyl radical.

3. Reaction of Cyclopentyl Bromide with $(CH_3)_3COCl$

See Table 76. The reported trans/cis ratio of 1-bromo-2-chloro-cyclopentanes corresponds to $\Delta\Delta G^{\ddagger} > 2.2$ kcal/mole. The comments in Sec. 5; IIB apply to these results with regard to the 2-bromocyclopentyl radical.

4. Addition of HBr to Cyclopentenes

Photoaddition [276] of HBr at 0-5° to what was reported to be 1-bromocyclopentene* was reported to give a 93-99% yield of "saturated products" (presumed to be 1,2-dibromocyclopentane) whose trans/cis ratio was reported to be 10-19. [†], [‡], [**]

Results of the addition of $BrCCl_3$, $CF_3CF_2CF_2CF_2I$, and $BrCH(CN)_2$ are given in Table 119. The copolymerization of indene and O_2 at 50°, followed by reduction, gave 62% 1,2-dihydroxyindane with $0.56 \leq$ trans/cis ≤ 2.1 [320]. In view of the lack of control experiments which demonstrate that the products are formed in the same ratios as from the direct reaction of cyclopentyl radical with addendum, we defer judgment on the meaning of these results with regard to the question of bridging. If the results are taken at face value, 2-bromocyclopentyl radicals can be slightly (maximum observed stereoselectivity corresponds to 1.6 kcal/mole) bridged [1; III(a)]. It is not possible to judge whether this small stereoselectivity is unusual because most of the entries in Table 119 are lower limits.

---

* Basis of assignment of structure: (1) prepared from what was reported to be trans-1,2-dibromocyclopentane [†]; (2) acceptable C and H analyses.

† Basis of assignment of structure of trans-1,2-dibromocyclopentane: prepared from Br_2 and cyclopentene.

‡ Basis of assignment of structure of cis-1,2-dibromocyclopentane: acceptable C and H analyses.

** " . . . these dibromides were checked for stability under reaction conditions. In no instance was there evidence for any interconversions between cis and trans isomers. "

TABLE 119

Addition of Radicals to Cyclopentenes

Substrate	Addendum	Reaction conditions	Reported product			
			Structure (% yield)	trans/cis	Controls	Ref.
Cyclopentene	$BrCCl_3$	$h\nu$	(cyclopentane, CCl_3, Br) (84)	$>5.2^a$	None	312
Cyclopentene	$CF_3(CF_2)_3I$	75–82°, 7 mole % AIBN	(cyclopentane, $(CF_2)_3CF_3$, I) (80)	$>4^a$	None	313, 314
Cyclopentene	$BrCH(CN)_2$	$h\nu$, $\leq 38°$, 5 hr, CH_2Cl_2 ^{b}	(cyclopentane, $CH(CN)_2$, Br) (34)	3.8^d	g	316
1-Methylcyclopentene	$BrCH(CN)_2$	$h\nu$, $\leq 38°$, 5 hr, CH_2Cl_2 ^{c}	(cyclopentane, $CH(CN)_2$, Br, CH_3) (44)	$2.6^{e,f}$	None	316
Indene	$BrCCl_3$	$h\nu$	(indane, CCl_3, Br) (72)	$>2.5^a$	None	312

^{a} Identification of products OK.

^{b} Cycloalkene/$BrCH(CN)_2$ = 0.69 – 22.

^{c} Cycloalkene/$BrCH(CN)_2$ = 1.0.

[d] What was reported to be the cis isomer "remained unchanged" upon treatment with Et_3N/CH_2Cl_2 at 0°. The crude reaction mixture gave a 51% yield of 2,2-dicyanobicyclo[3.1.0]hexane under these conditions.

[e] Basis of identification of trans [$CH(CN)_2$, Br]-product: (1) ir absorption at 2250 cm^{-1}; (2) acceptable Br and N analyses; (3) reaction with Et_3N/CH_2Cl_2 gave a 37% yield of 1-methyl-2,2-dicyanobicyclo[3.1.0] hexane.

[f] Basis of identification of cis [$CH(CN)_2$, Br]-product: (1) ir absorption at 2232 cm^{-1}; (2) acceptable C, H, and N analyses; (3) "remained unchanged" upon reaction with Et_3N/CH_2Cl_2.

[g] $\Phi_{30°} = 3400$. The dark reaction involved 5% conversion/hr and was completely inhibited by the addition of t-butylcatechol.

V.   β-BROMOCYCLOBUTYL RADICALS

Control of Stereochemistry

1.   Reaction of Silver 1,2-Cyclohexanedicarboxylates with Br_2

See Sec. 6; IIIC3 for general comments. Reaction of silver cis-1,2-cyclohexanedicarboxylate and silver trans-1,2-cyclohexanedicarboxylate with Br_2/CCl_4 at <65° was reported to yield 1,2-dibromocyclobutane, trans[*]/cis[†] = 5.6 and 2.5, respectively.[‡]

The relative amounts of reported [291, 322] products from the reaction

correspond to trans/cis = 4.0.[‡]

---

[*] Basis of identification of product: produced in unspecified yield by reaction of cyclobutane with Br_2 in $CHCl_3$.

[†] Basis of identification of product: produced in unspecified yield by photoreaction of what was reported to be 1-bromocyclobutene with HBr in isopentane at 0°.

[‡] No control experiments were reported.

[**] Basis of assignment of structure: (1) acceptable C, H, and Br analyses; (2) produced in 43% yield from $KMnO_4$ oxidation of cyclooctatetraene dibromide.

[||] Basis of identification of product: (1) number of bands in ir spectrum; (2) consistency of NMR spectrum with structure.

[***] Basis of identification of product: As in [||] + acceptable H analysis.

[††] Mixture gave acceptable C, H, and Br analyses.

Reaction of silver <u>trans</u>-1,2-cyclobutanedicarboxylate (46% racemic) with Br_2 in refluxing CCl_4 yielded 32% <u>trans</u>-1,2-dibromocyclobutane ($[\alpha]_D^{27}$ -6.0°) [323]. "It may be concluded that the reaction does not proceed entirely, if at all, through symmetrical intermediates such as

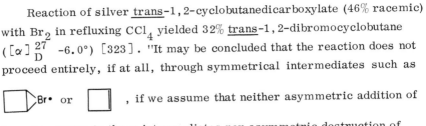 , if we assume that neither asymmetric addition of

bromine atoms to these intermediates nor asymmetric destruction of racemic dibromide in the reaction mixture occurs."

The results in this section are subject to the relevant comments in Sec. 6; IIIC3.

2. Addition of HBr to Cyclobutenes

Photoaddition [276] of HBr at 0-5° to what was reported to be 1-bromocyclobutene* was reported to give an 86-91% yield of "saturated products" (presumed to be 1,2-dibromocyclobutane) whose <u>trans</u>/<u>cis</u> ratio was 3.2 - 4.6.[†] In view of the lack of control experiments which demonstrate that the products are formed in the same ratios as from the direct reaction of cyclobutyl radical with addendum, we defer judgment on the meaning of these results with regard to the question of bridging. If the results are taken at face value, 2-bromocyclobutyl radicals can be slightly (maximum observed stereoselectivity corresponds to 0.8 kcal/mole) bridged [1; III(a)].

---

*Basis of assignment of structure: (1) prepared by reaction of <u>trans</u>-1,2-dibromocyclobutane with KOH; (2) boiling point agreed with that published for material which gave acceptable C and H analyses.

[†]"All of these dibromides were checked for stability under reaction conditions. In no instance was there any evidence for any interconversions between <u>cis</u> and <u>trans</u> isomers."

VI. 3-BROMO-2-NORBORNYL RADICALS

Control of Stereochemistry

1. Reaction of Norbornenes with $BrCCl_2CCl_2Br$

Photoreaction of norbornene with $BrCCl_2CCl_2Br$ in CCl_4 at 70° in the presence of 1 mole % AIBN yielded 94% 2,3-dibromonorbornane, trans/exo-cis = 32* [324]. (The stereochemistry of product formation from either the 3-endo- or the 3-exo-2-norbornyl radical cannot be obtained from the data reported.)

Photoreaction (275-W sunlamp[†]) of benzonorbornadiene with $BrCCl_2CCl_2Br$ in CCl_4 yielded 100% 2,3-dibromobenzonorbornadiene, trans/exo-cis = 8 [325]. (The stereoselectivity of product formation from either the 3-endo- or 3-exo-benzonorbornenyl-2 radical cannot be obtained from the data reported.)

In view of the lack of control experiments, including those relevant to whether the addition step is radical in nature, which demonstrate that the products are formed in the same ratio as from the direct reaction of the free 3-bromo-2-norbornyl radicals and the ambiguities mentioned above, we defer judgment on the meaning of these results with regard to the question of bridging.

2. Addition of HBr to Norbornenes

Results of the addition of HBr (or DBr) to norbornene, 2-bromo-, and 2-chloronorbornene are given in Table 120. Results of the addition of CH_2Br_2, $BrCH_2COOMe$, $BrCCl_3$, CCl_4, $CHCl_3$, $CFBr_3$, and $BrCH(CN)_2$ to norbornenes are in Table 121. In view of the lack of adequate proofs of structure of many of the products and, in all cases, of control experiments which demonstrate that the products are formed in the same ratio as from

---

[*] Basis of identification of products: NMR spectrum of reaction mixture corresponded to that published previously. Omission of the AIBN, the light, or both resulted in yields of 58, 40, and 0%, respectively.

[†] Use of a 22-W bulb or a low pressure uv source resulted in the production of no dibromides.

TABLE 120
Addition of HBr to Norbornenes

| Substrate | Reaction conditions | Reported products (% yield) | |
		2-exo, 3-exo	2-endo, 3-exo
Norbornene	DBr, hν, hexane, 0.4 mole % Bz$_2$O$_2$	D, Br (61)[a]	D, Br (19)[a]
2-Chloro- norbornene	HBr, hν, "cooling bath," pentane		
2-Bromo- norbornene	HBr, hν, pentane, 0°		
2-Bromo- norbornene	HBr, hν, pentane, -25°, <1 hr		
2-Bromo- norbornene	HBr, hν, pentane, -65[c]		

[a] Basis of identification of product: (1) Treatment of reaction mixture with Et$_2$MeCo$^-$K$^+$ yielded norbornene attached to whose 2-position was 39% D and 61% H; (2) qualitative arguments concerning "expected" preferred stereochemistry and isotope effects in the above reaction; (3) KMnO$_4$ oxidation of the product in (1) yielded 1,3-cyclopentanedicarboxylic acid containing no D; (4) the 2-exo-D, 3-endo-Br norbornene[b] was "estimated to be ca. 20% by v.p.c. and infrared analysis . . ." [We conclude from our examination of the details of the procedure that the method used was highly subjective.]

[b] Basis of assignment of structure: "Any possibility for formation of an endo-endo product . . . is apparently in opposition to the manifest preference of both the bromine radical attack step and the product-forming step to occur from the exo-direction."

[c] Identification of products OK.

[d] Irradiation (-25°, 2 hr, pentane) of 2-exo, 3-endo-dibromonor-bornane in the presence of HBr yielded 78% material which contained

2-exo, 3-endo	2-endo, 3-endo	Stereochemistry of product formation (trans/cis)[h] from		Controls	Ref.
		β-Br-endo	β-Br-exo		
D, Br (20)[b]	< 4%[k]	> 5	0.31	None	326
Br, Cl (?)[c]	Br, Cl (?)[c]	> 6[j]	< 0.07[j]	None	327–329
Br, Br (58)[c,e,i]	Br, Br (25)[c,e,i]	> 5[i] or > 0.4[i,l]	< 0.07[i] or ?[l]	None	329, 330
Br, Br (55)[c,f,i]	Br, Br (27)[c,f,i]	> 10[i] or > 0.5[i,l]	< 0.05[i] or ?[l]	d	329, 330
Br, Br (54)[c,g,i]	Br, Br (31)[c,g,i]	> 12[i] or > 0.5[i,l]	< 0.05[i] or ?[l]	None	329, 330

75% reactant and < 1% of the endo, endo-dibromide. Similarly, the endo, endo dibromide yielded 75% material which contained 73% reactant and 6% endo, exo-dibromide.

[e] 4-10% unidentified dibromides reported.

[f] 3-7% unidentified dibromides reported.

[g] 3-18% unidentified dibromides reported.

[h] Not structure of product.

[i] At the earliest stage of reaction for which data were reported.

[j] Based on our arbitrarily chosen limit of detectability of 5% for the unreported "other isomer."

[k] Authors' guess.

[l] Corresponds to formation of the 2-exo, 3-endo product via endo-attack of bromine.

the direct reaction of the free 3-bromo-2-norbornyl radicals, we defer
judgment on the meaning of these results with regard to the question of
bridging. If the results are naively taken at face value, endo- and exo-β-
bromonorbornyl radicals (at least some — see footnote 1 to Table 120 regarding this qualification) and a wide variety of exo-β-X norbornyl radicals
are moderately bridged [1; III(a)]; the exo-β-X radicals would show stereoselectivity (trans > cis addition) in the opposite sense from that of the exo-
β-Br radicals (cis > trans addition). The latter is in the opposite sense
from what is usually associated with bridged [1; I(a)-(c)] radicals.

VII.   3-BROMOALKYL RADICALS

A.   Energy

1.   Reaction of Alkyl Bromides with Br_2

See Sec. 6;IC3.

2.   Reaction of Alkyl Bromides with N-Bromosuccinimide

See Sec. 6;IC4.

3.   Reaction of Alkyl Bromides with Cl_2

See Sec. 6;IC5.

4.   Reaction of Alkyl Bromides with SO_2Cl_2

See Sec. 6;IC6.

5.   Reaction of Haloalkyl Bromides with Chromous Salts

See Sec. 5;IC9 for general comments. As shown in Table 83, a 3-Br
substituent exhibits an accelerating effect corresponding to $\Delta\Delta G^{\ddagger} = 2.1$
kcal/mole in the reaction of 3-substituted propyl halides with Cr^{II}en.
This result is subject to the discussion in Sec. 5; VIB8 and 6;IC7.

TABLE 121

Addition of Radicals to Norbornenes

Substrate	Addendum	Reaction conditions	Reported Products (% Yield)		Stereochem. of addition[a] (trans/cis) β-X-exo	Controls	Ref.
			2-exo,3-exo	2-endo,3-exo			
Norbornene	CH_2Br_2	Reflux, 0.8 mole % Bz_2O_2	Br, CH_2Br(23)[d]	Br, CH_2Br(18)[c]	0.8	None	331
Norbornene	$BrCH_2COOMe$	80°, 0.8 mole % Bz_2O_2	Br, CH_2COOMe(31)[f]	Br, CH_2COOMe(23)[e]	0.8	None	331
Norbornene	$BrCCl_3$	hν		Br, CCl_3(95)[g]	>20	None	312
Norbornene	$BrCCl_3$	Reflux, 0.4 mole % Bz_2O_2		Br, CCl_3(77)[ai]	>3	None	339
Norbornene	$CFBr_3$	Room temp., 2 mole % AIBN	Br, $CFBr_2$(?)[af]	Br, $CFBr_2$(?)[ag]	>30	None	337
Norbornene	CCl_4	Reflux, 0.4 mole % Bz_2O_2		Cl, CCl_3(91)[g]	>10	None	339
Norbornene	CCl_4	80°, Bz_2O_2	Cl, CCl_3(4)[an]	Cl, CCl_3(73)[ao]	18	ar	340

| Substrate | Addendum | Reaction conditions | Reported Products (% Yield) | | Stereochem. of addition[a] (trans/cis) β-X-exo | Controls | Ref. |
			2-exo, 3-exo	2-endo, 3-exo			
Norbornene	BrCH(CN)$_2$	hν, CH$_2$Cl$_2$, <37°	Br, CH(CN)$_2$(75)[ah]	Br, CH(CN)$_2$(21)[ah]	0.22	None	316
Benzonorbornadiene	BrCCl$_3$	75–80°, 1.6 mole % Bz$_2$O$_2$		Br, CCl$_3$(92)[g]	>12	None	332
Benzonorbornadiene	BrCCl$_3$	Bz$_2$O$_2$		Br, CCl$_3$(95)[ah]	>20	None	338
Benzonorbornadiene	CCl$_4$	Reflux, 0.4 mole % Bz$_2$O$_2$		Cl, CCl$_3$(100)[g]	>20[h]	None	332
Benzonorbornadiene	CCl$_4$	Bz$_2$O$_2$		Cl, CCl$_3$(45)[ah]	>0.8	None	338
Benzonorbornadiene	BrCCl$_3$	hν, reflux		Br, CCl$_3$(37)[k]	>0.6	None	333
(structure)	CFBr$_3$	80°, AIBN, 35 min	(structure) (?)[w]	(structure) (?)[g]	0.68	u	337

| | CFBr$_3$ | 80°, 2 mole % AIBN | (?)^x | (?)^y | 1.6 | None | 337 |
| (X=<u>endo</u>-OAc) | | | | | | | |

| | CFBr$_3$ | 80°, 3 mole % AIBN | — | (?)^{ad} | >40^h | None | 337 |
| (X=<u>exo</u>-OAc) | | | (?)^{ad} | | | | |

| | CFBr$_3$ | 80°, 3 mole % AIBN, 5 min | (?)^z | aa | 1.3 | v | 337 |
| (X=<u>endo</u>-OCOCF$_3$) | | | | aa | | | |

271

Substrate	Addendum	Reaction conditions	Reported Products (% Yield)		Stereochem. of addition[a] (trans/cis) β-X-exo	Controls	Ref.
			2-exo,3-exo	2-endo,3-exo			
(X = t-BuO – OCOCF$_3$)	CFBr$_3$	80°, 3 mole % AIBN, 5 min	—	(CFBr$_2$, H, Br, X) + (H, Br, CFBr$_2$, H, X) ae >40[h]	ae	v	337
(Me, OCOPh)	CFBr$_3$	80°, 4 mole % AIBN, 20 min	(CBr$_2$F, Br, Me, OCOPh) (?)[ab]	(CFBr$_2$, H, Br, Me, OCOPh) (?)[ac] + (H, Br, CFBr$_2$, Me, OCOPh, H) (?)[ac] ae 0.86	ae	u	337
(Me, Me)	CCl$_4$	Reflux, 5 mole % Bz$_2$O$_2$	(CCl$_3$, Cl, Me) (31)[ap]	(H, Cl, CCl$_3$, CH$_3$, CH$_3$) (26)[aq] 0.84		ar	340

272

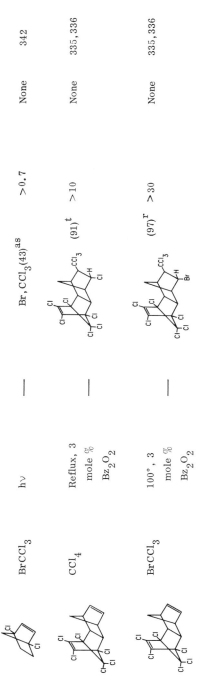

Substrate	Reagent	Conditions		Product		Additive	Ref.
[structure]	$BrCCl_3$	$h\nu$	—	$Br, CCl_3 (43)^{as}$	>0.7	None	342
[structure]	CCl_4	Reflux, 3 mole % Bz_2O_2	—	$(91)^t$	>10	None	335, 336
[structure]	$BrCCl_3$	$100°$, 3 mole % Bz_2O_2	—	$(97)^r$	>30	None	335, 336
Norbornadiene	$BrCCl_3$	$80–140°$, 0.6 mole % AIBN	—	$Br, CCl_3 (6.1)^q$	>0.1	None	335
(j) [structure]	$BrCCl_3$	$h\nu$, reflux	—	$(21)^n$	>0.3	None	333
[structure]	$BrCCl_3$	$h\nu$, reflux	—	$(25)^o$	>0.3	None	333
[structure]	$BrCCl_3$	$80°$, 2.5 mole % Bz_2O_2	—	$(44)^p$	$>0.8^{au}$	None	333, 334

273

Substrate	Addendum	Reaction conditions	Reported Products (% Yield) 2-exo,3-exo	Reported Products (% Yield) 2-endo,3-exo	Stereochem. of addition[a] (trans/cis) β-X-exo	Controls	Ref.
(i) [hexachloronorbornene structure, H, Cl, CCl_3]	$BrCCl_3$	$h\nu$, reflux	—	(40)[m] [structure with H, Cl, CCl_3, Cl, Cl, Cl, H, Br]	> 0.7	None	333
[hexachloronorbornene structure]	CCl_4	Reflux, 8 mole % Bz_2O_2	—	(70)[g] [structure with Cl, CCl_3, H, Cl, Cl, Cl]	> 15[h]	None	341
[bicyclic chlorinated structure]	$BrCCl_3$	80°, 3 mole % Bz_2O_2	—	(40)[g] [structure with Cl, CCl_3, H, Br, Cl]	> 0.7	None	341
[polycyclic hexachloro structure, Cl, Cl, Cl, Cl, Cl, Cl]	$CHCl_3$	Reflux, 3 mole % Bz_2O_2	—	(31)[t] [structure with CCl_3, H, Cl, Cl, Cl, Cl, Cl]	< 2	None	335, 336
[polycyclic structure with Cl, Cl, Cl, Cl, Cl, Br]	$CHCl_3$	Reflux, 3 mole % Bz_2O_2	—	(73)[r] [structure with CCl_3, H, Br, Cl, Cl, Cl, Cl]	< 0.4	None	335, 336

a Not structure of product.

b Crude reaction mixture gave acceptable C and H analyses.

c Basis of identification of product: (1) acceptable C and H analyses; (2) NMR spectrum — poorly-resolved triplet (J ~ 4.5 cps on which is superimposed a 1.5-cps coupling) at τ = 6.28 which is reduced to a doublet on irradiation at the τ = 7.56 absorption; (3) see b.

d Basis of identification of product: (1) NMR spectrum — absorption at τ = 5.77 (q, J = 7.5 and 2.5 cps) unaffected by irradiation at τ = 7.3–7.5 absorption; (2) see b.

e Basis of identification of product: (1) see b; (2) NMR spectrum — broadened triplet at τ = 6.20 (J ~ 4.3 cps) which coalesced to a poorly resolved doublet on irradiation at τ = 7.60 absorption; (3) acceptable C and H analyses.

f Basis of identification of product: (1) see b; (2) NMR spectrum — τ = 5.74 (q, J = 6.2 and 2.0 cps).

g Identification of product OK.

h Based on our arbitrarily chosen limit of detectability of 5% for the unreported "other isomer."

i Basis of assignment of structure: (1) acceptable C and H analyses; (2) prepared as one of two products of the Zn/HOAc reduction of the product of the reaction of 1,2,3,4,5-pentachlorocyclopentadiene with cis-1,2-dibromoethylene.

j Basis of assignment of structure: (1) acceptable C and H analyses; (2) prepared by Zn/HOAc reduction of the product of reaction of 1,2,3,4-tetrachlorocyclopentadiene with cis-1,2-dibromoethylene; (3) NMR spectrum — τ = 3.17 (d), 6.33 (d), 7.11 (sextet, J = 0.7, 5.2 cps).

[k] Basis of identification of product: (1) NMR spectrum — τ = 5.13 (q, J = 5.0 and 2.3 cps), 5.72 (d, J = 5.0 cps); (2) acceptable C and H analyses; (3) reaction with Bu^tOK/Bu^tOH yielded 55% of what was reported to be (I).

[l] Basis of assignment of structure: (1) acceptable C and H analyses; (2) ν = 1625 cm^{-1}; (3) NMR spectrum — τ = 5.15 (d, J = 2.0 cps).

[m] Basis of identification of product: (1) acceptable C and H analyses; (2) NMR spectrum — τ = 5.10 (d, J = 4.1 cps), 5.67 (d, J = 2.0 cps), 5.88 (q, J = 4.1 and 2.0 cps).

[n] Basis of identification of product: (1) acceptable C and H analyses; (2) NMR spectrum — τ = 5.31 (d, J = 2.8 cps), 5.96 (t, J = 2.8 cps), 6.95 (d, J = 9.3 cps), 7.35 (q, J = 2.8 and 9.3 cps).

[o] Basis of identification of product: (1) acceptable C and H analyses, (2) NMR spectrum — τ = 5.14 (d, J = 4 cps), 5.80 (d, J = 4 cps), 7.57 (s), 7.64 (s).

[p] Basis of identification of product: (1) acceptable C analysis; (2) NMR spectrum — τ = 5.44 (d, J = 5.0 cps), 5.68 (d, J = 5.0 cps); (3) 6.27 μ(s).

[q] Basis of identification of product: (1) NMR spectrum — τ = 3.71, 3.41; (2) reaction with hexachloro-

cyclopentadiene yielded 85% of what was reported to be [r]

r Basis of identification of product: (1) acceptable C and H analyses; (2) NMR spectrum — τ = 5.76 (J = 6 and 3.5 cps); (3) reaction with KOH/EtOH yielded 86% of what was reported to be[s]

s Basis of assignment of structure: acceptable C and H analyses.

t Basis of identification of product: (1) acceptable C and H analyses; (2) reaction with KOH/EtOH gave a 75% yield of what was reported to be

u "Negligible reaction" occurred in 1 hr in the absence of AIBN.

v "Negligible reaction" occurred in the absence of AIBN.

w Basis of identification of product: ^{19}F NMR spectrum of the crude reaction mixture contained a doublet assigned to FBr_2CCH.

x Basis of identification of product: (1) the broad absorption observed in the room temperature ^{19}F NMR spectrum of the crude reaction mixture was shown to be temperature-dependent in the manner shown for what was reported to be the exo, cis-adduct[w] in the anhydride system; (2) proton NMR spectrum of crude reaction mixture — τ = 5.45 (J = 7.5 and 2 cps).

y Basis of identification of product: ^{19}F NMR spectrum of crude reaction mixture contained doublets assigned to FBr_2CCH; (2) proton NMR spectrum of crude reaction mixture — τ = 6.44 (J = 23.5, 6.0, and 2 cps).

z Basis of identification of product: (1) see footnote x (1); (2) proton NMR spectrum of crude reaction mixture — τ = 5.47 (J = 7 and 2 cps).

aa Basis of identification of product: (1) see footnote y (1); (2) proton NMR spectrum of crude reaction mixture — τ = 6.73 (J = 6 and 2 cps).

ab Basis of identification of product: (1) see footnote x (1); (2) proton NMR spectrum of crude reaction mixture — τ = 5.5 (J = 7 cps). Note that there is no report of a 2-cps coupling only for that compound in which an exo–H has been replaced by a CH_3 group. Might the absorption cited here and in footnotes x, y, z, and aa correspond to an exo, rather than to an endo hydrogen?

ac Basis of identification of product: (1) see footnote y (1); (2) proton NMR spectrum of crude reaction mixture — τ = 6.5 (J = 6), 5.8 (m).

ad Basis of identification of product: (1) ^{19}F NMR absorptions of crude reaction mixture are sharp doublets; (2) proton NMR spectrum of crude reaction mixture — τ = 4.80 (m), 5.35 (m), 5.89 (m).

ae Basis of identification of product: (1) see footnote ad(1); (2) proton NMR spectrum of crude reaction mixture — τ = 4.46 (m), 5.03 (m), 5.83 (m).

af Basis of identification of product: see footnote y(1).

ag Basis of identification of product: (1) see footnote ad(1); (2) proton NMR spectrum of crude reaction mixture — τ = 5.87 (m), 7.37 (J = 24, 6, and 2 cps).

ah Basis of identification of product not mentioned.

ai Basis of identification of product: (1) acceptable C, Br, and Cl analyses; (2) reaction with KOH/MeOH yielded 54% of what was reported to be [aj]

aj Basis of assignment of structure: (1) acceptable C and Cl analyses; (2) 6.10 (s), 10.99 (s), 11.17 (s) μ;

(3) reaction with aq. AgNO₃ yielded 78% of what was reported to be [ak]

ak Basis of assignment of structure: (1) acceptable C and Cl analyses; (2) obtained from endo-2-Cl, exo-3-trichloromethylnorbornane as per sequence ai(2)-aj(3); reaction with HNO₃ yielded 77% of what was reported to be [al]

al Basis of assignment of structure: (1) acceptable C, H, Cl, and N analyses of semicarbazone; (2) 252, 345 mμ; (3) 5.73, 6.18, 8.31, 9.22, 10.98, 13.45 μ; (4) reaction with NaBH₄ yielded what was reported to be [am]

am Basis of assignment of structure: (1) acceptable C, H, and Cl analyses; (2) different from what was reported to be the exo-isomer. [ak]

an Basis of identification of product: (1) acceptable C, H, and Cl analyses; (2) consistency of NMR spectrum with structure.

ao Basis of identification of product: consistency of NMR spectrum with structure.

ap Basis of identification of product: (1) acceptable C, H, Cl, and molecular weight analyses; (2) consistency of NMR spectrum with structure; however, 5,5- vs. 6,6-diMe uncertain.

aq Basis of identification of product: (1) acceptable C, H, and Cl analyses; (2) see footnote ap(2).

ar "... there was no interconversion between products under either reaction or analytical (vpc) conditions."

as Basis of identification of product: (1) acceptable C and H analyses; (2) NMR spectrum — $\tau = 5.71$ (t, J = 2.5, at 2.5 cps, 1 H), 5.20 (q, J = 2.5 and 3.5 cps, 1 H).

at Note that the corresponding coupling constant in the 1,4-dechloro compound was reported [339] to be 6.1 cps.

au Could be β-X-endo.

6. Reaction of Haloalkyl Bromides with $Co(CN)_5^{3-}$

See Sec. 6;IC8 for general comments. The rate constant $[25.0°,$ $H_2O/MeOH = 20/80$ (v/v)] for reaction of $ICH_2CH_2CH_2Br^*$ is 16 times that for reaction of $CH_3CH_2CH_2I^{**}$ [343]. This small ($\Delta\Delta G^{\ddagger} = 1.6$ kcal/mole) accelerating effect is subject to the discussion in Sec. 6; IC8.

B. Control of Stereochemistry

1. Reaction of Haloalkyl Bromides with Chromous Salts

General comments are given in Sec. 5;IC9. The effects of solvent and of ligand concentration on the trans/cis ratios of 1, 2-dimethylcyclopropanes resulting from eliminative cyclization of dl- and meso-2, 4-dibromopentane are indicated by the data in Table 122. The lack of any stereospecificity provides no cause to consider $CH_3\dot{C}HCH_2CHBrCH_3$ to be bridged.

2. Reaction of Norbornenes with CH_2Br_2 and $CFBr_3$

Upon the future availability of data for additional systems the data in Table 121 may become useful in the present context.

VIII. 3-BROMOCYCLOHEXYL RADICALS

A. Energy

1. Reaction of Cyclohexyl Bromide with Cl_2

See Sec. 5; IIA1.

2. Reaction of Cyclohexyl Bromide with Br_2

See Sec. 6; IIIB1.

---

*Products: 92% cyclopropane.[†]

**Products: 37% propylene, 39% propane.[†]

[†]"The volatile organic products were identified, and their yields determined by v.p.c. or, in some cases, by n.m.r."

B. Control of Stereochemistry

1. Reaction of Cyclohexyl Bromide with Cl_2

See Table 71. The reported stereoselectivity of formation of 1,3-disubstituted product is small, corresponding to $\Delta\Delta G^{\ddagger} = 0.1$ kcal/mole, and provides no cause to consider the 3-bromocyclohexyl radical to be bridged.

2. Reaction of Cyclohexyl Bromide with Br_2

See Sec. 6;IIIB1. In view of the lack of knowledge of details of the mechanism (Sec. 5;ID1) and of adequate control experiments (Sec. 5;ID1), we must withhold judgment on whether these results are relevant to the question of bridging. If they are assumed to be so and are taken at face value, they indicate that the 3-bromocyclohexyl radical is at most weakly bridged [1;III(a)], the largest reported trans/cis ratio corresponding to $\Delta\Delta G^{\ddagger} = 0.4$ kcal/mole.

3. Reaction of Silver 1,3-Cyclohexanedicarboxylates with Br_2

See Sec. 6;IIIC3 for general comments. Riemann [289] has reported that silver trans-1,3-cyclohexanedicarboxylate reacted with Br_2/CCl_4 at 25° to give 13.8% trans* - and 13.6% cis* - 1,3-dibromocyclohexane. Similarly, reaction of the cis-dicarboxylate was reported to give 8.2% trans* - and 7.7% cis-dibromide.* Reaction of silver trans-1,3-cyclo-hexanedicarboxylate (acid $[\alpha]_D^{25} = -24°$) was reported to yield 16% inactive 1,3-dibromocyclohexane (trans/cis* = 1.06). In view of the comments made in Sec. 6;IIIC3, the lack of adequate control experiments, and the poor yields, it is not clear whether these results are relevant to the question of bridging (note that the bridged [1;I(a)] 3-bromocyclohexyl radical is achiral, while the unbridged [1;I(a)] one is chiral).

IX. 3-BROMOCYCLOPENTYL RADICALS

A. Energy

1. Reaction of Cyclopentyl Bromide with Br_2

See Sec. 6;IVB1.

---

*Trans vs. cis assigned based on comparison of order of elution from a silicone rubber VPC column with that previously reported for an ethylene glycol succinate column.

2. Reaction of Cyclopentyl Bromide with Cl_2 and $(CH_3)_3COCl$

See Sec. 6; IVB2.

B. Control of Stereochemistry

1. Reaction of Cyclopentyl Bromide with Cl_2

See Table 75. The observed stereoselectivity is very small and provides no cause to consider the 3-bromocyclopentyl radical to be bridged [1; III(a)].

2. Reaction of Cyclopentyl Bromide with $(CH_3)_3COCl$

See Table 76. The reported trans/cis ratio of 1-bromo-3-chlorocyclopentanes corresponds to $\Delta\Delta G^{\ddagger} = 0.5$ kcal/mole. In view of the lack of knowledge of details of the mechanism (Sec. 5; ID1) and the lack of adequate control experiments (Sec. 5; ID1), it should not be concluded from this small stereoselectivity that the 3-bromocyclopentyl radical is slightly bridged [1; III(a)].

X. 4-BROMOALKYL RADICALS

Energy

1. Reaction of Alkyl Bromides with Br_2

See Sec. 6; IC3.

2. Reaction of Alkyl Bromides with N-Bromosuccinimide

See Sec. 6; IC4.

3. Reaction of Alkyl Bromides with Cl_2

See Sec. 6; IC5.

XI. 4-BROMOCYCLOHEXYL RADICALS

A. Energy

Reaction of Cyclohexyl Bromide with Cl_2

See Sec. 5; IIA1.

TABLE 122

Reaction of 2,4-Dibromopentane with Chromous Salts [169]

Substrate	Solvent	en/Cr^{II}	$\bigtriangledown$ CH$_3$ CH$_3$ (trans/cis) [a]
meso	87/13(v/v)DMSO/H$_2$O	0	1.5
dl	87/13(v/v)DMSO/H$_2$O	0	1.5
meso	86/14(v/v)DMF/H$_2$O	2.2	1.0
dl	86/14(v/v)DMF/H$_2$O	2.2	1.0
meso	22/78(v/v)DMF/en	130	0.79
dl	22/78(v/v)DMF/en	130	0.79

[a] Room temperature.

B. Control of Stereochemistry

1. Reaction of Cyclohexyl Bromide with Cl$_2$

See Table 71. In view of the lack of adequate proofs of structure of the products, of knowledge of details of the mechanism (Sec. 5; ID1) and of adequate control experiments (Sec. 5; ID1), we must withhold judgment on whether these results are relevant to the question of bridging. If they are assumed to be so and are taken at face value, they indicate that the 4-bromocyclohexyl radical is at most weakly bridged [1; III(a)], the largest reported trans/cis ratio corresponding to $\Delta\Delta G^{\ddagger} = 0.3$ kcal/mole.

2. Reaction of Cyclohexyl Bromide with Br$_2$

See Sec. 6; IIIB1. The reported stereoselectivity of formation of 1,4-dibromocyclohexanes is nil and provides no cause to consider the 4-bromocyclohexyl radical to be bridged [1; III(a)].

XII. MISCELLANEOUS δ-BROMO RADICALS

Structure

The ESR spectra attributed to the o-bromophenoxymethyl and 2,4-dibromo-5-methoxyphenoxymethyl radicals on the basis of qualitative

consistency with the structures are described in Table 87. The "normal"
coupling constants provide no cause to consider the radicals to be bridged
[1; I(a)-(c)].

XIII. ε-BROMOALKYL RADICALS

Energy

Trahanovsky and Doyle [344] have made the possibly relevant observation that $Br(CH_2)_5CO_3Bu^t$ and $CH_3(CH_2)_4CO_3Bu^t$ decompose at 106° with
rate constants of 1.37×10^{-5} and 1.44×10^{-5} sec^{-1}, respectively. These
results would provide no cause to consider the 5-bromopentyl radical to be
bridged [1; II(a)].

XIV. 2-BROMOVINYL RADICALS

A. Structure

An ESR spectrum, attributed to the 3-bromo-3-hexenyl radical, was
obtained by observation of the reaction mixture resulting from the photolysis
at 77° K of 3-hexyne and HBr [175]; "the spectrum . . . was discernible as
a major five line spectrum with additional splittings that were not resolved.
The five line splitting was measured to be about 19 gauss, and the other
splittings appear to be separated by 4 gauss." "It would . . . be expected
that [an unbridged] . . . structure . . . would . . . result in a non-
equivalence of the two methylenes because of the proximity of the bromine
and therefore not give a major five line pattern." ESR spectra, attributed
to the 2-bromo-2-butenyl radical, were similarly observed starting with
2-butyne and HBr (or DBr); "the spectrum consists of seven equally spaced
lines with a hyperfine constant of 8.3 gauss." "In each case an analysis of
the hyperfine pattern strongly suggested a bridged type intermediate as the
initiating free radical." Although the spectrum attributed to $CH_3C \equiv CCH_2 \cdot$
(from $CH_3 \equiv CCH_3$, γ-rays or hν, 77° K) [345] is not very similar to that
reported above, we believe that the credibility of the assignments described
above is weakened by association with the contents of Secs. 6; IA, IIIA,
IVA.

B. Control of Stereochemistry

Addition of Br_2 and HBr to Alkynes

Results of the addition of Br_2 and of HBr to several acetylenes are
given in Tables 123 and 124, respectively. In view of the uncertain structures of some of the products, the lack of adequate control experiments in
all cases, the general confusion regarding the processes which are
occurring, and indications that some of the products may be unstable under
the reaction conditions, we must withhold judgment on whether these
results are relevant to the question of bridging.

$$\text{XV.} \quad -C\equiv C-\overset{\displaystyle .}{C}=C\overset{\displaystyle \nearrow Br}{\underset{\displaystyle \searrow}{}} \quad \text{RADICALS}$$

Control of Stereochemistry

Addition of HBr to Diacetylenes

Photoreaction[*] of $HC\equiv C-C\equiv CH$ with HBr in ether at room temperature
in the presence of hydroquinone was reported [534] to yield 19%
$HC\equiv C-CH=CHBr$ [†] (cis/trans) > 5).[‡] In view of the lack of control experiments and the uncertainty regarding the nature of the processes which are
occurring, we must withhold judgment on whether this report is relevant to
the question of bridging.

---

[*] ". . . the rate of . . . reaction increases noticeably when the
reaction mixture is irradiated with UV light."

[†] Basis of identification of product: (1) absorption at 3280, 3084,
3026, 3004, 2105, 1621, 1581, 980, 942, 717, and 708 cm^{-1}; (2) NMR
spectrum — τ = 6.4-6.5 (m), 3.16-3.82 (m, J = 8.44 cps).

[‡] Based on our visual inspection of the NMR spectrum.

TABLE 123

Addition of Br_2 to Acetylenes

Substrate	Reaction conditions	Reported product			Ref.
		Structure (% yield)	trans/cis	Controls	
HC≡CCH₂OH	hν, 25°, CCl₄	BrCH=CBrCH₂OH (74)	>2.8[c]	None	346
HC≡CCMe₂OH	hν, 25°, hexane	BrCH=CBrCMe₂OH (86)	0.10[d]	a,f,g	346
HC≡CCMe₂OH	hν, 25°, CCl₄	BrCH=CBrCMe₂OH (?)	0.14[d]	a,g	346
HC≡CCMe₂OH	hν, <35°, petroleum ether, 0.2 mole % Bz₂O₂	BrCH=CBrCMe₂OH (54)	0.084[d]	ai	349
HC≡CCMe₂OAc	hν, 25°, hexane	BrCH=CBrCMe₂OAc (77)	0.087[e]	h	346
(cyclohexyl) OH / C≡CH	hν, 25°, hexane	(cyclohexyl) OH / CBr=CHBr (82)	0/100[d]	b,i	346
(cyclohexyl) OH / C≡CH	hν, <35°, petroleum ether, 2.5 mole % Bz₂O₂	(cyclohexyl) OH / CBr=CHBr (85)	<0.2[d]	a,i	349
HC≡CH	-5 to -3°, dark, Bz₂O₂	BrCH=CHBr (?)	0.82°	j,u	347

Substrate	Reaction conditions	Reported product		trans/cis	Controls	Ref.
		Structure (% yield)				
$CH_3C{\equiv}CH$	-30 or $-60°$,[k] pentane or hexane or heptane,[l] $h\nu$	$CH_3CHBr{=}CHBr$ (50)		3.7[q]	m,r	347
$HC{\equiv}CC(CH_3)_3$	$h\nu$, hexane	$(CH_3)_3CCHBr{=}CHBr$ (91)		<0.1[p]	n,s	347
$HOCH_2C{\equiv}CCH_2OH$	$h\nu$, $55°$, CCl_4	$HOCH_2CBr{=}CBrCH_2OH$ (67 or 84)[v]		<0.5[aa]	w	348
$AcOCH_2C{\equiv}CCH_2OAc$	$h\nu$, hexane	$AcOCH_2CBr{=}CBrCH_2OAc$ (78)		<0.3[ab]	ah	348
$HOCMe_2C{\equiv}CCMe_2OH$	$h\nu$, CCl_4	$HOCMe_2CBr{=}CBrCMe_2OH$ (94 or 97 or 100)[v]		<0.3[ac]	x,af	348
$PhC{\equiv}CPh$	$h\nu$, hexane	$PhCBr{=}CBrPh$ (69)		>2[ad]	y	348
$HOOCC{\equiv}CCOOH$	$h\nu$, room temp., ether	$HOOCCBr{=}CBrCOOH$ (90)		0.10[ae]	z,ag	348

[a] For HOAc in daylight, HOAc in the dark, HOAc in the dark in the presence of LiBr, $CHCl_3$ in the dark, and MeOH in the dark, trans/cis ratios of 3.8, 4.9, 100/0, 0.70, and 0.19, respectively, were reported.

[b] For HOAc in the dark, HOAc in the dark in the presence of LiBr, $CHCl_3$ in the dark, and MeOH in the dark, trans/cis ratios of 19, 100/0, 1.0, and 0.56, respectively, were reported.

[c] Basis of identification of product: (1) acceptable Br analysis; (2) ir spectrum consistent with structure.

[d] Identification of product OK.

[e] Basis of identification of products: ir spectra consistent with assigned structures.

[f] Irradiation of what was reported to be the trans–isomer[e] in the presence of 1 drop of Br_2 in hexane was reported to give trans/cis = 0.23.[e]

[g] On standing for 2 months in the dark, what was reported to be the trans–isomer[e] gave a 69% yield of what was reported to consist of 69% cis–isomer. Under the same conditions the latter was "scarcely affected."

[h] Irradiation of what was reported to be the trans–isomer[e] was reported to give trans/cis = 0.28 (also Ref. 347).

[i] Irradiation of what was reported to be the trans–isomer[d] in hexane was reported to give trans/cis = 0.27, cis mp 68–70° (also Ref. 347). Storage in the refrigerator for 6 months gave a 62% yield of material, mp 66–70°.

[j] Reaction in HOAc in the dark was reported to yield trans/cis = 100/0.

[k] Different temperatures given in different parts of translated paper.

[l] Different solvents given in different parts of translated paper.

[m] Reaction in propionic acid in the dark in the presence of LiBr was reported to give trans/cis = 100/0.

[n] Reaction in HOAc in the dark in the presence of LiBr was reported to give trans/cis > 3.3.

[o] "The contents of isomeric dibromoethylenes in mixtures were determined from the intensities of the 579 cm^{-1} line, which is characteristic for the cis–isomer, and the 744 and 1241 cm^{-1} lines, which are characteristic for the trans–form."

[p] Basis of identification of product: (1) acceptable C, H, and Br analyses; (2) "the i.r. spectrum showed that there was essentially no trans–isomer in the product (complete absence of absorption at 900 cm^{-1})."

q Basis of identification of trans product: ir spectrum consistent with structure. Basis of identification of cis product: not mentioned.

r Irradiation of what was reported to be the trans-isomer[q] in the presence of 1 drop of Br_2 in hexane led to 32% of what was reported to be the cis-isomer.[q]

s Irradiation of what was reported to be the trans-isomer[p] in the presence of 1 drop of Br_2 in hexane gave a 98% yield of material which was reported to consist of 66% cis-isomer.[t]

t Basis of assignment of structure: ir spectrum consistent with structure.

u Irradiation of either what was reported to be the trans-isomer[o] or of what was reported to be the cis-isomer[o] in the presence of 1 drop of Br_2 in hexane was reported to give trans/cis = 0.82.

v Different yields given in different parts of translated paper.

w Reaction in MeOH in the dark, DMF in the dark, and DMF in the dark in the presence of LiBr was reported to give trans/cis < 0.5.

x Reaction in DMF in the dark was reported to give trans/cis > 3.

y Reaction in ether in the dark was reported to give trans/cis > 4.

z Reaction in MeOH in the dark and DMF in the dark was reported to give trans/cis = 3.7 and > 2, respectively.

aa Basis of identification of product: ir bands "in the region of 1600 cm^{-1}" and at about 800 cm^{-1}.

ab Basis of identification of product: (1) "intense absorption band in the region of 1600 cm^{-1} and bands of various intensities at 700 cm^{-1}"; (2) acceptable C, H, and Br analyses; (3) produced by reaction of what was reported to be cis-$HOCH_2C{\equiv}CCH_2OH^{aa}$ with Ac_2O.

ac Basis of identification of product: (1) acceptable C, H, and Br analyses; (2) "the spectra of disubstituted trans-dibromides have intense bands at 900 cm^{-1}, which are absent in the spectra of the cis isomers. The spectra of the cis dibromides have an intense absorption band in the region of 1600 cm^{-1} and bands of various intensities at 700 cm^{-1}, which are absent in the spectra of trans-dibromo olefins."

ad Basis of identification of product: mp agreed with that of authentic material.

ae Basis of identification of products: melting points reported.

af Irradiation at 20° of what was reported to be the trans-isomer,^{ac} mp 129–130°, in the presence of 1 drop of Br$_2$ in CCl$_4$ gave an 82% recovery of material, mp 128–129.5°.

ag Irradiation at 20° of what was reported to be the trans-isomer^{ae} in ether resulted in its recovery "in almost quantitative yield."

ah [319] Reaction of toluene and Br$_2$ at 0° in the dark resulted in "feeble evolution of HBr." Addition of 16 mole % AcOCH$_2$C≡CCH$_2$OAc resulted in an exothermic reaction with vigorous HBr evolution. The formation of 82% PhCH$_2$Br (characterized only by boiling point) was reported. ". . . in absence of . . . [the diacetate] a mixture of o- and p-bromotoluenes is the main product."

ai ". . . the bromination . . . is sharply suppressed in the dark."

TABLE 124

Photoaddition of HBr to Acetylenes

Substrate	Reaction conditions	Reported product		Controls	Ref.
		Structure (% yield)	cis/trans		
HC≡CCH₂Br	-78°, liquid HBr	BrCH=CHCH₂Br (89)	82[a,c,m]	b	350
HC≡CCH₃	Room temp., gas phase	BrCH=CHCH₃ (?)	44.0-5.68[d,f]	e	351
BrC≡CCH₃ [g]	-78°, liquid HBr	BrCH=CBrCH₃ (38)	0.18[i]	l	353,354
BrC≡CCH₃ [g]	-78°, pentane	BrCH=CBrCH₃ (90)	0.20[i]	l	353,354
BrC≡CC(CH₃)₃ [h]	-78°, liquid HBr	BrCH=CBrC(CH₃)₃ (75)	0/100[j]	k	353,354
BrC≡CCH₃ [h]	-78°, pentane	BrCH=CBrC(CH₃)₃ (89)	0.05[j]	k	353,354

[a] At 1.8% conversion. At 27.7% conversion, cis/trans = 24.

[b] "... the equilibrium value ... is calculated to be ca. 90% cis at -78°." "No addition or isomerization occurred in the absence of illumination. No reaction products with molecular weight higher than that of dibromopropene were detected."

[c] Basis of distinction of cis and trans: indices of refraction agreed with those reported for material whose stereochemistry was assigned on the basis of ir spectra.

[d] As % conversion varies from 0.25 to 11.0.

[e] In the dark cis/trans ≅ 110 at 0.15% conversion. The equilibrium ratio was stated to be 4.19 (see Ref. 352). "The mixture obeyed Dalton's law, indicating that π - complexing and other condensation reactions do not occur on mixing."

f Basis of identification of products: [352] ir spectra agreed with those of material whose proof of structure rested on ir spectra and method of preparation (from $CH_3CHBrCH_2Br$ and PhO^-).

g Basis of assignment of structure: (1) prepared from KOBr and $CH_3C\equiv CH$; (2) boiling point and density agreed with those reported for material whose Br analysis was acceptable and whose ir spectrum was consistent with the proposed structure.

h Basis of assignment of structure: (1) prepared from KOBr and $(CH_3)_3CC\equiv CH$; (2) acceptable C, H, and Br analyses.

i Assignment of structure of trans isomer based on ir spectrum. That of cis-isomer based on nothing.

j Basis of assignment of structure: cis — acceptable C, H, and Br analyses and ir spectrum; trans — ir spectrum.

k Reaction at -70° in the dark in the presence of Ph_2NH was reported to give cis/trans = 2.3. " . . . trans-1,2-dibromo-3,3-dimethyl1-1-butene passes almost quantitatively into its cis isomer under uv irradiation at room temperature . . . " "cis-1,2-dibromo-3,3-dimethyl-1-butene was irradiated . . . at -70° in the presence of HBr. . . . the cis-dibromide is not isomerized. On the other hand, under the same conditions pure trans-dibromopropene was converted to the extent of about 20% into the cis isomer."

l " . . . under uv radiation at room temperature . . . trans-1,2-dibromopropene gives a mixture of cis and trans forms in the proportion of 2:3."

m A similar stereoselectivity was reported for bromoallene as substrate.

Chapter 7

IODO-SUBSTITUTED RADICALS

I. β-IODOALKYL RADICALS

A. Energy

1. "Direct" Tests of Bridging

As illustrated by statements made concerning the thermochemistry of the following reactions, β-iodoalkyl radicals have frequently been treated, without justification, as if they were not bridged [1; II(a)]: I· + trans-CHI=CHI → $CHI_2\dot{C}HI$ [191, 357]; I· + cis-CHI=CHI → $CHI_2\dot{C}HI$ [191]; I· + cis-CHCl=CHCl → $CHCl\dot{C}HCl$ [191]; I· + cis-CH_3CH=CHCH$_3$ → $CH_3\dot{C}HICHCH_3$ [191]; I· + cis-PhCH=CHPh → $PhCHI\dot{C}HPh$ [191]; ICH_2CH_2· → I· + C_2H_4 [355]; I· + ICH_2CH_2I [355, 356]; CHI_2CH_2I → CHI_2CHI· + H· [357].

In order to support the validity of estimates of heats of formation of β-iodoalkyl radicals which implicitly do not consider them to be bridged [1; II(a)] it has been point out [355] that ". . ˙. the kinetics of the I-atom catalyzed isomerization of cis-trans butenes in the gas phase . . . are in excellent agreement with the thermochemical data [sic] on the iodo-radical . . . " The equilibrium constant of I· + CH_3CH=CHCH$_3$ → $CH_3CHI\dot{C}HCH_3$ was estimated, with the implicit assumption that $CH_3CHI\dot{C}HCH_3$ is not bridged [1; II(a)], followed by an estimate, similarly based, of the rate of internal rotation in $CH_3CHI\dot{C}HCH_3$. These estimates were combined to give an estimate of the observed rate constant which agreed acceptably with that determined experimentally. It should be noted, however, that bridging [1; II(a)] in the radical would "affect" the equilibrium constant and the rate constant for internal rotation in compensatory ways as far as a calculated overall rate constant is concerned. Successful calculation of the observed rate constant cannot be used as evidence in support of the estimated energy of formation of the intermediate radical.

2. Reaction of Alkyl Iodides with $CH_3\cdot$

Benson [356] "attempted to check [the idea that C_2H_5I or n-propyl iodide would be very good hydrogen donors] . . . by adding C_2H_5I or n-propyl iodide to a flask containing $CH_3\cdot$ generated from the pyrolysis of $(Bu^tO)_2$ . . . If these iodides were good donors, one would expect enhanced yields of CH_4 and olefin [In place of CH_4 one might find C_2H_6 (or C_3H_8) and olefin arising from I-atom transfer]. However, this was not observed. Instead, there is a very rapid exchange of I between CH_3 and C_2H_5, but no evidence for enhanced H abstraction by either C_2H_5 or CH_3 or n-propyl radicals."

3. Reaction of Haloalkyl Iodides with $Co(CN)_5^{3-}$

See Sec. 6; IC8 for general comments. Second-order rate constants for the reaction of $Co(CN)_5^{3-}$ with iodoalkyl halides are given in Table 125 and indicate a large accelerating effect of a β-I substituent. Even if the conclusion of Sec. 6; IC8 is ignored, the meaning of this result is still uncertain because of problems of the type enumerated in Sec. 6; IC7 and, in general, the content of Sec. 1; VB. It would be premature to consider this result to be relevant to the question of bridging or to conclude from it that β-iodoalkyl radicals are bridged [1; II(a)].

TABLE 125

Reaction of Haloalkyl Iodides with $Co(CN)_5^{3-a}$ [343]

Halide	k, $M^{-1} sec^{-1}$	Relative[j] $\Delta G^{\ddagger}$, kcal/mole
ICH_2CH_2I[b]	32[i]	3.7
CH_3CH_2I[d]	0.059	0
$ICH_2CH_2CH_2I$[e]	0.50[i]	1.3
$CH_3CH_2CH_2I$[f]	0.043	-0.2
$I(CH_2)_4I$[g]	0.15[i]	0.6
$I(CH_2)_5I$[h]	0.06[i]	0.0

[a] 20/80 (v/v) H_2O/MeOH, 25.0°.

[b] Products: 105% C_2H_4.[c]

[c] "The volatile organic products were identified, and their yields determined by VPC or, in some cases, by NMR."

[d] Products: 12% C_2H_4, 13% C_2H_6.[c]

[e] Products: 100% cyclopropane.[c]

[f] Products: 37% propylene, 39% propane.[c]

[g] Products: 67% 1-butene, 16% trans-2-butene, 5% cis-2-butene.[c]

[h] Products: 76% 1-pentene, 5% cis-2-pentene.[c]

[i] The rate constant has been divided by 2.

[j] Corresponds to "corrected"[i] rate constants where applicable.

B. Control of Stereochemistry

Addition of I_2 to Olefins

Photoaddition* of I_2 to cis- and trans-2-butene (-42°, propane, 30 min) was reported to yield dl[†]- and meso[†]-2,3-diiodobutane, respectively, of

---

[*] "a comparable dark reaction for several hours does not result in extensive [emphasis added] loss of iodine."

[†] Heating of the meso- and dl-diiodides to room temperature regenerated, in unspecified yield, the original olefins with unchanged stereochemical purities; this reaction was "accelerated by illumination."
"Two lines of evidence indicate these stereospecific addition reactions occur with a trans geometric requirement. (a) Dehalogenation of the diiodides with zinc dust in methanol at -78° regenerates the olefin from which it had been prepared. A similar procedure, applied to the known 2,3-dibromobutanes, generates olefins by a stereoselective trans elimination. (b) The dl- and meso-2,3-dichloro- and -dibromobutane isomer pairs show similar characteristic differences in δ for the proton magnetic resonance absorptions of the CH_3 and the CH groups. The same differences occur in the spectra of the diastereomeric diiodides. If the analogy is applicable, the meso- and dl-forms are identified. Methods (a) and (b) are in agreement . . ."

$> 90\%$ stereochemical purity in unspecified yield [358] (see also Ref. 494). In view of the lack of control experiments that demonstrate that the products are formed only via those routes which produce them in the same ratio as does the direct reaction of the free β-iodoalkyl radical with I_2, we defer judgment on the meaning of these results with regard to the question of bridging.

It has been pointed out [355] that "in decomposition of ICH_2CH_2I which has been carefully studied in both solution and gas phase, the rate-determining step has been shown to be the attack of I atom on ICH_2CH_2I with an activation energy of 11 ± 1 kcal. However if the products are C_2H_4I and I· , the endothermicity of the reaction is 18 kcal . . ." It is implied that there is something inherently wrong with such a situation but, actually, there is no general requirement that $\Delta H^{\ddagger} \geq \Delta H$ (the transition state, i.e., the position of maximum free energy along the reaction coordinate, need not correspond to the position of maximum enthalpy along the reaction coordinate) and the implication, which is implicitly based upon an unbridged [1; II(a)] ICH_2CH_2· , turns out to be correct* only because the activation energy of the reverse reaction (ICH_2CH_2· + I_2 → I· + ICH_2CH_2I) is known (most probably extremely close to 0 if the iodoethyl radical is not bridged — see footnote g to Table 62). A concerted reaction, leading to I· + C_2H_4 + I_2, without the intermediacy of ICH_2CH_2·, is favored [355]. It is stated that "consideration of detailed balancing now requires that if the decompositions are concerted processes, then the additions are also concerted processes . . ." Since the mechanism "involving the formation of the ICH_2CH_2· radical (Skell's mechanism)" is of different kinetic order in iodine than "a concerted addition of I_2 catalyzed by I·," detailed balancing by itself does not impose the requirement that the addition reaction, as studied, proceed along the reverse of the path allegedly followed by the separately studied elimination reaction. It is only because of the relative concentrations of iodine involved in the studies cited by Benson et al. and

---

*It is not at all clear, considering the uncertainties that really should be attached to the 11- and 18-kcal figures and whether the reaction mechanism is as well understood as is implied, that the enthalpy of activation and of reaction are significantly different.

that of Skell and Pavlis that the conclusion of Benson et al. is valid within the framework of their assumptions.

The "concerted addition of I_2 catalyzed by I" can be formulated in more detail so as to become operationally equivalent to " . . . [a] modification of the chain process [which] involves complexing of the addendum and olefin prior to addition," a mechanism considered in another context [34]. This model is consistent with the report [358] that "these stereospecific addition reactions occur with a _trans_ geometric requirement"; its credibility is not weakened by reports of complex formation between I_2 and 1-octene at room temperature [359]; 2-methylbutene-1 at 0° [493], room temperature [359]; cyclohexene at 0° [493], room temperature [359, 363, 364], 15° [360], and 25° [360, 365]; cyclopentene at 0° [493] 15° [360], and 25° [360]; cycloheptene at 0° [493], 15° [360], and 25° [360]; _cis_-cyclooctene at 15° and 25° [360]; methylenecyclobutane at 15° and 25° [360]; methylenecyclopentane at 15° and 25° [360]; methylenecyclohexane at 15° and 25° [360]; methylenecycloheptane at 15° and 25° [360]; norbornene at 25° [360]; $(CH_3)_2C=CHC(CH_3)_3$ at 10° [361] and room temperature [363]; propylene at 0° [493], -196° [362], and -43° [362]; _cis_-2-butene at -196° [362], -43° [362], 0° [493], and room temperature [364]; _trans_-2-butene at -196° [362], -43° [362], 0° [493], and room temperature [364]; and 1-butene, 2-butene, $(CH_3)_2CHCH=CH_2$, $(CH_3)_2C=CH_2$, _cis_-2-pentene, _trans_-2-pentene, $CH_3CH=C(CH_3)_2$, and $CH_3CH_2CH=C(CH_3)_2$ at 0° [493].

If $CH_3CH\dot{I}CHCH_3$ were considered to be bridged [1; III(a)], whether it is bridged [1; I(a)-(c)] would still be an open question, the issue being one of competition between rotation about the central bond and reaction with I_2. As in Sec. 6; ID1, estimates of rate constants for internal rotation in the gas phase at -42° of 1×10^7 and $\sim 7 \times 10^9$ sec^{-1} may be obtained. Using the results of a study of the kinetics of the gas phase iodine-catalyzed _cis-trans_ isomerization of 2-butene and implicitly assuming that $CH_3CH\dot{I}C\dot{H}CH_3$ is not bridged [1; II(a)], Benson et al. [191] have estimated the activation parameters for internal rotation in $CH_3CH\dot{I}C\dot{H}CH_3$; from these a rate constant of 4×10^7 sec^{-1} at -42° can be obtained. Based on the data in footnote g to Table 62 we estimate the rate constant for reaction with I_2 to be 2×10^9 M^{-1} sec^{-1} at -42°. In view of the magnitude of $[I_2]$, reaction with I_2 could easily be faster than internal rotation. It is important to keep in mind the inappropriateness of our models, our very uncertain estimates of the rate constants corresponding to them, and our application of estimates derived from data obtained for systems in the gas phase to a reaction in solution.

II. γ-IODOALKYL RADICALS

A. Structure

The products of the decomposition of deuteriated 4-iodobutyryl peroxide are given in Table 126. The 1- and 3-carbon atoms have not become equivalent to a significant extent at any time during the history of any of the products reported. If 1-chloro-3-iodopropane and 1,6-diiodohexane, in particular, are formed via the pathways usually considered for $RCO_2 \rightarrow_2 \xrightarrow{CCl_4} RCl + R\text{-}R$, a 3-iodopropyl radical is not bridged [1; I(a)].

B. Energy

1. Decomposition of 4-Iodobutyryl Peroxide

Drury and Kaplan [366] have made the possibly relevant observation that $ICH_2CH_2CH_2CO_2\rightarrow_2$ and $CH_3CH_2CH_2CO_2\rightarrow_2$ decompose at 85.1° (0.010 M in CCl_4) with rate constants of 6.5×10^{-5} and 15×10^{-5} sec^{-1}, respectively. These results would provide no cause to consider the 3-iodopropyl radical to be bridged [1; II(a)].

2. Reaction of Alkyl Iodides with $CH_3 \cdot$

See Sec. 7; IA2.

3. Reaction of Haloalkyl Iodides with $Co(CN)_5^{3-}$

See Sec. 6; IC8 for general comments. Second-order rate constants for the reaction of $Co(CN)_5^{3-}$ with iodoalkyl halides are given in Table 125 and indicate a moderate accelerating effect of a 3-I substituent. This result is subject to the comments in Sec. 7; IA3.

TABLE 126

Decomposition of 0.010 M $(ICH_2CH_2CD_2CO_2)_2$[a] in CCl_4 at 95.0° [366]

Product isolated from reaction mixture	Distribution of deuterium	Product isolated from Ph₃SnH-treated reaction mixture	Distribution of deuterium[b]
1-Chloro-3-iodopropane	97.7 ± 0.1% α to Cl	1-Chloropropane	97.8 ± 0.1, 97.9 ± 0.1% α to Cl
1,6-Diiodohexane	95.8 ± 2% in methylenes	n-Hexane	97.6 ± 1.3% in methylenes
(3-Iodopropyl)-4-iodobutyrate	?	n-Propyl butyrate	94.5 ± 0.9, 96.1 ± 0.4% α to carbonyl; 96.6 ± 0.7, 96.3 ± 0.7% α to O.

[a] 96.3 ± 1.3% D α to carbonyl; 99.5 weight % pure.

[b] Multiple entries represent the results of independent reactions and analyses.

4. Reaction of Haloalkyl Iodides with Chromous Salts

See Sec. 5; IC9 for general comments. As shown in Table 83, a 3-I substituent exhibits an accelerating effect corresponding to $\Delta\Delta G^{\ddagger} =$ 2.0 kcal/mole in the reaction of 3-substituted propyl halides with Cr^{II}en. This result is subject to the discussion in Sec. 5; VIB8 and 6; IC7.

5. Reaction of Haloalkyl Iodides with $PhN=NCPh_3$

See Sec. 5; IC9 and Table 60. The "acceleration" "caused by" I would correspond to $\Delta\Delta G^{\ddagger} = 0.3$ kcal/mole. This result is subject to the comments in Sec. 6; IC9.

III.   δ-IODOALKYL RADICALS

A.   Energy

Reaction of Haloalkyl Iodides with $Co(CN)_5^{3-}$

See Sec. 6; IC8 for general comments. Second-order rate constants for the reaction of $Co(CN)_5^{3-}$ with iodoalkyl halides are given in Table 125 and indicate a small accelerating effect of a 4-I substituent. This result is subject to the comments in Sec. 7; IA3.

B.   Energy-Structure

Reaction of Haloalkyl Iodides with Sodium Naphthalene

The reaction of RX with sodium naphthalene and the attendant competition

$$R\cdot \ + \ (\text{naphthalene})^{\overline{\cdot}} \xrightarrow[k_2]{\ k_1\ } \begin{array}{l} R\text{-naphthalene}^{-} \\ \\ R^{-} \end{array}$$

has been studied [367]. For primary alkyl halides, $k_1/k_2 = 0.9 \pm 0.2$, and for $ICH_2CH_2CH_2CH_2I$, $k_1/k_2 = 0.96 \pm 0.08$. The conclusion drawn was that "not only are unusual [bridged] structures . . . not required to account for these results, they are contraindicated. If the 4-iodobutyl radicals were bridged, . . . k_1/k_2 equal to that for simple primary alkyl radicals would have been unexpected, since this ratio apparently reflects

electronic factors." Setting aside the question of whether identity of behavior in only one particular competitive situation is sufficient to rule out
significant difference in structure, this result at most indicates that those
4-iodobutyl radicals which react with sodium naphthalene are not bridged.
It is not relevant to the question of where a possible equilibrium between
bridged and unbridged 4-iodobutyl radicals would lie and is not incompatible
with a preferred bridged structure.

IV.   ε-IODOALKYL RADICALS

Energy

Reaction of Haloalkyl Iodides with $Co(CN)_5^{3-}$

See Sec. 6; IC8 for general comments. Second-order rate constants
for the reaction of $Co(CN)_5^{3-}$ with iodoalkyl halides are given in Table 125
and indicate no accelerating effect of a 5-I substituent. This result provides no cause to consider the 5-iodopentyl radical to be bridged [1; II(a)].

Chapter 8

SULFUR-SUBSTITUTED RADICALS

I. β-THIYL-ALKYL RADICALS

A. Structure

ESR spectra assigned to β-thio-alkyl radicals on the basis of qualitative consistency with the structures are summarized in Tables 127 and 128. The "normal" α-H coupling constants and g-values provide no cause to consider these radicals to be bridged [1; I(a)-(c)].

Spectra of material resulting from the photolysis of RSSR and $\overset{\backslash}{\underset{/}{C}}=\overset{/}{\underset{\backslash}{C}}$ are summarized in Table 129. The α-H coupling constants are all "normal." Based primarily on the "anomaly" that " . . . the values of a_β for $CH_3SCH_2CH_2\cdot$ at the lowest temperatures are below . . . the approximate low-temperature limit for a primary alkyl radical," it is argued that there are "distortions in the structures of these radicals primarily at the β carbon atom . . . Concomitant with this distortion the sulfur atom is placed closer to the . . . trigonal center." We believe this argument to be subject to the discussion in the last paragraph of Section 5; IA. In addition, "a . . . homoconjugative interaction between the p_z orbital [at the trigonal center] and the sulfur atom is qualitatively consistent with the large fluorine coupling constants observed in the $CF_3SCH_2CH_2\cdot$ radical." Note the absence of observable coupling by CH_3 in $CH_3SCH_2CH_2\cdot$ and the 2.00-G fluorine coupling constant observed in the $CF_3OCH_2CH_2\cdot$ radical (Table 13).

The data presently at hand are not inconsistent with slightly bridged [1; I(c)] structures for RS-$\overset{|}{C}$-$\overset{|}{C}\cdot$ and might, with some difficulty ("normal" α-H coupling constants, no observable CH_3 coupling), be shown to be consistent with such structures. On the other hand, there is no strong indication that any such radical is bridged [1; I(a)-(c)]; $RSCH_2CH_2\cdot$ are not bridged [1; I(a)] with equivalent geminal hydrogens.

TABLE 127

ESR Spectra Attributed to β-Thio-Alkyl and Other Radicals

Radical	Method of generation	Coupling constant, G			Ref.
		α-H	β-H	g	
$CH_3\dot{C}HSSCH_2CH_3$	a	16.8	20.6		368
$CH_3\dot{C}HSCH_2CH_3$	a	16.8	19.8		368
$CH_3CH_2CH_2\dot{C}HSSCH_2CH_2CH_2CH_3$	a	17.3	18.3		368
$CH_3CH_2CH_2\dot{C}HSCH_2CH_2CH_2CH_3$	a	16.9	17.2		368
$\cdot CH_2SSCH_3$	a	17.34			473
$CH_3\dot{S}CHCH(SCH_3)CH_2OH$	b	16.6	9.0	2.0049	483
$CH_3CH_2\dot{S}CHCH(SCH_2CH_3)CH_2OH$	b	16.6	8.4	2.0048	483
$Pr^i\dot{S}CHCH(SPr^i)CH_2OH$	b	16.5	7.6	2.0047	483
$Bu^t\dot{S}CHCH(SBu^t)CH_2OH$	b	16.4	9.4	2.0046	483
$HOOCCH_2\dot{S}CHCH(SCH_2COOH)CH_2OH$	b	16.8	8.6	2.0046	483

[a] Photolysis of RSSR and $(Bu^tO)_2$.

[b] Solution containing H_2O_2 and $HC\equiv CCH_2OH$ mixed in a flow system at 15-25° with a solution containing $TiCl_3$, H_2SO_4, and RSH.

B. 1,2-Thiyl Migrations

See the first two paragraphs of Sec. 5;IB for a general discussion. Results (only those products possibly indicative of a thiyl shift) of the reaction of $(Bu^tO)_2$ with several sulfides are given in Table 130.

C. Control of Stereochemistry

1. Addition of RSX to Olefins

Data are given in Table 131. Those results corresponding to the absence of HBr from the reaction system show a pattern of no stereospecificity. While, because of the lack of adequate control experiments, it would be

TABLE 128

ESR Spectra Attributed to $RS-\overset{X}{\underset{|}{\overset{|}{C}}}-\overset{|}{\underset{|}{C}}\cdot$

Radical	Method of generation	α-H	β-H	N	g	Ref.
$CH_3CH_2SCH_2\dot{C}HCOOH$	a	19.6	11.7		2.0031	483
$(CH_3)_2CHSCH_2\dot{C}HCOOH$	a	19.6	11.9		2.0031	483
$HOOCCH_2SCH_2\dot{C}HCOOH$	a	19.5	12.0		2.0031	483
$CH_3CH_2SCH(COOH)\dot{C}HCOOH$	b	19.9	8.2		2.0032	483
$(CH_3)_2CHSCH(COOH)\dot{C}HCOOH$	b	20.0	8.0		2.0032	483
$(CH_3)_3CSCH(COOH)\dot{C}HCOOH$	b	19.9	7.2		2.0032	483
$HOOCCH_2SCH(COOH)\dot{C}HCOOH$	b	20.0	9.0		2.0033	483
$CH_3CH_2SCH(CH_3)\dot{C}HCOOH$	c	19.0	5.5		2.0032	483
$(CH_3)_2CHSCH(CH_3)\dot{C}HCOOH$	c	19.0	5.3		2.0032	483
$(CH_3)_3CSCH(CH_3)\dot{C}HCOOH$	c	19.9	5.3		2.0031	483
$HOOCCH_2SCH(CH_3)\dot{C}HCOOH$	c	19.4	5.6		2.0031	483
$HOCH(COOH)\dot{C}HCOOH$	b	20.9	12.7		2.0033	483
$HOCH(COOH)\dot{C}HCOOH$	d	21.1	12.7			55
$HOCH(COOH)\dot{C}HCOOH$	g	20.8	12.5			484
$HOOCCH_2C(OH)(COOH)\dot{C}HCOOH$	l	20.9			2.0032	490
$CH_3\dot{C}HCOOH$	i	20.2	25.0			484
$CH_3\dot{C}HCOOH$	l	20.3			2.0033	490
$HOCH_2\dot{C}HCOOH$	f	20.5	27.6			484
$Cl\dot{C}HCOOH$	e	20.9				55
$HOCH(CH_3)\dot{C}HCOOH$	h	20.6	15.6			484
$HOOCCH_2\dot{C}HCOOH$	m	20.9				489
$(CH_3)_2CHSCH_2\dot{C}HCN$	j	19.9	14.5	3.4	2.0028	483
$HOOCCH_2SCH_2\dot{C}HCN$	j	20.4	14.6	3.4	2.0031	483
$HOCH_2\dot{C}HCN$	j	20.2	28.6	3.6	2.0029	483
$HOCH_2\dot{C}HCN$	k	20.1	28.2	3.5		484

[a] Solution containing H_2O_2 and CH_2=CHCOOH mixed in a flow system at 15-25° with a solution containing $TiCl_3$, H_2SO_4, and RSH.

[b] As in a, except <u>cis</u>-HOOCCH=CHCOOH instead of CH_2=CHCOOH.

[c] As in a, except <u>trans</u>-CH_3CH=CHCOOH instead of CH_2=CHCOOH.

[d] As in a, except RSH omitted and <u>cis</u>- or <u>trans</u>-HOOCCH=CHCOOH instead of CH_2=CHCOOH.

[e] As in d, except $ClCH_2COOH$ instead of HOOCCH=CHCOOH.

[f] As in a, except RSH omitted.

[g] As in b, except RSH omitted.

[h] As in c, except RSH omitted.

[i] As in d, except CH_3CH_2COOH instead of HOOCCH=CHCOOH.

[j] As in a, except CH_2=CHCN instead of CH_2=CHCOOH.

[k] As in j, except RSH omitted.

[l] R· generated by photolysis of aqueous RH at 31°.

[m] Solution containing $HOOCCH_2CH_2COOH$ and H_2O_2 mixed in a flow system at 25° with one containing Fe^{2+} / EDTA.

improper to conclude from them that β-thio-alkyl radicals are not bridged [1; III(a)], the results provide no cause to consider them to be bridged [1; III(a)]. In the presence of HBr, 2-butene adds CH_3SD with a stereo-specificity corresponding to $\Delta\Delta G^{\ddagger} > 1.2$ kcal/mole. This result is subject to a discussion of the type included in Sec. 6; ID8.

2. Reaction of $CH_3CHOHCH(SCH_3)CH_3$ with $(Bu^tO)_2$

See Sec. 6; ID10 and Table 111.

II. β-THIYL-CYCLOHEXYL RADICALS

Control of Stereochemistry

Additions to Cyclohexenes

Results of the addition of CH_3COSH, H_2S, CH_3SH, $PhCH_2SH$, PhSH, and NCS-SCN to a variety of cyclohexenes are given in Table 132. In view of the lack of adequate proofs of structure in some cases and, in all cases,

TABLE 129

ESR Spectra Attributed to $RS-\overset{|}{\underset{|}{C}}-\overset{|}{\underset{|}{C}}\cdot$ [473]

Radical	Temperature	Coupling Constant, G		
		α	β	$CH_3(CF_3)$
$CH_3SCH_2CH_2\cdot$	$-70(-124)°$	21.60(21.68)	14.89(13.50)	"Unresolved"
$(CH_3)_3CSCH_2CH_2\cdot$	$-73(-110)°$	21.65(21.65)	16.87(15.97)	
$CF_3SCH_2CH_2\cdot$	$-83(-127)(-162)°$	22.03(21.96)(21.82)	13.82(12.81)(11.89)	4.34(4.42)(4.47)
$CH_3SCH_2\dot{C}HCH_3$	$-30(-81)°$	21.26(21.36)	13.89(12.94)	24.25(24.13)
$(CH_3)_3CSCH_2\dot{C}HCH_3$	$-84(-133)°$	21.38(21.45)	13.81(12.68)	24.10(24.08)
$CH_3SCH_2\dot{C}HCH_2SCH_3$	$+10(-41)°$	20.99(21.15)	15.42(14.64)[a]	

[a] Methylene groups equivalent.

TABLE 130

Reaction of Sulfides with t-Butyl Peroxide

Substrate	Reaction conditions	Reported product (% yield)	Controls	Ref.
PhC(CH$_3$)(SBu)$_2$ [b]	Reflux, CHCl$_3$, 23 mole % (Bu^{t}O)$_2$	PhCH(SBu)CH$_2$SBu (40) [c]	None	369
PhC(CH$_3$)(SBu)$_2$ [b]	PhCl, 5 mole % (Bu^{t}O)$_2$	PhCH(SBu)CH$_2$SBu (76)	a	369
(CH$_3$)$_2$C(S-p-tolyl)$_2$ [f]	Reflux, PhCl, 40 mole % (Bu^{t}O)$_2$	CH$_3$CH(S-p-tolyl)CH$_2$S-p-tolyl (40) [c]	d	370
(CH$_3$)$_2$C(SBu)$_2$ [f]	Reflux, PhCl, 5 mole % (Bu^{t}O)$_2$	CH$_3$CH(SBu)CH$_2$SBu (75) [c]	e	371
PhC(CH$_3$)(SPh)$_2$ [c]	Reflux, PhCl, 24 mole % (Bu^{t}O)$_2$	PhCH(SPh)CH$_2$SPh (24) [g]	h	372
(CH$_3$)$_2$C(SPh)$_2$ [f]	Reflux, PhCl, 40 mole % (Bu^{t}O)$_2$	CH$_3$CH(SPh)CH$_2$SPh (19) [c]	i	373

[a] No PhCH(SBu)CH$_2$SBu was formed in the absence of (Bu^{t}O)$_2$.

[b] Basis of assignment of structure: (1) acceptable C and H analyses; (2) prepared in 30% yield from PhCOCH$_3$ and BuSH/HCl.

[c] Assignment of structure OK.

[d] Omission of peroxide resulted in 94% recovery of starting material.

[e] Omission of peroxide resulted in 85% recovery of starting material.

[f] Basis of assignment of structure not mentioned.

g Basis of assignment of structure: (1) acceptable C and H analyses; (2) ir spectrum contained no absorption considered characteristic of a CH_3 group; (3) melting point same as that of material prepared from $PhCH=CH_2$ and $PhSSPh/I_2$.

h Omission of peroxide resulted in a 35% yield of the product. Reaction in the presence of 100 mole % of what was reported to be $PhSC(CH_3)=CH_2$ resulted in a 4% yield of what was reported to be $PhSCH(CH_3)CH_2SPh$ (identified from VPC retention time).

i Omission of peroxide resulted in 92% recovery of starting material.

TABLE 131

Addition of RSX to Olefins

Olefin	Addendum	Reaction conditions	Reported product			Ref.
			Structure (% yield)	threo/erythro (dl/meso)	Controls	
cis- or trans-3-hexene	$(SCN)_2$	20–30°, benzene	EtCH(SCN)CH(SCN)Et (30)[c]	a	b	374
cis- or trans-2-butene	$(SCN)_2$	20–30°, benzene	MeCH(SCN)CH(SCN)Me (?)[c]	a	b	374
cis- or trans-stilbene	$(SCN)_2$	20–30°, benzene	PhCH(SCN)CH(SCN)Ph (?)[c]	a	b	374
cis- or trans-2-butene	CH_3SD	−70°	$CH_3CH(SCH_3)CHDCH_3$ (?)	a	e	376
trans-2-butene	CH_3SD, DBr	−78°	$CH_3CH(SCH_3)CHDCH_3$ (?)	< 0.05[d]	e	376
cis-2-butene	CH_3SD, DBr	−78°	$CH_3CH(SCH_3)CHDCH_3$ (?)	> 20[d]	e	376
cis- or trans $CH_3ClC{=}CHCH_3$[f]	CH_3COSH	hν, −78°	$CH_3CH(SCOCH_3)CHClCH_3$ (58)	12[g]	h	268

[a] Not mentioned; ratio stated to be the same starting from cis- or trans-olefin.

[b] "reaction is strongly promoted by light, is influenced by the presence of peroxides."

[c] Basis of identification of products not mentioned.

[d] Basis of identification of product: Reaction of CH_3SNa with threo- or erythro-3-deuterio-2-bromobutane was assumed to proceed with inversion.

[e] "reaction conditions producing only minor amounts of olefin isomerization" were used.

[f] Basis of distinction of cis vs. trans: ir spectrum.

[g] Basis of identification of products: reaction with OH^-/H_2O gave a 59-63% mixture of episulfides which, upon reaction with PhLi, gave 2-butene (cis/trans = 12) in unspecified yield.

[h] Recovered olefin when reactant was trans[f] had an ir spectrum "essentially identical" to that of trans-olefin. [f] Recovered olefin when reactant was cis[f] had an ir spectrum which showed <3% isomerization.

TABLE 132

Addition of RSX to Cyclohexenes

Cyclohexene	RSX	Reaction conditions	Reported products — Structure (% yield)	$\dfrac{\text{trans/cis}}{\text{(RS, X)}}$	Controls	Ref.
Cyclohexene	NCS–SCN		(SCN, SCN) (?)[av]	"Stereospecifically trans"	None	374
	PhSH	$(Bu^tO)_2$, steam bath	(SPh) (50)[az]	> 1	None	379
	CH_3COSH	hν	(SAc) (85)[az]	4.9	None	379
	CH_3COSH		(SAc) (94)[az]	> 16	None	380
	CH_3COSH	86°, AIBN, hexane[b]	(SAc) (?); (SAc) (?)[ad]	?; 0.26	None	309
	CH_3COSH	86°, AIBN, hexane[c]	(SAc) (?); (SAc) (?)[ad]	?; 0.12	None	309
	CH_3COSH	86°, AIBN, hexane[d]	(SAc) (?); (SAc) (?)[ad]	?; 0.19	None	309
	CH_3COSH	86°, AIBN, hexane[e]	(SAc) (?); (SAc) (?)[ad]	?; 0.21	None	309

Substrate	Reagent	Conditions	Products	Ratio	Notes	Ref.
(olefin)	CH₃COSH	40–55°, hν	SAc (66)[az]; SAc (17)[az]	>4; <0.8	None	378
Cl-cyclohexene	H₂S	−60 to −80°, hν, neat[ay]	Cl–SH (65)[bd]	6.1	bc	383
Cl-cyclohexene	H₂S	−60 to −80°, hν, neat[ba]	Cl–SH (65)[bd]	9.0	bc	383
Cl-cyclohexene	H₂S	−60 to −80°, hν, neat[bb]	Cl–SH (64)[bd]	12.3	bc	383
Cl-cyclohexene	CH₃COSH	~50°, hν[ay]	Cl–SAc (80)	1.9	None	383
Cl-cyclohexene	CH₃COSH	~50°, hν[ba]	Cl–SAc (84)	2.7	None	383
Cl-cyclohexene	PhSH	~50°, hν[ay]	Cl–SPh (49)[bd]	>1	None	384
Cl-cyclohexene	H₂S	~−50°, hν	Cl–SH (?)[ad,ae]; Cl–SH (?)[ad,ae]	59[j]; 0.51[j]	None	309
Cl-cyclohexene	CH₃SH	−60°, hν	Cl–SCH₃ (?)[ab,ah]; Cl–SCH₃ (?)[ab,aj]	45[k]; 0.5[k]	None	309
Cl-cyclohexene	CH₃SH	10°, hν	Cl–SCH₃ (?)[ab,ah]; Cl–SCH₃ (?)[ab,aj]	20[k]; 0.55[k]	None	309

Cyclohexene	RSX	Reaction conditions	Reported products — Structure (% yield)		$\dfrac{trans/cis}{(RS, X)}$	Controls	Ref.
(Cl-cyclohexene)	CH_3SH	hν, pentane	(Cl, SCH$_3$ product) (~81)[ax];	(Cl, SCH$_3$ product) (~9)[ax]	>35; ~0.2	None	377
(Cl-cyclohexene)	CH_3COSH	−60°, hν, hexane[a]	(Cl, SAc product) (?)[j,ad];	(Cl, SAc product) (?)[j,ad]	74; 1.4	None	309
(Cl-cyclohexene)	CH_3COSH	63°, AIBN, hexane[a]	(Cl, SAc product) (?)[j,ad];	(Cl, SAc product) (?)[j,ad]	12.3; 1.3	None	309
(Cl-cyclohexene)	CH_3COSH	65°, AIBN, hexane[f]	(Cl, SAc product) (?)[j,ad];	(Cl, SAc product) (?)[j,ad]	13.6; 1.3	None	309
(methylcyclohexene)	CH_3COSH	−60°, hν, hexane[e]	(SAc product) (?);	(SAc product) (?)[as,at]	?; 0.11	None	309
(methylcyclohexene)	CH_3COSH	5°, hν, hexane[c]	(SAc product) (?);	(SAc product) (?)[as,at]	?; 0.15	None	309
(methylcyclohexene)	CH_3COSH	5°, hν, hexane[d]	(SAc product) (?);	(SAc product) (?)[as,at]	?; 0.17	None	309
(methylcyclohexene)	CH_3COSH	5°, hν, hexane[e]	(SAc product) (?);	(SAc product) (?)[as,at]	?; 0.22	None	309

Substrate	Reagent	Conditions	Product 1	Product 2	Ratio		Ref.
	CH_3COSH	86°, AIBN, hexane[b,c]	SAc (?);	SAc (?)[as,at]	?; 0.13	None	309
	CH_3COSH	86°, AIBN, hexane[d]	SAc (?);	SAc (?)[as,at]	?; 0.27	None	309
	CH_3COSH	86°, AIBN, hexane[e]	SAc (?);	SAc (?)[as,at]	?; 0.23	None	309
	H_2S	5°, hν, hexane	SH Cl (?)[x,y];	SH Cl (?)[z,aa]	36^j; 0.89^j	None	309
	CH_3SH	5°, hν, hexane	SCH_3 Cl (?)[p,af];	SCH_3 Cl (?)[q,ag]	16^k; 0.8^k	None	309
	CH_3SH	86°, AIBN, hexane	SCH_3 Cl (?)[p,af];	SCH_3 Cl (?)[q,ag]	7.0^k; 1.1^k	None	309
	$PhCH_2SH$	10°, hν, hexane	SCH_2Ph Cl (?)[ac,ak];	SCH_2Ph(?) Cl [ac,al]	7.1^l; 1.4^l	None	309
	$PhCH_2SH$	86°, AIBN, hexane	SCH_2Ph Cl (?)[ac,ak];	SCH_2Ph(?) Cl [ac,al]	5.1^l; 2.8^l	None	309
	$PhSH$	80°, Bz_2O_2, neat	SPh Cl (m)[am,an,ar];	SPh (m) Cl [ao,ar]	40; 3.5	None	309

Cyclohexene	RSX	Reaction conditions	Reported products		trans/cis (RS,X)	Controls	Ref.
			Structure (% yield)				
(Cl-cyclohexene)	CH_3COSH	$-60°$, $h\nu$, hexane[a]	(SAc, Cl) $(?)^{n,r}$;	(SAc, Cl) $(?)^{t,w}$	49.2; 1.9	None	309
(Cl-cyclohexene)	CH_3COSH	$5°$, $h\nu$, hexane[b]	(SAc, Cl) $(?)^{n,r}$;	(SAc, Cl) $(?)^{t,w}$	11.4; 2.0	None	309
(Cl-cyclohexene)	CH_3COSH	$5°$, $h\nu$, hexane[c]	(SAc, Cl) $(?)^{n,r}$;	(SAc, Cl) $(?)^{t,w}$	12.9; 2.1	None	309
(Cl-cyclohexene)	CH_3COSH	$5°$, $h\nu$, hexane[d]	(SAc, Cl) $(?)^{n,r}$;	(SAc, Cl) $(?)^{t,w}$	13.0; 2.0	None	309
(Cl-cyclohexene)	CH_3COSH	$5°$, $h\nu$, neat[e]	(SAc, Cl) $(?)^{n,r}$;	(SAc, Cl) $(?)^{t,w}$	14.0; 1.8	None	309
(Cl-cyclohexene)	CH_3COSH	$37°$, AIBN, pentane[a]	(SAc, Cl) $(?)^{n,r}$;	(SAc, Cl) $(?)^{t,w}$	12.4; 2.4	None	309
(Cl-cyclohexene)	CH_3COSH	$40°$, AIBN, pentane[f]	(SAc, Cl) $(?)^{n,r}$;	(SAc, Cl) $(?)^{t,w}$	15.5; 2.4	None	309
(Cl-cyclohexene)	CH_3COSH	$45°$, AIBN, CS_2[f]	(SAc, Cl) $(?)^{n,r}$;	(SAc, Cl) $(?)^{t,w}$	8.4; 1.7	None	309

					Ref.
(Cl-cyclohexene) + CH$_3$COSH	46°, AIBN, CS$_2$ [a]	(?)^{n,r} ; (?)^{t,w}	8.6; 2.1	None	309
(Cl-cyclohexene) + CH$_3$COSH	63°, AIBN, hexane [g]	(?)^{n,r} ; (?)^{t,w}	6.3; 1.9	None	309
(Cl-cyclohexene) + CH$_3$COSH	63°, AIBN, MeOH [f]	(?)^{n,r} ; (?)^{t,w}	7.7; 1.8	None	309
(Cl-cyclohexene) + CH$_3$COSH	64°, AIBN, MeOH [a]	(?)^{n,r} ; (?)^{t,w}	6.9; 1.8	None	309
(Cl-cyclohexene) + CH$_3$COSH	65°, AIBN, hexane [f]	(?)^{n,r} ; (?)^{t,w}	8.0; 1.8	None	309
(Cl-cyclohexene) + CH$_3$COSH	66°, AIBN, MeOH [h]	(?)^{n,r} ; (?)^{t,w}	8.2; 2.0	None	309
(Cl-cyclohexene) + CH$_3$COSH	67°, AIBN, PhCl [f]	(?)^{n,r} ; (?)^{t,w}	11.6; 1.9	None	309
(Cl-cyclohexene) + CH$_3$COSH	67°, AIBN, PhCl [a]	(?)^{n,r} ; (?)^{t,w}	9.9; 1.9	None	309
(Cl-cyclohexene) + CH$_3$COSH	68°, AIBN, hexane [h]	(?)^{n,r} ; (?)^{t,w}	12.5; 2.2	None	309

	RSX	Reaction conditions	Reported products — Structure (% yield)	trans/cis (RS, X)	Controls	Ref.
Cyclohexene	CH_3COSH	86°, AIBN, heptane[i]	(?)[n,r] ; (?)[t,w]	6.8; 2.1	None	309
	CH_3COSH	86°, AIBN, heptane[f]	(?)[n,r] ; (?)[t,w]	8.2; 2.2	None	309
	CH_3COSH	98°, AIBN, PhCl[f]	(?)[n,r] ; (?)[t,w]	6.2; 1.8	None	309
	CH_3COSH	105°, AIBN, PhCl[a]	(?)[n,r] ; (?)[t,w]	7.1; 2.1	None	309
	CH_3COSH	106°, AIBN, heptane[i]	(?)[n,r] ; (?)[t,w]	6.6; 2.6	None	309
	CH_3COSH	106°, AIBN, heptane[f]	(?)[n,r] ; (?)[t,w]	8.1; 2.4	None	309
(decalin)	NCS-SCN		(?)[av]	"2a, 3a… _trans_"	None	374
Cholestene	CH_3COSH	20°, hν, CCl_4	(97)[aw]	>32	None	381

(103)[aw]

				None	381
	CH₃COSH	20°, hν, CCl₄	large	None	381
Cholesteryl acetate	CH₃COSH	20°, hν, CCl₄	>16	None	381
	CH₃COSH	Steam bath	>1.1	None	382

[Note: In all footnotes, "trans"- and "cis"-isomer refer to the products of trans- and cis-addition, respectively.]

a-i [RSX] = 1.0, 0.045, 0.45, 4.5, 13.5, 5.9, 0.1, 10, and 0.59, respectively.

j Basis of identification of _trans_-isomer: saponification followed by methylation gave what were reported

to be and in the same ratio as the acetyl compounds in the reaction mixture.

k Analyzed by NMR integration of SCH₃ singlets.

l Analyzed by NMR integration of SCH₂Ph singlets.

m Total yield of isomers = 35%.

[n] Basis of identification of trans-isomer: (1) produced in 82% yield by reaction of what was reported to

be with Ac_2O; (2) acceptable C, H, and S analyses; (3) absorptions at 710 and 735 cm^{-1} (considered

to be indicative of equatorial Cl); (4) NMR τ = 5.42 – 6.22 (2H, poorly resolved multiplet); (5) survived in

unspecified yield treatment of the crude product mixture with NaOEt/EtOH, then H_3O^+, then Ac_2O/py; (6) treat-

ment of the crude product mixture with NaOEt/EtOH followed by MeI resulted in the recovery of only what were

reported to be + in the same ratio as the acetyl compounds in the reaction mixture.

[o] Basis of assignment of structure: (1) acceptable C, H, and Cl analyses; (2) footnote n(1).

[p,q] Basis of identification of trans-isomer: NMR analysis of mixture of two sulfides obtained by saponifica-

tion and methylation of the crude reaction mixture resulting from + CH_3COSH.

[r] Basis of identification of cis-isomer: (1) produced in 61% yield by reaction of what was reported to be

with AcCl; (2) acceptable C, H, and Cl analyses; (3) NMR τ = 6.00 (1H, $W_{1/2}$ = 8 cps^u), 5.60

(1H, $W_{1/2}$ = 6 cps^u); (4) absorption at 680 cm^{-1}; (5) does not survive treatment of crude product mixture with

NaOEt/EtOH, then H_3O^+, then Ac_2O/py.

[s] Basis of assignment of structure: (1) prepared in 81% yield from KSCN + ; (2) acceptable C, H,

and S analyses.

[t] Basis of identification of <u>trans</u>-isomer: (1) NMR $\tau = 6.38$ (1H, $W_{1/2} = 21$ cps[v]), 5.50 (1H, $W_{1/2} = 8$ cps[u]); (2) absorption at 675 and 740 cm^{-1} ai; (3) acceptable C, H, and S analyses; (4) see footnote n(4); (5) see footnote n(6).

[u] Indicative of equatorial H.

[v] Indicative of axial H.

[w] Basis of identification of cis-isomer: (1) absorption at 725 cm^{-1}; (2) NMR $\tau = 6.15$-6.75 (2H, complex multiplet); (3) see footnote r(5).

[x] Basis of identification of trans-isomer: comparison of retention time of acetyl derivative with that of material described in footnote o.

[y] Basis of identification of cis-isomer: as in footnote x, except footnote r.

[z] Basis of identification of trans-isomer: as in footnote x, except footnote t.

[aa] Basis of identification of cis-isomer: as in footnote x, except footnote w.

[ab] Basis of identification of trans-isomer: NMR spectrum agreed with that of authentic material.

[ac] Basis of identification of trans-isomer: as in footnote p, q except PhCH$_2$I used.

[ad] Basis of identification of products: none.

[ae] Basis of identification of trans-isomer: comparison of VPC retention time of acetyl derivative with that of material described in footnote j.

af Basis of identification of cis-isomer: (1) prepared by reaction of [structure] with MeSNa followed by SOCl$_2$/py; (2) acceptable C, H, and Cl analyses; (3) "the n.m.r. and i.r. spectra . . . supported the assigned diaxial configuration."

ag Basis of identification of cis-isomer: when crude reaction mixture heated, NMR band attributed to [ah] [structure with Cl, SCH$_3$] grew at expense of band assigned to cis-isomer.

ah Basis of identification of cis-isomer: (1) prepared by reaction of [structure] with MeS$^-$ followed by SOCl$_2$/py; (2) acceptable C, H, and Cl analyses; (3) absorption at 685 cm^{-1}. [ai]

ai Considered to be indicative of axial Cl.

aj Basis of identification of cis-isomer: its NMR band identified from fact that, when crude reaction mixture heated, the band attributed to the material described in footnote af grew at its expense.

ak Basis of identification of cis-isomer: as in footnote af(1),(3), using PhCH$_2$SNa, in a mixture along with the material described in footnote q.

al Basis of identification of cis-isomer: produced by heating the material described in footnote q.

am Basis of identification of trans-isomer: (1) prepared in 52% yield by reaction of [structure: OTs, Cl] with [structure: SO$_2$Ph] ; (4) PhSLi; (2) acceptable C, H, and Cl analyses; (3) the corresponding sulfone was converted to see footnote ar.

an Basis of identification of cis-isomer: (1) prepared as in footnote af(1),(2), except PhSNa used; (2) Ni(Ra) desulfurization of intermediate hydroxy sulfide gave 25% [structure: cyclohexane bearing OH and tert-butyl] ; (3) see footnote ar.

ao Basis of identification of cis-isomer: (1) prepared in 40% yield by reaction of what was reported to be[ap] [structure: cyclohexane bearing OTs and Cl] with PhSLi; (2) acceptable C, H, and Cl analyses; (3) "... n.m.r. spectrum ... consistent with its configuration assignment"; (4) reaction with H$_2$O yielded what was reported to be[aq] [structure: cyclohexane bearing OH and SPh] ; (5) converted to [structure: cyclohexene bearing SO$_2$Ph and tert-butyl] ; (6) see footnote ar.

ap Basis of assignment of structure: for corresponding alcohol see footnote j to Table 117.

aq Basis of assignment of structure: (1) reaction with Ni(Ra)/EtOH gave 75% [structure: cyclohexane bearing OH and tert-butyl] ; (2) p-nitrobenzoate gave acceptable C, H, and S analyses.

ar Crude reaction mixture solvolyzed under conditions known to preserve the material described in footnote am. After separation of that material, 73% of the material described in footnote aq and 27% remained.

as Basis of identification of cis-isomer: (1) NMR $\tau = 7.07$ ($W_{1/2} = 25$ cps^{v}); (2) see also footnote at.

at Basis of identification of products: (1) crude reaction mixture gave acceptable C and H analyses; (2) on reaction with Ni(Ra) crude reaction mixture was reported to be 90%

and 10%

au

; (3) on reaction with NaOEt/EtOH followed by Ni(Ra) crude reaction mixture gave results as in 2.

au Material considered to be cis had no absorption at 620–640 cm^{-1}, while that considered to be trans did have absorption there.

av Basis of identification of product not mentioned.

aw Basis of identification of product: (1) acceptable C and H analyses; (2) molecular rotations.

ax Basis of identification of trans-isomer OK; that of cis-isomer not mentioned.

ay RSX/Olefin = 1.

az Identification of products OK.

ba RSX/Olefin = 18.

bb RSX/Olefin = 56.

bc No reaction occurred in the absence of light. "Trans-2-chlorocyclohexanethiol was introduced at the start of a typical addition and the products were isolated in the usual manner. The observed composition of the isolated 1:1 adduct (25.2% trans) agrees well with that which would be expected (29% trans) if fractionation does not occur during the reaction or isolation."

bd Basis of identification of products: trans — OK, except 1,1-disubstituted cyclohexane not rigorously excluded; cis — OK.

of control experiments which demonstrate that the products are formed in the same ratio as from the direct reaction of the free 2-thiocyclohexyl radical, we defer judgment on the meaning of these results with regard to the question of bridging. If the results are taken at face value, 2-thiyl-cyclohexyl radicals can be moderately bridged [1; III(a)] (largest stereoselectivity observed corresponds to $\Delta \Delta G^{\ddagger}$ = 2.0 kcal/mole) while 2-trichloro-methyl, perfluoroalkyl, or dicyanomethylcycloalkyl radicals (Table 118) are not bridged [1; III(a)]. Within this framework of superficiality, one can speculate about this difference in "behavior" in terms of bridging [1; I(a)-(c)] only by thiyl groups or in terms of a combination of conformational factors in the substituted cyclohexyl radical and its transition state leading to product. A great deal of experimental work and discussion have been brought to bear on this question. At one extreme, much of it has been premature because the rudiments of the mechanism have not been established and the significance of the relative amounts of products (indeed, sometimes the nature of the products) remains unknown; at the other, all of this effort may be more relevant to the question of the conformations of cyclohexyl radicals and preferred modes of attack on them than to the question of bridging.

III.   β-THIYL-CYCLOPENTYL RADICALS

Control of Stereochemistry

Additions to Cyclopentenes

Results of the co-oxidation of indene with PhSH and CH_3COSH and of the addition of CH_3COSH to 1-methylcyclopentene are given in Table 133. In view of the lack of control experiments which demonstrate that the products are formed in the same ratios as from the direct reaction of the cyclopentyl radical, we defer judgment on the meaning of these results with regard to the question of bridging. If the results are taken at face value, 2-thiyl-cyclopentyl radicals can be slightly (maximum observed stereoselectivity corresponds to 1.8 kcal/mole) bridged [1; III(a)]. It is not possible to judge whether this small stereoselectivity is unusual because most of the entries in Table 119 are lower limits.

TABLE 133

Addition of RSX to Cyclopentenes

Cyclopentene	RSX	Reaction conditions	Reported product		trans/cis	Controls	Ref.
			Structure (% yield)				
Indene	PhSH/O_2	20°, benzene	[structure with SPh, OH]	(67)[a,c]	> 2	None	385
Indene	PhSH/O_2	20°, benzene	[structure with SOPh(and SPh), OH]	(77)[b,c]	> 20	None	385
Indene	CH_3COSH/O_2	5°, heptane	[structure with SCOCH$_3$, OH(and OOH)]	(53)[d]	> 1.9	None	386
Indene	CH_3COSH/O_2	Heptane, 10 mole % Bu^tNH_2	[structure with SCOCH$_3$, OH(and OOH)]	(74)	> 18	None	386
[cyclopentene structure]	CH_3COSH	hν	[structure with SCOCH$_3$]	(80)[e]	0.41	None	379

[a] After reaction with LiAlH$_4$.

[b] After adequate time allowed for formation of sulfoxide; followed by reaction with LiAlH$_4$.

^{c} Basis of identification of products: assignment of gross structure OK. Distinction of cis and trans rests on the assumption that the major product (64/14 mixture) of the LiAlH$_4$ reduction of

is the trans-isomer.

^{d} Basis of identification of products: assignment of gross structure OK. The NMR spectra of trans–OH, OH; OH, OCHO; OCH$_3$, OCHO; OH, Br; Br, Br; and OH, NH$_2$ compounds all show an octet for the 3-CH$_2$ group. Those of the cis–OH, OH; OH, OCHO; OCH$_3$, OCHO; and OCOCH$_3$, OCHO compounds show a doublet or two nearly superimposable doublets. The distinction of cis and trans in the present system rests on these analogies.

^{e} Identification of products OK.

Photoreaction of ⬡ and CH_3SD^* was reported to give trans/cis[†]
$(SCH_3, D) = 4.3\text{-}10$, in unspecified yield** [510]. In view of the uncertainty in the identification of the products and the lack of control experiments, we defer judgment on the meaning of these results with regard to

the question of bridging. If the results are taken at face value, ⬡-SCH₃

radicals can be slightly bridged (reported stereoselectivity corresponds to
0.6-1.4 kcal/mole) [1; III(a)].

IV. 3-THIYL-2-NORBORNYL RADICALS

Control of Stereochemistry

Addition of Thiols to Norbornenes

Results of the addition of phenyl and p-tolyl thiol to some norbornenes
are given in Table 134. Results of the addition of CH_2Br_2, $BrCH_2COOMe$,
$BrCCl_3$, CCl_4, $CHCl_3$, $CFBr_3$, and $BrCH(CN)_2$ to norbornenes are given
in Table 127. In view of the lack of control experiments which demonstrate
that the products are formed in the same ratio as from the direct reaction
of free 3-thiyl-2-norbornyl radicals with addendum, we defer judgment on
the meaning of these results with regard to the question of bridging. If the
results are taken at face value, exo-β-thiyl-norbornyl radicals are at least
slightly bridged [1; III(a)]. Note that the exo-β-X radicals would show

---

*> 98% D.

[†] The absorption at $\tau = 6.96$ [footnote [‡] (3)] was assigned to the
proton α to SCH_3 and the 5-cps coupling taken to be due to the cis-proton
α to D.

[‡] Basis of assignment of structure (per-H compound): (1) prepared
in 50% yield by reaction of (cis + trans)- ⬡-Cl and CH_3SK in acetone;
(2) absorption at 3070, 3040, 3000, 1020 cm^{-1}; (3) NMR spectrum —
$\tau = 9.38\text{-}9.97$, $8.88\text{-}9.00$ (6H), 7.94, 6.96 (d of d, J = 5, 2 cps).

** No control experiments were reported.

stereoselectivity (trans > cis addition) in the opposite sense from that of the exo-β-thiyl radicals (cis > trans addition). The latter is in the opposite sense from what is usually associated with bridged [1; I(a)-(c)] radicals.

V. 3-THIYL-2-BICYCLO[2.2.2]OCTYL RADICALS

Control of Stereochemistry

Addition of Thiols to Bicyclo[2.2.2]octenes

The addition of p-tolyl thiol to I in refluxing hexane gave a 71% yield of what was reported to be II, trans-addition/cis-addition = 0.3* [387]. This report provides no cause to consider the presumed intermediate radical to be bridged [1; III(a)].

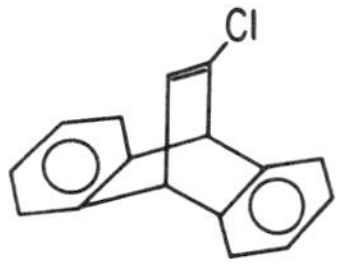

(I) 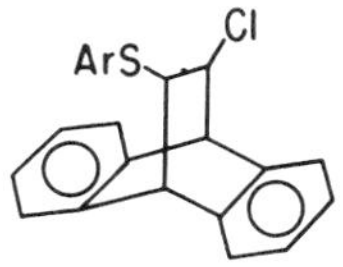   (II)

VI. 3-THIYL-CYCLOBUTYL RADICALS

Control of Stereochemistry

Addition of Thiols to Bicyclobutanes

Results of the addition of MeSH and $(CH_3)_3CSH$ to CH_3——CN and —CN are given in Table 135. Any stereoselectivity is nil and provides no cause to consider 3-thiyl-cyclobutyl radicals to be bridged [1; III(a)].

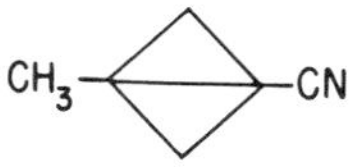

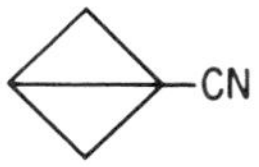

---

*Identification of trans-isomer based on strong presumptive evidence; that of cis-isomer based on acceptable C and H analyses.

TABLE 134

Addition of Thiols to Norbornenes

Norbornene	Thiol	Reaction conditions	Reported product		Controls	Ref.
			Structure (% yield)	Stereochemistry of addition:[d] trans/cis		
Norbornene	PhSD	Diethylperoxydicarbonate, cyclohexane	2-exo-SPh,3-exo-D(48)[b]	<0.1[a]	None	331
	PhSH	"Exothermic"	2-exo-SPh,3-endo-Me(71)[c]	<0.4	None	331
	p-MeC$_6$H$_4$SH	<40°, hexane	(85)[e]	<0.2	None	387

[a] Based on our arbitrarily chosen limit of detectability of the "other isomer" of 10%.

[b] Basis of identification of product: (1) acceptable C, H and S analyses; (2) NMR τ = 6.94 (q, J = 7.5, 1.5 cps) [2-exo-thiophenoxynorbornane has τ = 6.92 (octet, J = 7.5, 4.7, 1.5 cps)].

[c] Basis of identification of product: (1) acceptable C and H analyses; (2) NMR τ = 7.44 (q; J = 5.0, 2.0; 1H), 8.95 (d; J = 6.8; 3H).

[d] Not structure of product.

[e] Identification of product based on strong presumptive evidence.

TABLE 135

Addition of RSH to R'—⟨▷⟩—CN

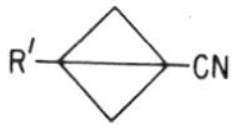

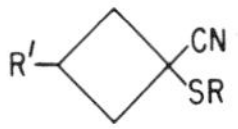

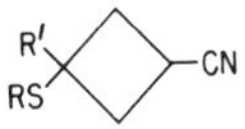

R'	R	Reaction conditions	Products				Controls	Ref.
			R'—⟨▷⟩—CN, SR % Yield	cis/trans	R'/RS—⟨▷⟩—CN % Yield	cis/trans		
CH$_3$	CH$_3$	8°, hν	72	1.0	18	1.0	None	388
CH$_3$	(CH$_3$)$_3$C	hν	63	1.2 or 0.8	14	1.2 or 0.8	None	388
CH$_3$	(CH$_3$)$_3$C	Reflux, 6 mole % AIBN	80	1.2 or 0.8	17	1.2 or 0.8	None	388
H	(CH$_3$)$_3$C	Reflux, 6 mole % AIBN			50	1.1 or 0.9	None	414

VII. δ-THIYL-ALKYL RADICALS

Structure

The proton ENDOR spectrum at -80°
of the material, presumed to be I, resulting from Zn/PhEt reduction of II indicated
a CH_3 coupling constant of 0.116 G and two
equally populated methylene group environ-
ments (coupling constants = 0, 0.60 G) [389]. By inference, the spin
density on sulfur is small.

VIII. 2-THIYL-VINYL RADICALS

Control of Stereochemistry

1. Addition of Disulfides to Acetylenes

Results of the addition of dimethyl and dibutyl disulfide to propyne and
1-heptyne are given in Table 136. In view of indications that some of the
products may be unstable under the reaction conditions and the fact that the
largest observed stereoselectivity corresponds to $\Delta\Delta G^{\ddagger} = 0.8$ kcal/mole,
these results provide no cause to consider the 2-thiyl-vinyl radicals to be
bridged [1; III(a)].

2. Addition of Thiols to Acetylenes

Results of the addition of thiols to several acetylenes are given in
Table 137. In view of the uncertain structures of some of the products,
the lack of adequate control experiments in all cases, and indications that
some of the products may be unstable under the reaction conditions, we
must withhold judgment on whether these results are relevant to the question of bridging.

TABLE 136

Addition of Disulfides to Acetylenes [390]

Acetylene	Disulfide	Reaction conditions	Product			
			Structure (% yield)	trans/cis	Controls	
$CH_3C \equiv CH$	CH_3SSCH_3	$20°$, h$\vee$	$CH_3C(SCH_3)=CHSCH_3$ (>80)	4.3	None	
$CH_3C \equiv CH$	BuSSBu	$20°$, h$\vee$	$CH_3C(SBu)=CHSBu$ (>80)	4.0	None	
$C_5H_{11}C \equiv CH$	CH_3SSCH_3	$20°$, h$\vee$	$C_5H_{11}C(SCH_3)=CHSCH_3$ (>80)	3.3	a	
$C_5H_{11}C \equiv CH$	CH_3SSCH_3	$120°$, $(Bu^tO)_2$	$C_5H_{11}C(SCH_3)=CHSCH_3$ (>80)	3.3	None	

a Trans/cis decreased steadily from 4.3 at 2 hr irradiation to 2.8 at 48 hr irradiation. Irradiation in the presence of CH_3SSCH_3 for 2 hr of $C_5H_{11}C(SCH_3)=CHSCH_3$, trans/cis = 61 and 0.18, yielded trans/cis = 3.1 and 2.8, respectively.

TABLE 137

Addition of Thiols to Acetylenes

Acetylene	Thiol	Reaction conditions	Reported products		Controls	Ref.
			Structure (% yield)	cis/trans		
HC≡CCOOH	PhCH$_2$SH	steam bath, 2 mole % ascaridole	PhCH$_2$SCH=CHCOOH (>30)[a]	> 0.4	None	391
HC≡CPh	PhSH	0°, hν, 2 hr, heptane[f]	PhSCH=CHPh (?)[d]	> 20	g	395
HC≡CPh	PhSH	Ice-bath to room temp., with or without Ph$_2$NH	PhSCH=CHPh (97)[b]	< 0.03	None	392
HC≡CPh	p-MeC$_6$H$_4$SH	0°, hν, heptane[h]	PhCH=CHSAr (?)[d]	9	i	395
HC≡CPh	p-MeC$_6$H$_4$SH	Ice-bath to room temp., with or without Ph$_2$NH	PhCH=CHSAr (94)[b]	2.9	None	392
HC≡CPh	p-MeC$_6$H$_4$SH	1.6 Mole % Bz$_2$O$_2$, reflux, cyclohexane	PhCH=CHSAr (80)[p]	0.92	None	396
HC≡CPh	CH$_3$SH	0°, hν, neat[h]	PhCH=CHSCH$_3$ (?)[d]	6.7	j	395

		Conditions	Product	Ratio		Ref.
HC≡CPh	EtSH	0°, hν, 1 hr, heptane[h]	PhCH=CHSEt (?)[d]	7.3	k	395
HC≡CPh	EtSH	80°, hν, 2 hr, heptane[l]	PhCH=CHSEt (?)[d]	3.3	o	395
HC≡CPh	BuSH	0°, hν, 1 hr, heptane[h]	PhCH=CHSBu (?)[d]	9.0	m	395
HC≡CPh	BuSH	80°, hν, heptane[l]	PhCH=CHSBu (?)[d]	2.4	None	395
HC≡CPh	CH$_3$COSH	0°, hν, heptane[h]	PhCH=CHSCOCH$_3$ (?)[d]	4.0	n	395
HC≡CSEt	EtSH	"Free radical conditions"	EtSCH=CHSEt (?)[c]	cis "main product"	None	393
HC≡CSEt	EtSH	hν, heptane	EtSCH=CHSEt (65)[c]	>1.9	None	401
HC≡CCOOCH$_3$	MeSH	hν	CH$_3$SCH=CHCOOCH$_3$ (35)[d]	0.1	None	394
HC≡CCOOCH$_3$	EtSH	hν	EtSCH=CHCOOCH$_3$ (21)[d]	0.25	None	394
HC≡CCOOCH$_3$	PhSH	hν	PhSCH=CHCOOCH$_3$ (44)[d]	0.7	None	394
HC≡CCOOCH$_3$	CH$_3$COSH	55°, 2 mole % AIBN	CH$_3$COSCH=CHCOOCH$_3$ (67)[d]	0.72	e	394
HC≡Cmesityl	mesitylSH	Reflux, cyclohexane, 2 mole % Bz$_2$O$_2$	mesitylSCH=CHmesityl (55)[d]	>1.2	None	396
HC≡CC$_4$H$_9$	CH$_3$COSH	q	C$_4$H$_9$CH=CHSCOCH$_3$ (5-75)[d]	4	r	397

Acetylene	Thiol	Reaction conditions	Reported products		Controls	Ref.
			Structure (% yield)	cis/trans		
HC≡CCF$_3$	H$_2$S	X-rays	CF$_3$CH=CHSH (73)[s]	1.5 or 0.7	None	398
HC≡CCH$_3$	EtSH	68°, AIBN	CH$_3$CH=CHSEt (56)[t]	1.3	None	399
HC≡CCH$_2$CH$_3$	EtSH	70°, 0.2 mole % AIBN	CH$_3$CH$_2$CH=CHSEt (28)[u]	~1	None	399
PhC≡C(CH$_2$)$_3$SH[v]		Reflux, hν, pentane	(83)[w]	< 0.2	None	400

[a] Basis of identification of products: gross structure OK. Cis distinguished from trans on the basis of its lower melting point (144-5° vs. 162-3°) and longer uv λ (2840 vs. 2750 Å).

[b] Basis of identification of product: melting point.

[c] Basis of identification of product not mentioned.

[d] Identification of products OK.

[e] Reaction at room temperature in the presence of 2 mole % hydroquinone yielded 68% CH$_3$COSCH=CHCOOCH$_3$, cis/trans = 3. Reaction without AIBN or hydroquinone showed "similar rates and product distribution" to that with AIBN.

[f] Acetylene/thiol = 22.

[g] At acetylene/thiol = 10, 1.1, 1.0, and 0.05, cis/trans = 9, 1.3, 1.2 (41% yield), and 0.19, respectively. In the absence of light, at acetylene/thiol = 1.0, cis/trans = 2.0. PhCH=CHSPh (cis/trans > 20), irradiated at 0° in heptane for 1 hr in the presence of 8 mole % PhSH, gave cis/trans = 0.8.

h Acetylene/thiol = 10.

i In the absence of light, at acetylene/thiol = 1.0, cis/trans = 2.7.

j At acetylene/thiol = 1.0, cis/trans = 4.7 (93% yield).

k At acetylene/thiol = 1.0, cis/trans = 4.0 (97% yield) in the presence of light and 5.5 in the absence of light. PhCH=CHSEt (cis/trans = 3.8), irradiated at 0° in the presence of 0.4 mole % EtSH, gave cis/trans = 3.2.

l Acetylene/thiol = 1.0.

m At acetylene/thiol = 1.0, cis/trans = 2.8 (95% yield). In the presence of 4 mole % cumene hydroperoxide the conversion of acetylene was 71% that in its absence (+hν). PhCH=CHSBu (cis/trans = 3.3), irradiated neat at 0° for 1 hr in the presence of ~5 mole % BuSH, gave cis/trans = 2.7.

n In the presence of 4 mole % cumene hydroperoxide the conversion of acetylene was 125% that in its absence (+hν).

o PhCH=CHSEt (cis/trans = 3.8), at 80° in heptane for 5 hr in the presence of 0.4 mole % EtSH gave cis/trans = 2.9.

p Basis of identification of products: agreement of melting points with those of authentic samples.

q Acetylene/thiol varied from 1 to 530.

r $C_4H_9CH=CHSCOCH_3$, equilibrated at 21-25° (hν, ~1 mole % PhSH) and 90°, was cis/trans = 1.

"Approximately 5 mole % of thiolacetic acid was added to a sample of monoadduct containing 74% cis isomer. After 4 days at 5°, the mixture contained 71% cis isomer and the thiolacetic acid could not be detected by VPC.

s Assignment of gross structure OK.

t Basis of identification of products: gross structure by agreement of VPC retention time with that of an authentic sample. Cis and trans not distinguished.

u Basis of identification of products: (1) acceptable C and H analyses; (2) ir spectrum.

v Basis of assignment of structure: By alkaline hydrolysis of thiouronium salts prepared from the corresponding halides (60–65% yield). > 95% pure by VPC (the heterocyclic products are the major impurities).

w Basis of assignment of structure: "... par analyse élémentaire et spectroscopique [ir, uv, NMR], et certains cycles ont été préparés par d'autres méthodes."

IX. $-C \equiv C - \overset{\bullet}{C} = C \overset{S^{\diagup}}{\diagdown}$ RADICALS

Control of Stereochemistry

Addition of Thiols to Diacetylenes

Results of the addition of CH_3CH_2SH to two diacetylenes are given in Table 138. In view of the small stereoselectivities reported, the lack of adequate proofs of structure of products and of adequate control experiments, these results provide no cause to consider the title radicals to be bridged [1; III(a)].

X. $\overset{\diagdown}{\diagup}C = C - \overset{\bullet}{C} = C \overset{S^{\diagup}}{\diagdown}$ RADICALS

Control of Stereochemistry

Addition of Thiols to Vinylacetylenes

Reaction of what was reported to be cis-$HC \equiv CCH = CHSBu*$ with BuSH at 70-80° in the presence of AIBN was reported [393, 403] to yield 75% $BuSCH = CHCH = CHSBu$ [†] (cis-cis/cis-trans = 0.2).

XI. THIYLMETHALLYL RADICALS

Structure

The ESR spectrum of material resulting from the photolysis of $CH_2 = CHCH = CH_2$ and CH_3SSCH_3 is summarized in Table 139. The CH_3S group does not significantly affect the spin distribution in the allyl radical.

---

*Basis of assignment of structure: (1) prepared from BuSK and $HC \equiv C - C \equiv CH$; (2) acceptable C, H, and S analyses.

[†]Basis of identification of products: (1) acceptable C, H, and S analyses; (2) "the geometric structure . . . follows from the structure of [the] sulfones, which is in turn determined with the aid of dipole moments, UV- and IR-spectra, and also by quantitative cis-cis and cis-trans isomerization into trans-trans by the action of I_2."

TABLE 138

Addition of Thiols to Diacetylenes [402]

Diacetylene	Thiol	Reaction conditions	Reported product		
			Structure (% yield)	cis/trans	Controls
$HC{\equiv}CC{\equiv}CSiMe_3$	EtSH	65°, 2-4 mole % $(Bu^tO)_2$	$Me_3SiC{\equiv}C\text{-}CH{=}CHSEt$ (16)[b]	~9	d
$EtC{\equiv}CC{\equiv}CSiMe_3$	EtSH	105°, 2-4 mole % $(Bu^tO)_2$	$Me_3SiC{\equiv}C\text{-}CH{=}C(Et)SEt$ (65)[c]	~0.3[a]	d,e

[a] Based on our visual inspection of published NMR spectrum.

[b] Basis of identification of product: (1) ir spectrum — 2137(s) (considered characteristic of conjugated triple bond at Si), 1560 cm^{-1} (considered characteristic of conjugated double bond at S); no absorption at 2050, 3080, 3300 cm^{-1} (considered characteristic of cumulene, terminal vinyl, terminal ethynyl); (2) NMR spectrum — (H_2O external standard) -4.80(s), 3.72(t), 2.11(q), -0.61 to -0.36 (J = 10 cps), -1.73, -1.48 (J = 10 cps) + weak signals nearby with J = 16 cps.

[c] Basis of identification of product: (1) ir spectrum — bands at 2128(s), 1571 cm^{-1} (see b); none at 2050, 3080, or 3300 cm^{-1} (see b); (2) NMR spectrum (see b) — -0.56(s), -0.13(s); (3) cis vs. trans — treatment with I_2 led to material whose NMR spectrum had no absorption at -0.46 and, at -0.14, a larger one than before.

[d] "When the already prepared adducts were heated, no cis-trans isomerization was observed."

[e] Reaction in the presence of hydroquinone: ". . . under mild conditions cis isomers were formed, but the reaction rate was low. Under more severe conditions the rate was higher, but a mixture of cis and trans isomers was formed with a predominance of the former if the reaction temperature did not exceed 85-90°."

TABLE 139

ESR Spectra Attributed to

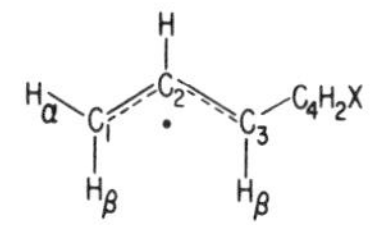

X	Temperature	Coupling constants (G)						Ref.
		$H_{1\alpha}$	$H_{1\beta}$	H_2	$H_{3\beta}$	H_4	
CH_3S	$-66(-105)°$	14.51(14.47)	13.53(13.53)	3.81(3.75)	13.26(13.24)	9.20(8.75)	473
H	$-110°$	14.78	13.83	3.85	13.83	16.43	475
CH_3	$-133°$	14.85	13.89	3.85	13.58	14.18	473
$(CH_3)_3CO$	$-67\left(\begin{array}{c}-127\\ \text{or} -137\end{array}\right)°$	14.68(14.83)	13.60(13.74)	3.93(3.88)	13.60(13.74)	16.47(16.97)	473,475

XII. β-SULFONYLALKYL RADICALS

Control of Stereochemistry

Photoaddition of $PhSO_2I$ to cis- or trans-2-butene in CCl_4 at 0-26° was reported [404] to give a 101% crude yield of the same mixture [†] of threo- and erythro-$PhSO_2CH(CH_3)CHICH_3$. [*]

XIII. 3-SULFONYL-2-NORBORNYL RADICALS

Control of Stereochemistry

Results of the addition of p-$CH_3C_6H_4SO_2Cl$ to norbornenes are given in Table 140. All that is known about any stereoselectivity is that its lower limit is small.

---

[*] Basis of identification of product: 7% of the product gave an acceptable halogen analysis.

[†] " . . . i.r. spectra of the recovered olefins showed no detectable cis-trans isomerization."

TABLE 140

Addition of $\underline{p}$-CH$_3$C$_6$H$_4$SO$_2$Cl to Norbornenes [405]

Nornornene	Reaction conditions	Reported product		
		Structure (% yield)	trans/cis	Controls
Norbornene	85–90°, 5 mole % Bz$_2$O$_2$	$\underline{endo}$-2-Cl, $\underline{exo}$-3-SO$_2$Ar[a] (64)	>1.8	None
Norbornene	120–130°, hν	$\underline{endo}$-2-Cl, $\underline{exo}$-3-SO$_2$Ar[a] (68)	>2.1	None
	152–155°, 10 mole % (Bu^{t}O)$_2$	(32+13)[a,b]	>1	None

[a] Identification of product based on strong presumptive evidence.

[b] Melting point 215–245° and 265–268°, respectively.

Chapter 9

CYANO-SUBSTITUTED RADICALS

I. β-CYANOALKYL RADICALS

A. Structure

ESR spectra, which have been assigned to β-cyanoalkyl radicals on the basis of qualitative consistency with the structures, are summarized in Table 141. The "normal" α-H coupling constants attributed to $HO\dot{C}HCH_2CN$, $\cdot CH_2CH(CH_3)CN$, and $\cdot CH_2CH_2CN$ would provide no cause to consider them to be bridged [1; I(a)-(c)]; $\cdot CH_2CH_2CN$ would not be bridged [1; I(a)] with equivalent geminal hydrogens.

B. Energy

1. Reaction of Alkyl Cyanides with Cl_2

The products resulting from the chlorination of $CH_3CH_2CH_2CH_2CN$ (Table 27), $(CH_3)_2CHCH_2CN$ (Table 30), and $CH_3CH_2CH_2CN$ (Table 28) have been reported. The comments in Sec. 5; IC2 apply to these results with regard to β-, γ-, and δ-cyanoalkyl radicals.

Data potentially useful in the sense discussed in paragraph 2 of Sec. 5; IC3 have been obtained: the relative reactivities per molecule toward Cl_2 of $CH_3CH_2CH_2Cl$, $CH_3CH_2CH_2CH_2Cl$, and $CH_3CH_2CH_2CN$ have been reported [137] to be (1.00), 1.26, and 0.60, respectively. No control experiments were reported.

2. Reaction of Alkyl Cyanides with $PhICl_2$ and Cl_2O

See Sec. 5; IC4,5 for general comments. The products resulting from the chlorination of $CH_3CH_2CH_2CN$ have been reported (Tables 40 and 42).

The comments in Sec. 5;IC2 apply to these results with regard to β- and γ-cyanoalkyl radicals.

3. Reaction of Alkyl Cyanides with $(CH_3)_3COCl$

The products resulting from the chlorination of $CH_3CH_2CH_2CN$ have been reported(Table 46). The comments in Sec. 5;IC2 apply to these results with regard to β- and γ-cyanoalkyl radicals.

Data useful in the sense discussed in paragraph 2 of Sec. 5;IC3 have been obtained: the relative reactivities toward $(CH_3)_3COCl$ of $CH_3CH_2CH_2CN$, $CH_3CH_2CH_2CH_2Cl$, and $CH_3CH_2CH_2OAc$ have been reported [138] to be 0.18, 1.48, and 0.64, respectively. If these results, subject to the comments in Sec. 5;IC2, are taken at face value, they permit the conclusion that there is no extraordinary stabilization of the transition states derived from $CH_3CH_2CH_2CN$.

4. Reaction of Alkyl Cyanides with Cl_3CSO_2Cl, Cl_3CSCl, SCl_2, and S_2Cl_2

The products resulting from the chlorination of $(CH_3)_2CHCH_2CN$ have been reported (Table 142). The comments in Sec. 5;IC2 apply to these results with regard to β- and γ-cyanoalkyl radicals.

C. Control of Stereochemistry

It has been reported [406] that liquid phase photobromination (1 M Br_2) of $CH_3CH_2CH(CH_3)CH_2CN$, $[\alpha]_{5481} = +7.98°$, yielded $CH_3CH_2CBr(CH_3)CH_2CN$, $[\alpha]_{5481} = +1.69°$, and that "... [the reaction] proceeds with high selectivity at the tertiary carbon." We find evaluation of this result difficult because the optical purity, chemical purity, and % yield of the product were not reported and because it has not been established that the reaction proceeds via attack on free $CH_3CH_2CH(CH_3)CH_2CN$ to give free $CH_3CH_2\overset{\bullet}{C}(CH_3)CH_2CN$ (structure - ?) followed by reaction with an achiral bromine-containing substance to give $CH_3CH_2CBr(CH_3)CH_2CN$ which is not asymmetrically destroyed. For these reasons we defer judgment on the meaning of this result with regard to the question of bridging [1; III(a)].

If additional information becomes available in the future and if the radical is judged to be bridged [1; III(a)], then whether it is bridged [1; I(a)-(c)] would still be an open question. For example, an initially produced

TABLE 141

ESR Spectra Attributed to β-Cyanoalkyl and Other Radicals

Radical	Method of generation	Coupling constants (G)			Temperature	Ref.
		C_1-H	C_2-H	N		
n-Octyl	a	22.1	28.3		-40°	56
·CH_2CH_2CN	a	23.0	28.4		-40°	56
·$CH_2CH(CH_3)CN$	b	23.2	25.1	—	25°	158
$HOCHCH_2OH$	b	15.4 ± 0.4	20.6 ± 0.4			73
$HOCHCH_2Cl$	b	15.4 ± 0.4	20.6 ± 0.4			73
$HOCHCH_2CN$	b	15.5 ± 0.1	16.5 ± 0.1	"Absent"		73

See also Table 13.

[a] Photoreaction of RBr, $(Bu^tO)_2$, and Et_3SiH.

[b] R· generated by mixing in a flow system a solution containing H_2SO_4 and $TiCl_3$ with a solution containing H_2SO_4, RH, and H_2O_2.

TABLE 142

Chlorination of $(CH_3)_2CHCH_2X$

X	Chlorinating agent	Reaction conditions	% Yield of reported monochlorinated substrate	Distribution of reported monochlorinated substrates (%)			Controls	Ref.
				1,1	1,2	1,3		
CN	Cl_3CSCl[a]	0°, hν, neat	≦90	1.5[e]	76[f]	22.5[g]	None	514
CN	Cl_3CSCl[a]	35°, hν, neat	≦90	2[e]	71[f]	27[g]	None	514
CN	Cl_3CSCl[a]	35°, hν, 50% CCl_4	≦90	2[e]	71[f]	27[g]	None	514
CN	Cl_3CSCl[a]	35°, hν, 8 M in PhH	≦90	1[e]	95[f]	4[g]	None	514
CN	Cl_3CSCl[a]	70°, hν, neat	≦90	3[e]	63[f]	34[g]	None	514
$OCOCH_3$	Cl_3CSCl[b]	20°, hν, neat	?	5–11[d]	58–61[d]	31–35[d]	c	515
CN	Cl_3CSO_2Cl[b]	30°, hν, neat	?	1[e]	92[f]	7[g]	None	514
CN	Cl_3CSO_2Cl[b]	60°, hν, neat	?	1[e]	85[f]	14[g]	None	514
CN	Cl_3CSO_2Cl[b]	90°, hν, neat	85	1.5[e]	75[f]	23.5[g]	None	514
CN	SCl_2[b]	30°, hν, neat	80	1[e]	86[f]	13[g]	h	514
CN	SCl_2[a]	30°, hν, PhH	?	0	100[f]	0	h	514
CN	SCl_2[a,b]	60°, hν, neat	?	2[e]	70[f]	28[g]	h	514
CN	S_2Cl_2[b]	30°, hν, neat	85	1[e]	84[f]	15[g]	None	514

[a] 40 mole %.
[b] 20 mole %.

[c] No reaction occurred in the absence of light.   Lower yields were obtained in the presence of hydro-quinone.   Reaction accelerated by Bz_2O_2.

[d] Identification of products OK.

[e] Basis of assignment structure: prepared in unspecified yield by reaction of $(CH_3)_2CHCH(OH)CN$ with PCl_5 [513].

[f] Basis of assignment of structure: (1) prepared in 70% yield by reaction of $(CH_3)_2C=CHCN$ with HCl; (2) reaction with refluxing 10% NaOH gave $(CH_3)_2C=CHCN$ in unspecified yield [513].

[g] Basis of assignment of structure: none [513].

[h] No reaction occurred in the absence of light.   Reaction retarded by air, β-naphthylamine, or hydroquinone.

$CH_3CH_2\overset{\cdot}{C}(CH_3)CH_2CN$ radical which is planar about the trisubstituted carbon atom and which has a "normal" C-C-CN angle, i.e., is not bridged [1; I(c)], would become statistically achiral only on a time scale large compared to the period of rotation about the $\overset{\cdot}{C}-CH_2CN$ bond. In other words, the rotamers of $CH_3CH_2\overset{\cdot}{C}(CH_3)CH_2CN$ which result directly from $CH_3CH_2CH(CH_3)CH_2CN$ are chiral and their racemization is accomplished in part by rotation about the $\overset{\cdot}{C}-CH_2CN$ bond; if such rotation is not much faster than reaction with Br_2, the $CH_3CH_2CBr(CH_3)CH_2CN$ will be optically active even though the radical is not bridged [1; I(a)-(c)]. (Of course, if the radical is not planar about the trisubstituted carbon atom, an additional process becomes necessary for the interconversion of enantiomers.) In order to establish the viability of this alternative, it is necessary to know the rates of rotation about the $\overset{\cdot}{C}-CH_2CN$ bond and of reaction of the radical with Br_2. Data which might be helpful in estimating the former are given in Table 61. Based on these data, we estimate the potential energy barrier to rotation in two models for unbridged $CH_3CH_2\overset{\cdot}{C}(CH_3)CH_2CN$, $CH_3CH_2CH(CH_3)CH_2CN$ and $R_2C=CRCH_2CN$,* to be 4.5 or ~ 2.3 kcal, respectively. If a pre-exponential factor of $\sim 10^{12}$ is used [191, 206, 239] rate constants at 0° (assumed) in the gas phase of 3×10^8 and 6×10^9 sec^{-1}, respectively, are obtained. Unfortunately, there is an even less secure basis for an estimate of the rate constant for reaction of unbridged $CH_3CH_2\overset{\cdot}{C}(CH_3)CH_2CN$ with Br_2. Available information is collected in Table 62. It is seen that the two rates ($[Br_2] = 1$) could, but may not, be of similar magnitude. We cannot emphasize too strongly that all of these estimates of rate constants for internal rotation and for reaction with Br_2 are extremely crude. We have of necessity chosen highly approximate models and have estimated their properties by use of highly approximate methods. Our purpose was only to determine whether, based on the very limited data available, the alternative "explanation" presented is a priori unreasonable. It does not appear to be so.

---

*See the corresponding footnote in Sec. 5; ID1.

II.   γ- CYANOALKYL RADICALS

Energy

1. Reaction of Alkyl Cyanides with Cl_2

   See Sec. 9:IB1.

2. Reaction of Alkyl Cyanides with $PhICl_2$, Cl_2O, and $(CH_3)_3COCl$

   See Sec. 9;IB2,3.

3. Reaction of Alkyl Cyanides with Cl_3CSO_2Cl, Cl_3CSCl, SCl_2, and S_2Cl_2

   See Sec. 9;IB4.

III.   δ- CYANOALKYL RADICALS

Energy

Reaction of Alkyl Cyanides with Cl_2

   See Sec. 9;IB1.

IV.   γ- CYANOCYCLOBUTYL RADICALS

Control of Stereochemistry

Addition of Thiols to Cyanobicyclobutanes

   See Sec. H6ai.

Chapter 10

VINYL-SUBSTITUTED RADICALS

In studying bridging by vinyl groups one has a degree of flexibility
unavailable in other systems: β-vinylalkyl radicals with a bridged
[1; I(a) or (c)]=CH- group are species well known under the name of
cyclopropylcarbinyl radicals.* Therefore, questions can be asked which
are much more intimate and subtle than could realistically be asked with
regard to other systems.

I. β-VINYLALKYL RADICALS

A. Structure

Can an unbridged [1; I(a)-(c)] β-vinylalkyl radical exist? The ESR
spectrum of $CH_2=CHCH_2CH_2\cdot$ and those assigned to allylcarbinyl radicals
on the basis of qualitative consistency with the structures are summarized
in Table 143. All of these spectra are "typical of those expected for a
normal radical" [171] and provide no cause to consider the allylcarbinyl
radical to be bridged [1; I(a)-(c)].

---

*Note that, in the application of definitions 1; I(a)-(c) to radicals of
the type $\overset{4}{C}H_2=\overset{3}{C}H\overset{2}{C}H_2\overset{1}{C}H_2\cdot$, the definition of the bridging (?) group is
arbitrary, =CH- and $CH_2=CH-$ being reasonable possibilities. If =CH-
is chosen, definition 1; I(a) refers to the equality of the C_1-C_3 and C_2-C_3
distances and definition 1; I(c) refers to the $C_1-C_2-C_3$ angle. If $CH_2=CH-$
is chosen, definition 1; I(a) refers to the distance from C_1 and C_2 to
some ill-defined point(s) on the line between C_3 and C_4 and definition
1; I(c) (limited to C_1-C_2-X angles) refers to the $C_1-C_2-?$ angle, ex-
tremes being the $C_1-C_2-C_3$ and $C_1-C_2-C_4$ angles.

355

Can more than one bridged [1; I(a) or (c)] β-vinylalkyl radical exist? The ESR spectrum of the cyclobutyl radical is summarized in Table 144 and is consistent with the structure of a strongly bridged [1; I(c)] β-vinyl-alkyl radical. The ESR spectrum of ▷–CH$_2$· and those assigned to cyclopropylcarbinyl radicals on the basis of qualitative consistency with the structures are summarized in Table 145. These spectra are consistent with structures either bridged [1; I(a)] or strongly bridged [1; I(c)], but different from that of the cyclobutyl radical.

B. Structural Isomers of β-Vinylalkyl Radicals

Can β-vinylalkyl radicals exist in both bridged [1; I(a)-(c)] and unbridged [1; I(a)-(c)] forms? See Sec. 10; IA for relevant ESR data.

1. Photochlorination of Hydrocarbons

Results of the photochlorination of $CH_3CH_2CH=CH_2$, cyclobutane, and methylcyclopropane are given in Table 146. It appears as if a different radical system is produced from each of these C_4H_8 isomers. In each case, that RCl which is closest in structure to its RH precursor is produced, a result consistent with, but not supporting, the idea that the β-vinylalkyl radical derived from (1) cyclobutane is moderately bridged [1; I(c)], (2) methylcyclopropane is bridged [1; I(a) or (c)], and (3) 3-butene is unbridged [1; I(a)-(c)] or at most moderately bridged [1; I(c)].

2. Decarbonylation of Allylcarbinylcarboxaldehydes

Montgomery and Matt [412, 413] have demonstrated the coexistence of at least two transient species derived from allylcarbinylcarboxaldehydes, one of which is capable of maintaining the stereochemical integrity of the original double bond while the other is not.

"The di-t-butyl peroxide-initiated radical-chain decarbonylations of 3-methyl-trans-4-hexenal and 2-methyl-trans-4-hexenal have been investigated in order to secure information regarding the possible intermediacy of cyclopropylcarbinyl-type radicals in homoallylic free radical rearrangements. The virtually exclusive hydrocarbon products from both aldehydes (chlorobenzene, 129.6°) were trans- and cis-2-hexene, 4-methyl-trans-2-pentene, and 4-methyl-cis-2-pentene. The distribution of olefinic products

TABLE 143

ESR Spectra Attributed to Allylcarbinyl Radicals

Radical	Method of generation	Temperature	Coupling constants (G)				Ref.
			α	β	γ	δ	
$CH_2=CHCH_2CH_2\cdot$	a	$-133°$	22.23	29.71	0.63	0.35	171
$CH_2=CHCH_2CH_2\cdot$	b	$-105°$	22.17	28.53	0.61	0.35	431
$CH_2=CHCH_2CH_2\cdot$	c	$-90°$	22.17	28.53	0.61	0.35	432
$CH_2=CHCH_2CH_2\cdot$	d	$>-100°$					432
$CH_3CH=CHCH_2CH_2\cdot$	c	$-114°$	(22.3, 22.3)	(29.3, 30.9)	(0.7, 0.7)	(–, 0.7)	433
$(CH_3)_2C=CHCH_2CH_2\cdot$	c	$-73°$	21.98	30.05	0.67		433
$\triangle CH=CHCH_2CH_2\cdot$	c	$-78°$	(22.06, 22.06)	(28.37, 29.86)	(0.63, 0.61)	(–, 0.61)	433
$CH_2=C(CH_3)CH_2CH_2\cdot$	c	$-70°$	22.18	26.86			433
$HOCH=CHCH_2CH_2\cdot$	c	$-48°$	(21.96, 21.96)	(28.26, 29.64)	(0.69, 0.61)	(–, 0.61)	433

[a] Electron-irradiation of ethylene.

[b] Photolysis of $CH_2=CHCH_2CH_2CO_2^{\frac{1}{2}}{}_2$ in cyclopropane.

[c] $R'R''C=CRCH_2CH_2\cdot$ produced by photolysis of $\triangle CHR'R''$ containing $(Bu^tO)_2$.

[d] Photolysis of $\triangle CH_2CO_2^{\frac{1}{2}}{}_2$.

TABLE 144

ESR Spectrum of the Cyclobutyl Radical

Method of generation	Temperature	Coupling constants (G)			Ref.
		α	β	γ	
a	-80°	21.20	36.66	1.12	171
b	-90°	21.30	36.77	1.09	432
c	-85°	21.30	36.77	1.09	174

[a] Electron-irradiation of cyclobutane.

[b] Photolysis of cyclobutanecarbonyl peroxide.

[c] Photolysis of $(Bu^tO)_2$ in cyclobutane.

was examined as a function of aldehyde concentration. Measurements were made on aldehyde solutions varying in concentration from ~6 to 0.094 M. From these data it has been concluded that substituted cyclopropylcarbinyl radicals are important decarbonylation intermediates and, further, that these radicals lie along the reaction coordinate for rearrangement." [412]

"The di-t-butyl peroxide initiated radical chain decarbonylations (chlorobenzene, 130°) of 4-pentenal, 4-pentenal-2,2-d_2, and cis-4-pentenal-5-d_1 have been examined in order to secure information regarding conceivable rearrangements of the allylcarbinyl radical. The virtually exclusive hydrocarbon product from 4-pentenal was 1-butene. Decarbonylation of 4-pentenal-2,2-d_2 afforded 1-butene in which the deuterium labels were distributed between the 3 and the 4 positions. The extent of deuterium scrambling was determined for a series of reactions in which the initial aldehyde concentration was varied from 0.50 to 6.0 M. The ratio of deuterons in the 4 position to those in the 3 position increased monotonically from 0.96 to 1.49 as the 4-pentenal-2,2-d_2 concentration was increased from 0.50 to 6.0 M. These observations suggest that vicinal vinyl group migration takes place in the allylcarbinyl radical. Decarbonylation of a 1.0 M solution of cis-4-pentenal-5-d_1 yielded a 1.03:1.00 mixture of cis-1-butene-1-d_1 and trans-1-butene-1-d_1. The ratio of olefins increased to 1.42:1.00

TABLE 145

ESR Spectra Attributed to Cyclopropylcarbinyl Radicals

Radical	Method of generation	Temperature	Coupling constants (G)			Ref.
			α	β	γ	
$\triangle\!\!-\!CH_2\cdot$	a	$-150°$	20.74	2.55	(2.98, 2.01)	432
$\triangle\!\!-\!CH_2\cdot$	b	$<-140°$	20.74	2.55	(2.98, 2.01)	432
$\triangle\!\!-\!\dot{C}HCH_3$	a	$-153°$	22.3	2.9	(1.9, 1.1)	433
$\triangle\!\!-\!\dot{C}(CH_3)_2$	a	$-150°$		2.27	(1.69, 1.27)	433
$\triangle\!\!\!\!\!-\!\!CH_2\cdot$	a	$-153°$	19.50	4.04	(2.27, 1.75)	433
$\triangle\!\!-\!\!\begin{smallmatrix}CH_2\cdot\\CH_3\end{smallmatrix}$	a	$-131°$	20.60		(3.51, 1.95)	433

^{a} R· produced by photolysis of $RH/(Bu^tO)_2$.

^{b} Photolysis of cyclopropaneacetyl peroxide.

L. KAPLAN

TABLE 146

Photochlorination (Cl_2) of C_4H_8

Substrate	Reaction conditions	Relative amount of			Controls	Ref.
		$CH_2=CHCH_2CH_2Cl$	▷CH_2Cl	□Cl		
$CH_3CH_2CH=CH_2$	Liquid, -9°	1.0 (5–6% yield)	—	—	None	157
□	Liquid	—	—	1.0	None	157
▷CH_3	0°, $CFCl_3$	<0.1	1.0	<0.1	None	411
□	Gas	—	—	1.0 (73% yield)	None	409, 410
▷CH_3	Gas	1.0 (35% yield)	1.0 (34% yield)	<0.15 (<5% yield)	None	409, 410
▷$^{13}CH_3$	Gas	0.5	1.0	—	None	409, 410

when the aldehyde concentration was increased to 7.3 M. From these results, it has been argued that cyclopropylcarbinyl radicals are important intermediates in the decarbonylation of 4-pentenal and, further, that they lie along the reaction coordinate for 1,2-vinyl group migration." [413]

Their "allylcarbinyl" radical may be unbridged [1; I(a)-(c)] or moderately bridged [1; I(a)-(c)]: their "cyclopropylcarbinyl" radical is a bridged [1; I(a) or (c)] β-vinylalkyl radical.

3. Addition of Thiols to Vinylcyclopropanes

Photoreaction at 5° of CH_3SH and $CH_2=C(CH_3)$ ◁ was reported [416] to give a mixture of $CH_3SCH_2CH(CH_3)$ ◁ and $CH_3SCH_2C(CH_3)=CHCH_2CH_3$, the "% rearranged product"* increasing monotonically to > 88% as the $CH_3SH/CH_2=C(CH_3)$ ◁ ratio decreased. If taken at face value, this result is consistent with the coexistence and at-least-one-way interconversion of at least two radicals, one which gives allylcarbinyl product and another which gives mainly cyclopropylcarbinyl product.

4. Decomposition of Peroxides and Peresters

"For the decomposition [at 125-131°] of $Ph_2C=CHCH_2CH_2CO_3Bu^t$ in $Et_3SnH/\underline{n}$-octane mixtures, the ratio Ph_2CH ◁ /$Ph_2C=CHCH_2CH_3$ is independent of the tin hydride concentration." [417] This result is consistent with an intermediate radical system whose components do not interconvert or a system which is kinetically equivalent to a single radical which gives both allylcarbinyl and cyclopropylcarbinyl product on reaction with Et_3SnH.

---

* "the percent rearranged material was determined [in the crude product mixture] from the integrated resonance peak of the single vinyl proton of the rearranged addition product . . . compared with that of the methyl thiol protons . . . ."

Results of the photolysis of >95% pure allylcarbinylcarbonyl, cyclo-
butylcarbonyl, and cyclopropylcarbinylcarbonyl peroxide and the corresponding t-butyl peresters are given in Table 147. It appears as if a
different radical system is produced (at least in the cage) from each substrate. The cyclobutanecarbonyl peroxide and perester yield only cyclobutyl products, from which it may be inferred that the intermediate β-
vinylalkyl radical is moderately bridged [1; I(c)]. The other two systems
yield product "fingerprints" which are both similar and different, a result
consistent with partial "leakage" between two different radical systems.

C. 1,2-Vinyl Migrations

See the first two paragraphs of Sec. 5; IB for a general discussion.

1. Decarbonylation of Allylcarbinylcarboxaldehydes

The occurrence of 1,2-vinyl migrations has been demonstrated in
several systems [412, 413, 434]. In addition to those described in Sec.
10; IB2, they have been observed during the decarbonylation of 3-methyl-4-
pentenal and 2-methyl-4-pentenal: "The principal hydrocarbon product
from the di-t-butyl peroxide initiated decarbonylation of 3-methyl-4-pentenal
in chlorobenzene at 129.6° was 1-pentene, an olefin of rearranged carbon
skeleton. 3-Methyl-1-butene, the simple decarbonylation product, was
formed in low yield. Under comparable reaction conditions 2-methyl-4-
pentenal afforded a similar mixture of 1-pentene and 3-methyl-1-butene.
1-Pentene/3-methyl-1-butene ratios, which were found to vary with initial
aldehyde concentration, were determined for chlorobenzene solutions
(about 6 M to 0.094 M) of 3-methyl-4-pentenal and 2-methyl-4-pentenal
that were carried to only a few percent reaction. . . . A detailed reaction
mechanism is proposed in which the decarbonylations and accompanying
rearrangements of the two aldehydes proceed in part through common free
radical intermediates. In this mechanism 1-pentene and 3-methyl-1-butene
are formed from classical homoallylic radicals. These intermediates are
interconverted by 1,2-vinyl group migration . . ." [434]

2. Addition of Radicals to Acetylenes

The photolysis of $(CH_3)_2CHCHO$ and $(CH_3)_3CCHO$ in acetylene at 137°
yielded $CH_3CH_2CH_2CH=CH_2$ and $(CH_3)_2CHCH_2CH=CH_2$, respectively [435].

3. Decomposition of Peresters

Decomposition of $Ph_2C=CHCH_2CD_2CO_3Bu^t$ in cyclohexane at 125°

yielded with 54% of the deuterium at C_a and the remainder

at C_b, C_c, C_d, and C_e; decomposition in $Et_3SnH/\underline{n}$-octane at 125° yielded

$$Ph_2C\overset{b}{=}\overset{a}{C}-\overset{}{C}-C-H$$

with the deuterium distributed equally between C_a and

C_b [417]. These results are in accord with those in Sec. 10; IC1.

D. Energy (Radical Formation from Allylcarbinyl Precursors)

Decomposition of Peroxides

$CH_2=CHCH_2CH_2CO_2)_2$ and $PhCH=CHCH_2CH_2CO_2)_2$ have been studied; results are given in Table 148. The former shows no enhancement in rate over that of a model compound, $CH_3CH_2CH_2CO_2)_2$. The latter decomposes 4-5 times faster ($\Delta\Delta G^{\ddagger} = 1.0$ kcal/mole at 60°) than a model compound, $PhCH_2CH_2CH_2CH_2CO_2)_2$; we do not consider this small difference to be significant, especially in view of the discussion in Sec. 1; VB. These results provide no cause to consider the allylcarbinyl radical to be bridged [1; II(a)].

E. Energy (Radical Formation from Cyclopropylcarbinyl Precursors)

1. Decomposition of Peroxides

Cyclopropaneacetylperoxide (Table 149) decomposes significantly faster ($\Delta\Delta G^{\ddagger} \cong 2$ kcal/mole) than a model compound, $(CH_3)_3CCH_2CH(CH_3)CH_2CO_2)_2$. In view of the abnormal product distribution (footnote c to Table 149) and the discussion in Sec. 1; VB, it would be premature to conclude from this result that cyclopropaneacetyl peroxide decomposes to bridged [1; II(a)] radicals.

2. Reaction of Hydrocarbons with $(CH_3)_3COCl$

Walling and Fredricks [411] have reported the relative reactivities (per H) of methyl C-H's (vs. neopentane) of methyl- and 1,1-dimethylcyclopropane toward $(CH_3)_3COCl$ to be 3.8-6.8 and 3.3-6.3, respectively, at 0°

TABLE 147

Photolysis of $C_4H_7CO_2)_2$ and $C_4H_7CO_3Bu^t$ (2537 Å, 30°) (418)

Substrate	Solvent	Reported products (% yield)[a]			
		CH_3CH_2 $CH=CH_2$	□	$CH_2=CHCH=$ CH_2[b]	□
$CH_2=CHCH_2CH_2CO_2)_2$	Decalin	36	—	< 2	—
□$CO_2)_2$	Decalin	—	26	—	10
▷$CH_2CO_2)_2$	Decalin	20	—	< 0.2	—
$CH_2=CHCH_2CH_2CO_2)_2$	Pentane	19	—	~ 2	—
□$CO_2)_2$	Pentane	—	19	—	10
▷$CH_2CO_2)_2$	Pentane	18	—	0.2	—
$CH_2=CHCH_2CH_2CO_3Bu^t$	Decalin	22	—	Trace	—
□CO_3Bu^t	Decalin	—	17	—	22
▷$CH_2CO_3Bu^{t}$ [c]	Decalin	20	—	Trace	—

[a] It was strongly implied that the products were determined by VPC retention time only.

[b] "unstable under reaction conditions"

[c] Perester "recovered after photolysis was carried to varying degrees of completion showed no evidence of rearrangement."

$CH_2=CHCH_2CH_2\overline{)}_2$	◇–◇	$CH_2=CH(CH_2)_3\triangleleft$ + $\triangleright CH_2\overline{)}_2$
48	—	—
—	40	—
2	—	38
48	—	—
—	42	—
17	—	31

TABLE 148

Rates of Decomposition of $(RCH_2CH_2CO_2)_2$

R	Temperature, °C	Solvent	Concentration, M	Rate constant, sec^{-1}	Ref.
$CH_2=CH$	60.1	Benzene, 10% methyl methacrylate	0.05–0.065	1.044×10^{-5}	52
CH_3	60.1[b]	Benzene	0.065	1.01×10^{-5}	419
$CH_2=CH$	70	Benzene, 10% methyl methacrylate	0.05–0.065	4.06×10^{-5}	52
CH_3	70[c]	Benzene	0.065	4.2×10^{-5}	419
$CH_2=CH$	77	Benzene, 10% methyl methacrylate	0.05–0.065	10.33×10^{-5}	52
CH_3	77[c]	Benzene	0.065	11×10^{-5}	419
$PhCH=CH$	60	CCl_4/10% styrene	0.026	3.47×10^{-5} a	420
$PhCH_2CH_2$	60[b]	CCl_4/10% styrene	0.026	0.715×10^{-5} d	420
$PhCH=CH$	70	CCl_4/10% styrene	0.027	10.6×10^{-5} a	420
$PhCH_2CH_2$	70	CCl_4/10% styrene	0.026	2.76×10^{-5} d	420

a "Kinetics followed through 3 half-live . . ."
b Extrapolated.
c Interpolated.
d " . . . followed through more than three half-lives."

TABLE 149

Rates of Decomposition of $(RCO_2)_2$

R	Temperature, °C	Solvent	Concentration, M	Rate constant, sec^{-1}	Ref.
▷–CH$_2$[a]	56.7	CCl$_4$	0.09	26 x 10^{-5} [b,c]	421, 422
(CH$_3$)$_3$CCH$_2$CH(CH$_3$)CH$_2$	60	Benzene	0.16	1.4 x 10^{-5}	423
⬡–CH$_2$	56.7[d]	CCl$_4$	0.06	4.7 x 10^{-5} [e,f]	421, 422

[a] 85–100% pure.

[b] The presence of 0.06 M I$_2$ had no effect on the rate constant at 56.5°. The presence of 100 mole % pivalic acid had no effect on the rate constant at 44.5°.

[c] Products at 78°: 132% CO$_2$, 56% ester, 7% acid, <8–9% RCl.

[d] Interpolated.

[e] The presence of 0.18 M I$_2$ had no effect on the rate constant at 64.3°.

[f] Products at 70°: 177% CO$_2$, 30% ester, 2.5% acid, 132% ⬡–CH$_2$Cl.

and 2.1-3.7 and 2.8-4.1, respectively, at 68°. These small accelera-
tions (largest possible $\Delta\Delta G^{\ddagger}$ = 1.0 kcal/mole), especially in view of the
discussion in Sec. 1; VB and the lack of adequate control experiments,
provide no cause to consider the product radical to be bridged [1; II(a)].

3. Decomposition of Ketals

Reaction of $CH_3CH(OR)(OBu^n)$ with 10 mole % $(Bu^tO)_2$ at 130° resulted
in a CH_3COOBu/CH_3COOR product ratio of 3.9 for R = cyclopropylcarbinyl
[424]. This small preference ($\Delta\Delta G^{\ddagger}$ = 1.1 kcal/mole), especially in
view of the discussion in Sec. 1; VB and the lack of adequate control experiments, provides no cause to consider the product radical to be
bridged [1; II(a)].

4. Reduction of Ketones by 2-Butanol

A mixture of two ketones was allowed to react with 50 mole % $(Bu^tO)_2$
and 3000 mole % 2-butanol. The relative amounts of residual ketones were
considered to be indicative of the "relative reactivities of . . . ketones
toward hydrogen atom addition." [425,426] Results are given in Table
150. Those for the PhCO⊲/PhCOCHMe$_2$ system would correspond to a
2.8 kcal/mole difference and those for the ▷-co⊲/⬡=O and
▷-coch$_3$/⬡=O systems differences of 0.2 and 0.1 kcal/mole, respectively. There is no reason to expect the relative amount of RCOR' in the
residual ketone to reflect inversely the relative stability of, say,
RĊ(OH)R'. In view of the uncertainty regarding the significance of the
ratio of residual ketones, the discussion in Sec. 1; VB, the lack of adequate
control experiments, and indications that said ratio may not reflect the
relative stability of RĊ(OH)R', * we do not consider the one indication
(2.8 kcal/mole difference) of abnormal behavior by ▷-COPh to indicate
that PhĊ(OH)⊲ is bridged [1; II(a)].

---

*See footnotes c, d, and e to Table 150.

5. Additions to Double Bonds

Results of the photoaddition of $BrCCl_3$, PhSH, and $\underline{n}$-$C_5H_{11}SH$ to

CH_2=CMe$\triangleleft$ and CH_2=CMeCHMe$_2$ are given in Table 151. The largest

ratio of residual olefins corresponds to 1.0 kcal/mole. In view of the uncertainty regarding the significance of the ratio of residual olefins, the lack of information regarding the products, the discussion in Sec. 1; VB, the lack of adequate control experiments, and the manner in which the experiments were conducted (footnotes a, b, and c to Table 151), we do not consider the ≤ 1.0 kcal/mole differences to be indicative of a bridged [1; II(a)] cyclopropylcarbinyl radical.

6. Decomposition of Azo Compounds

Rate constants for the decomposition of a series of substituted azo-methanes are given in Table 152. The replacement of CH_3 by cyclopropyl resulted in a large increase in rate of decomposition while replacement of $(CH_3)_2CHCH_2$ by cyclopropyl resulted in a small increase, corresponding to $\Delta\Delta G^{\ddagger} = 0.9$ kcal/mole. "The rate enhancement could arise (a) from resonance stabilization of a transition state leading to a cyclopropylcarbinyl radical or (b) from a concerted ring cleavage with the driving force originating from a relief of ring strain. * If the interaction is a conjugative one it is possible that transition state polarization introduces enough resemblance to the cyclopropyl carbonium ion to invalidate conclusions based on a supposed resemblance of the transition state to a free radical. . . . Since the rate acceleration seen in radical reactions that develop cyclopropylcarbinyl radicals are so much smaller than those seen in ioniza-tions leading to related carbonium ions, only a small increase in the amount of positive charge on the central atom on going to the transition state could account for the observed effect[†]." [428]

---

*See footnote e to Table 152. These data, if considered relevant, would be consistent with the proposition that cyclopropyl ring cleavage is concerted with initial homolysis.

[†] See footnote d to Table 152 for data indirectly related to this question.

TABLE 150

Reduction of Ketones by 2-Butanol

R	R'	Reaction conditions	Residual (RCOR'/PhCOCHMe$_2$) $\dfrac{\text{2-butanol}^a}{(\text{Bu}^t\text{O})_2{}^b}$ $\rightarrow$ Residual (RCOR' + PhCOCHMe$_2$)	Controls	Ref.
			RCOR' + PhCOCHMe$_2$		
△	Ph	125°	36^c	None	425
□	Ph	125°	11	None	425
⬠	Ph	125°	1.8	None	425
⬡	Ph	125°	1.2	None	425
△CH$_3$	Ph	125°	70^d	None	425
△Ph	Ph	125°	515	None	425
△Ph	Ph	125°	450	None	425
CH$_3$	Ph	125°	7.6	None	425

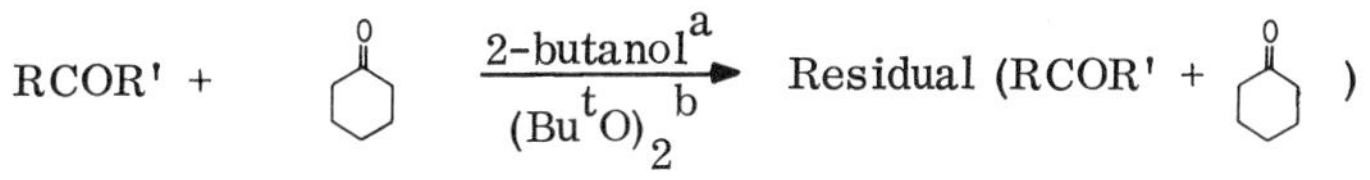

$$\text{RCOR'} + \text{(cyclohexanone)} \xrightarrow[\;(\text{Bu}^t\text{O})_2{}^b\;]{\;2\text{-butanol}^a\;} \text{Residual (RCOR'} + \text{cyclohexanone)}$$

R	R'	Reaction conditions	Residual (RCOR'/cyclohexanone)	Controls	Ref.
△	△	130°	1.29[e]	None	426
△	CH$_3$	130°	1.22	None	426

[a] 3000 mole %.

[b] 50 mole %.

[c] Products at 130° when PhCOCHMe$_2$ omitted: 45% PhCOCH$_2$CH$_2$CH$_3$, 19% PhCHOH △ .

[d] Products when PhCOCHMe$_2$ omitted: 1.7% PhCOCHMeCH$_2$CH$_3$, 15% PhCOCH$_2$CH$_2$CH$_3$, 31% PhCOCH$_2$CH$_2$CH$_2$CH$_3$, △ CHOHPh. (CH$_3$)

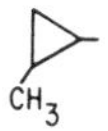

[e] Products when cyclohexanone omitted: 66% △ COCH$_2$CH$_2$CH$_3$.

TABLE 151

$$\text{CH}_2\text{=CMe}\triangle + \text{CH}_2\text{=CMeCHMe}_2 \xrightarrow[\text{h}\nu,\ 5°]{\text{addendum}} \text{Residual}\ (\text{CH}_2\text{=CMe}\triangle + \text{CH}_2\text{=CMeCHMe}_2)\ [416]$$

Addendum	Residual ($\text{CH}_2\text{=CMeCHMe}_2/\text{CH}_2\text{=CMe}\triangle$)[c]	Controls
BrCCl₃[a]	4.4	None
PhSH[b]	6.6	None
n-C₅H₁₁SH[b]	5.3	None

[a] In excess over olefin.

[b] Thiol/olefin ≅ 2.

[c] "The reactions were stopped in all cases before either of the two olefins was completely consumed."

TABLE 152

Decomposition of $R_1-\overset{\overset{\displaystyle R_2}{|}}{C}-N=N-\overset{\overset{\displaystyle R_2}{|}}{C}-R_1$
$\qquad\quad\underset{\displaystyle R_3}{\,}\qquad\qquad\underset{\displaystyle R_3}{\,}$

R_1	R_2	R_3	Reaction conditions	Rate constant, sec^{-1}	$\Delta G^{\ddagger}$, kcal/mole	Controls	Ref.
△	CH_3	CN	80.2°, toluene	$(25, 33)^{a,b} \times 10^{-4}$			286, 427
$(CH_3)_2CHCH_2$	CH_3	CN	80.2°, toluene	$(10,^a 7.1)^b \times 10^{-4}$			286, 427
$(CH_3)_3C$	CH_3	CN	80.2°, toluene	$(0.77, 1.09)^{a,b} \times 10^{-4}$			286, 427
CH_3	CH_3	CN	80.2°, toluene	1.5×10^{-4}			430
△	△	CN	80.2°, toluene	347×10^{-4} a		e	429
CH_3	CH_3	CH_3	135°, diphenyl ether/isoquinoline (90/10)		35.6^a		428, 429
△	CH_3	CH_3	135°, diphenyl ether/isoquinoline (90/10)		32.7^a		428, 429
△	△	CH_3	135°, diphenyl ether/isoquinoline (90/10)		30.6		428, 429

R_1	R_2	R_3	Reaction conditions	Rate constant, sec^{-1}	$\Delta G^{\ddagger}$, kcal/mole	Controls	Ref.
△	△	CH(CH$_3$)$_2$	135°, diphenyl ether/ isoquinoline (90/10)		30.8		428, 429
△	△	△	135°, diphenyl ether/ isoquinoline (90/10)		29.1	c, d	428, 429

[a] Extrapolated.

[b] Two stereoisomers.

[c] Rate constants at 118.50° of 3.77, 3.60, 3.30, and 4.45 x 10^{-4} were obtained in diphenyl ether/iso-quinoline (90/10), decalin/isoquinoline (90/10), isoquinoline, and cumene/isoquinoline (90/10), respectively.

[d] "A solution of the azo compound in 1,4-cyclohexadiene [not diphenyl ether] . . . was heated at 105° [not 135°] . . . The decomposition yielded 86% △─)$_2$C=CHCH$_2$CH$_3$ and 2% △─)$_3$C ·⫫$_2$. No tricyclopropylethane, within the limits of detection by glpc (0.25%), was produced. Hexacyclopropylethane is stable under the reaction conditions." "Irradiation of 1,1,1,1′,1′,1′-hexacyclopropylazomethane at 0° gives only hexacyclopropyl-ethane . . . At, or above, room temperature irradiation . . . gives olefinic products."

[e] Decomposition in the presence of a "slight excess" of "hexaphenylethane" yielded 53% of what was reported to be △─)$_2$C(CN)─CPh$_3$ and 25% △─)$_2$C(CN) ·⫫$_2$. The effect on the rate constant of variation of solvent was not reported.

F. Control of Stereochemistry

Reaction of ═△Br with Bu$_3$SnH yielded 75-80% ═△-Br ,

cis/trans = 2.2 [436].   The small stereoselectivity provides no cause to consider this allylcarbinyl radical to be bridged [1; III(a)].

II. 2-NORBORNENYL AND METHYLENENORBORNYL RADICALS

A. Structural Isomers of 2-Norbornenyl Radicals

Can 2-norbornenyl radicals exist in both bridged [1; I(a)-(c)] and unbridged [1; I(a)-(c)] forms?

1. Decarbonylation of 5-Norbornene-2-carboxaldehyde

Reaction of the title compound with 33 mole % (Bu^{t}O)$_2$ in chlorobenzene at 132° yielded 9% (norbornene + nortricyclene), the norbornene/nortricyclene ratio being only slightly sensitive to [aldehyde]. No control experiments were reported. These results, if taken at face value, would be consistent with an intermediate radical system whose components do not interconvert or a system which is kinetically equivalent to a single radical which gives both norbornenyl and nortricyclyl products.

2. Reaction of Norbornene with (CH$_3$)$_3$COCl

Photoreaction [165] of norbornene with (CH$_3$)$_3$COCl at 25° in CCl$_4$

yielded 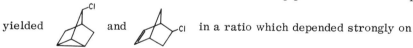 in a ratio which depended strongly on

[(CH$_3$)$_3$COCl] in a way which, if taken at face value (no control experiments were reported), would be consistent with the coexistence and at-least-one-way interconversion of at least two radicals, one which gives mainly nortricyclyl product and another which gives mainly norbornenyl product.

3. Addition of Thiols to Olefins

Reaction of with PhSH ("exothermic," temperature rose to

~100°) yielded a total of 33% of what were reported to be , *

, † and ‡ [437]. "The ratio of components

does not vary when the reaction is diluted." No control experiments were
reported. These results, if taken at face value, would be consistent with
an intermediate radical system whose components do not interconvert or a
system which is kinetically equivalent to a single radical which gives both
norbornenyl and nortricyclyl products.

4. Addition of PhSO$_2$X to Norbornadiene

Results are summarized in Table 153. If those for X = Cl are dis-
counted, these results, if taken at face value, would be consistent with the
coexistence and at-least-one-way interconversion of at least two radicals
which give what were reported to be norbornene and nortricyclene products
in different ratios.

5. Reaction of Halonorbornenes with Organotin Hydrides

Reaction of with Ph$_3$SnH at 38.5° in benzene resulted in a

ratio which varied only slightly with [Ph$_3$SnH]

([halide] ≥ [hydride]) [439].

Reaction of or with Bu$_3$SnH resulted in a

norbornene/nortricyclene product ratio which was independent of starting

---

* Basis of identification of product: (1) acceptable C and H analyses;
(2) NMR — τ = 5.1, 3.97, 7.69 (t, J = 9.9 cps). Note the absence of
additional coupling to what is presumed to be the <u>endo</u>-C$_6$-hydrogen.

† Basis of identification of product: NMR — τ = 4.73, 4.99, 8.65(s).

‡ Basis of identification of product: NMR — τ = 5.09, 5.34, 8.70.

halide under each of the following conditions [440]: (X = Br, 18-22°, neat), (X = Br, 100°, neat), (X = Br, 90°, xylene), and (X = Cl, 90°, AIBN). " . . . unreacted halides did not undergo interconversion under the reaction conditions. The hydrocarbons did not undergo interconversion under the reaction conditions, either." These results are consistent with an intermediate radical system which is kinetically equivalent to a single radical which gives both norbornenyl and nortricyclyl products.

B. 1,2-Vinyl Migrations*
See the first two paragraphs of Sec. 5; IB for a general discussion.

1. Reaction of Halonorbornenes with Organotin Hydrides

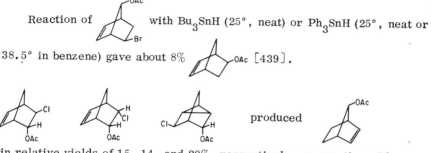

Reaction of [structure] with Bu$_3$SnH (25°, neat) or Ph$_3$SnH (25°, neat or 38.5° in benzene) gave about 8% [structure] [439].

[structures] produced [structure]

in relative yields of 15, 14, and 20%, respectively, on reaction with

Bu$_3$SnH (pentane, AIBN, hν, 25°). Similarly, [structure] and [structure] produced [structure] in relative yields of 5% † [441].

---

*Only those products possible indicative of a 1,2-vinyl shift are mentioned.

†"Each of the starting materials . . . was treated individually with tri-n-butyltin chloride under conditions similar to those employed in their reduction by tri-n-butyltin hydride. Similar experiments were carried out with each of the products . . . in the presence of tri-n-butyltin hydride and chloride . . . No isomerization had occurred."

TABLE 153

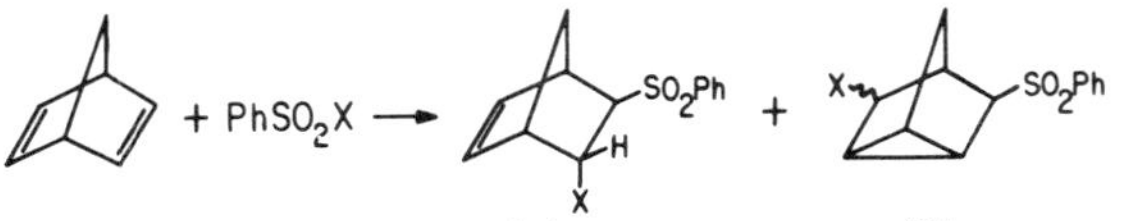

$$+ \text{PhSO}_2\text{X} \longrightarrow \quad (A) \quad + \quad (B) \qquad [438]$$

X	Reaction conditions[a]	A/B (% yield)	Controls
Cl	hν, Bz$_2$O$_2$, neat	0.05 (7.3)[b,d]	None
Cl	hν, Bz$_2$O$_2$[c]	0.07 (11.4)[b,d]	None
Br	hν, neat	0.5 (90)[b,e]	None
Br	hν[c]	0.15 (93)[b,e]	None
I	"exothermic," neat	5 (68)[f]	None
I	"exothermic"[c]	2.7 (68)[f]	None

[a] PhSO$_2$X/norbornadiene = 1:1.

[b] Basis of identification of A not mentioned.

[c] Reagents dissolved in 10 volumes of chlorobenzene.

[d] Basis of identification of B: (1) absorption at 12.3-12.5(m) μ (characteristic of nortricyclene); (2) did not decolorize KMnO$_4$/acetone; (3) acceptable C and H analyses.

[e] Basis of identification of B: see footnote d(1),(3).

[f] Basis of identification of B not mentioned. Basis of identification of A: (1) no absorption at 12.3-12.5 μ [see footnote d(1)]; (2) acceptable C and H analyses.

2. Addition of HBr to Norbornadienes and Methylenenorbornenes

Reaction of 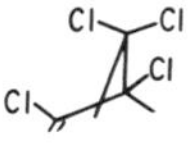 with HBr (60°, 1 mole % AIBN) was reported

[442-447] to yield 30% 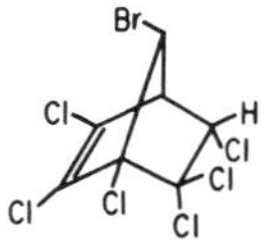 .* No control experiments were

reported.

_____

*Basis of identification of product: (1) acceptable C and Cl analyses; (2) absorption at 1618 cm^{-1}.

Reaction of 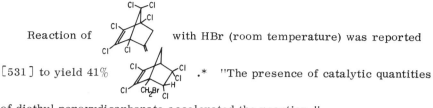 with HBr (room temperature) was reported

[531] to yield 41% .* "The presence of catalytic quantities

of diethyl peroxydicarbonate accelerated the reaction."

3. Addition of BrCCl₃ to Norbornadienes

Reaction of 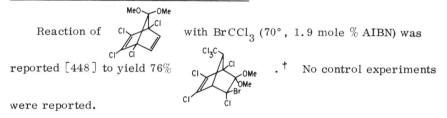 with BrCCl₃ (70°, 1.9 mole % AIBN) was

reported [448] to yield 76% .† No control experiments

were reported.

4. Addition of Thiols to Norbornadienes, Methylenenorbornenes, and
   Tetracyclenes

Results are summarized in Table 154.

---

*Basis of identification of product: (1) acceptable C and H
analyses; (2) NMR spectrum — τ = 4.99 (s), 6.33 (s), 7.25 (d, J =
10 cps), 7.55 (d, J = 10 cps); (3) absorption at 1610 cm^{-1}.

†Basis of identification of product: (1) acceptable C and H
analyses; (2) absorption at 1600 cm^{-1}; (3) NMR spectrum — τ = 6.31
(d, J = 0.8 cps, 1H), 5.80 (d, J = 0.8 cps, 1H), 6.27 (s), 6.38 (s).

TABLE 154

Addition of Thiols to Norbornadiene and Its Isomers

Substrate	Thiol	Reaction conditions	Reported product (% yield)	Controls	Ref.
	EtSH	20°	(42–relative)[a]	None	449
	EtSH	60°	(55–relative)[a]	None	449
	EtSH	80°, 0.5 mole % AIBN	(51)[a]	None	449
	PhSH	100°, 0.8 mole % AIBN	(<72)[b]	None	449
	CH$_3$COSH	"Exothermic"	(2–5)[l] + (~25)[m]	None	442–447

Reactant	Reagent	Conditions	Product	Catalyst	Ref.
(EtS/Cl structure)	EtSH	70°, 0.7 mole % AIBN	(13)[c]	None	449
(PhS/Cl structure)	PhSH	70°, 0.7 mole % AIBN	(14)[d]	None	449
(EtS/Cl structure)	EtSH	70°, 0.7 mole % AIBN	(7)[e] + –SEt (33)[f]	None	449
(SPh/Cl structure)	PhSH	70°, 0.7 mole % AIBN	(15)[g]	None	449
(MeS/Cl structure)	MeSH	60°, 0.2 mole % AIBN	(45)[ao]	None	452
(MeS/Cl structure)	MeSH	70–75°, 1 mole % AIBN	(48)[u]	j,k	442–447

Substrate	Thiol	Reaction conditions	Reported product (% yield)	Controls	Ref.
	EtSH	15–20°	(11)[s]	None	442–447
	EtSH	60°, 0.2 mole % AIBN	(50)[ak]	None	452
	EtSH	70–75°, 1 mole % AIBN	(38)[s]	h, k	442–447
	Bu^t SH	70–75°, 1 mole % AIBN	(47)[x]	k	442–447
	PhCH$_2$SH	Room temp., diethylperoxydicarbonate, "exothermic"	(20)[aw]	None	334

Reactant	Reagent	Conditions	Product	Yield	Ref.
(structure)	PhSH	Room temp., diethylperoxydicarbonate, "exothermic"	PhS— (structure) (28)^{as}	None	334
(structure)	PhSH	70–75°, 1 mole % AIBN	PhS— (structure) (21)^{y}	None	442–447
(structure)	p-MeC₆H₄SH	Room temp., diethylperoxydicarbonate, "exothermic"	ArS— (structure) (23)^{au}	None	334
(structure)	p-BrC₆H₄SH	Room temp., diethylperoxydicarbonate, "exothermic"	ArS— (structure) (33)^{av}	None	334
(structure)	MeOOCCH₂SH	Room temp., diethylperoxydicarbonate, "exothermic"	MeO₂CCH₂S— (structure) (21)^{ax}	None	334
(structure)	CH₃COSH	60°, 1 mole % AIBN	CH₃COS— (structure) (56)^{p}	None	442–447

Substrate	Thiol	Reaction conditions	Reported product (% yield)	Controls	Ref.
(structure)	MeSH	60°, 1.1 mole % AIBN	(74)[af]	None	451
(structure)	EtSH	60°, 0.3 mole % AIBN	(45)[ae]	None	451
(structure)	MeOOCCH₂SH	hν, ether	(16)[ag]	None	451
(structure)	CH₃COSH	hν, ether	(22)[ah]	None	451
(structure)	PhCOSH	hν, ether	(71)[aj]	None	451
(structure)	MeSH	70°, 1.1 mole % AIBN	(22)[ad]	None	448

				None	448
	EtSH	70°, 0.9 mole % AIBN	(19)[ac]	None	448
(z)	Bu^{t}SH	75–80°, AIBN, hexane	(32)[ab]	None	450
	EtSH	80°, 5.2 mole % AIBN	(72)[bb]	None	530
	PhSH	80°, 5.2 mole % AIBN	(44)[bc]	None	530

[a] Assignment of structure OK

[b] Basis of identification of product: (1) absorption at 1607 cm^{-1}; (2) NMR spectrum — τ = 6.87 (q, J = 3.5, 2.4 cps), 5.25 (octet, J = 3.5, 8.1, 3.5 cps), 7.21 (q, J = 8.1, 13.2 cps), 8.10 (octet, J = 3.5, 2.4, 13.2 cps), 6.45 (t, J = 2.4, 2.4 cps).

[c] Basis of identification of product: (1) absorption at 1614 cm^{-1}; (2) acceptable C and H analyses; (3) NMR spectrum — τ = 6.72 (q, J = 4.0, 1.8 cps), 5.44 (q, J = 4.0, 2.8 cps), 5.94 (d, J = 2.8 cps), 6.37 (d, J = 1.8 cps).

d Basis of identification of product: (1) absorption at 1606 cm^{-1}; (2) acceptable C and H analyses; (3) NMR spectrum — $\tau = 6.65$ (q, $J = 3.5$, 1.9 cps), 5.43 (q, $J = 3.5$, 2.6 cps), 5.88 (d, $J = 2.6$ cps), 5.99 (d, $J = 1.9$ cps).

e Basis of identification of product: (1) absorption at 1612 cm^{-1}; (2) acceptable C and H analyses; (3) NMR spectrum — $\tau = 6.43$ (octet, $J = 3.6$, 1.6, 0.6 cps), 5.05 (q, $J = 3.6$, 8.8 cps), 5.23 (q, $J = 8.8$, 0.6 cps), 6.67 (d, $J = 1.6$ cps); (4) [530] Reaction with H_2O_2/HOAc gave what was reported to be the sulfone[ay] in 75% yield.

f Basis of identification of product: (1) absorption at 1607 cm^{-1}; (2) acceptable C and H analyses; (3) NMR spectrum — $\tau = 6.61$ (q, $J = 2.7$, 1.9 cps), 4.97 (q, $J = 2.7$, 7.2 cps), 5.15 (d, $J = 7.2$ cps), 6.43 (d, $J = 1.9$ cps); (4) [530] Reaction with H_2O_2/HOAc gave what was reported to be the sulfone[ba] in 70% yield.

g Basis of identification of product: (1) absorption at 1607 cm^{-1}; (2) acceptable C and H analyses, (3) NMR spectrum — $\tau = 6.65$ (q, $J = 3.4$, 2.0 cps), 4.93 (q, $J = 3.4$, 8.4 cps), 5.17 (d, $J = 8.4$ cps), 6.17 (d, $J = 2.0$ cps); (4) [530] Reaction with H_2O_2/HOAc gave what was reported to be the sulfone[az] in 75% yield.

h A 10-fold decrease in [EtSH] increased the relative yield by 1.2-fold.

i A 10-fold decrease in [EtSH] increased the relative yield by 1.6-fold.

j A 13-fold decrease in [EtSH] increased the relative yield by 2.0-fold.

k "an alkylthiol and . . . [substrate] do not react at all at room temperature, nor upon heating if a sufficient amount of free radical initiator, . . . [AIBN], is not present. . ."

l "The NMR spectra were investigated for all adducts, for the sulfones obtained from them, the sulfoxides, the thiols, and the methylsulfonates synthesized from the isomeric thiolacetates. . ."

[m] Basis of identification of product: (1) acceptable C, H, Cl, and S analyses; (2) absorption at 1600 and 1713 cm^{-1}; (3) NMR — J_{H_7} = 2.5 cps; (4) treatment with Ni(Ra)/MeOH followed by Ni(Ra)/H$_2$/MeOH yielded 45% of what was reported to be[n]

[n] Basis of assignment of structure: (1) acceptable C, H, and Cl analyses; (2) different from what was reported to be the 1,4-isomer.[o]

[o] Basis of assignment of structure: melting point same as that of authentic sample.

[p] Basis of identification of product: (1) acceptable C, H, Cl, and S analyses; (2) absorption at 1612, 1710 cm^{-1}; (3) NMR spectrum — τ = 7.89(s), 6.32 (q, J = +3.80, +1.87 cps), 5.37 (d, J = +1.87 cps), 4.46 (d, J = +3.80 cps); (4) treatment with Zn/HOAc yielded 45% of what was reported to be[q]

(5) reaction with CH$_3$CO$_3$H followed by CH$_2$N$_2$ yielded 68% of what was reported to be[r]

[q] Basis of assignment of structure: (1) acceptable Cl and S analyses; (2) absorption at 1580, 1610, and 1712 cm^{-1}; (3) NMR spectrum — τ = 7.73 (s), 6.32 (q, J = +3.75, +1.95 cps), 5.39 (d, d, J = +1.95, -0.30 cps), 3.22 (d, d, J = +3.75, -0.30 cps).

[r] Basis of assignment of structure: (1) acceptable C, H, Cl, and S analyses; (2) absorption at 1185, 1377, 1617 cm^{-1}; (3) footnote 1.

[s] Basis of identification of product: (1) acceptable C, H, Cl, and S analyses; (2) absorption at 1612 cm^{-1};

(3) footnote l; (4) reaction with CH_3CO_3H/ether yielded 85% of what was reported to be [t]

[t] Basis of assignment of structure: (1) acceptable C, H, Cl, and S analyses; (2) absorption at 1050, 1060, and 1612 cm^{-1}; (3) footnote l.

[u] Basis of identification of product: (1) acceptable C, H, Cl, and S analyses; (2) absorption at 1622 cm^{-1};

(3) reaction with CH_3CO_3H/ether was conducted so as to yield either 95% of what was reported to be [v]

or 90% of what was reported to be [w]

[v] Basis of assignment of structure: (1) acceptable C, H, Cl, and S analyses; (2) absorption at 1148, 1344, and 1610 cm^{-1}; (3) footnote l.

[w] Basis of assignment of structure: (1) acceptable C, H, Cl, and S analyses; (2) absorption at 1050, 1070, and 1627 cm^{-1}; (3) footnote l.

[x] Basis of identification of product: (1) acceptable C, H, Cl, and S analyses; (2) absorption at 1610 cm^{-1}; (3) footnote l.

[y] Basis of identification of product: (1) acceptable C, H, Cl, and S analyses; (2) absorption at 1612 cm^{-1}; (3) footnote l.

[z] Basis of assignment of structure: (1) prepared in 14% yield by reaction of hexamethylcyclopentadiene with HOOCC≡CCOOH followed by Cu chromite/quinoline at 200-210°; (2) footnote aa.

[aa] "All the compounds obtained were characterized by elemental analysis, n.m.r. and IR spectra; their purity and the percentage in the reaction mixture were determined by t.l. and g.l. chromatography."

[ab] Basis of identification of product: (1) NMR spectrum — τ = 8.19 (J = 2.93 cps), 7.67 (J = 1.05, 2.93 cps), 7.26 (J = 1.05 cps); (2) footnote aa.

[ac] Basis of identification of product: (1) NMR spectrum — τ = 5.68 (d, J = 3.7 cps), 6.82 (q, J = 3.7, 1.7 cps), 6.61 (d, J = 1.7 cps); (2) acceptable C, H, and S analyses; (3) absorption at 1610 cm^{-1}; (4) conversion in 87% yield to what was reported to be the sulfone (acceptable C and H analyses, unchanged after Bu^{t}OK/Bu^{t}OH reflux).

[ad] Basis of identification of product: (1) NMR spectrum — τ = 5.68 (d, J = 3.8 cps), 6.80 (q, J = 3.8, 1.5 cps), 6.61 (d, J = 1.5 cps); (2) acceptable H analysis; (3) absorption at 1609 cm^{-1}; (4) conversion in 80% yield to what was reported to be the sulfone (acceptable C and H analyses, unchanged after Bu^{t}OK/Bu^{t}OH reflux).

[ae] Basis of identification of product: (1) acceptable H and S analyses; (2) absorption at 1600 cm^{-1}; (3) NMR spectrum — τ = 7.09(s), 4.91 (s, 1H), 7.31 (d, J = 10.5 cps), 7.60 (d, J = 10.5 cps).

[af] Basis of identification of product: (1) acceptable C and H analyses; (2) absorption at 1600 cm^{-1}; (3) NMR spectrum — τ = 7.11 (s), 4.98 (s, 1H), 7.31 (d, J = 10.5 cps), 7.61 (d, J = 10.5 cps).

[ag] Basis of identification of product: (1) acceptable S analysis; (2) absorption at 1600 cm^{-1}; (3) NMR spectrum — τ = 6.80 (s), 4.85 (s, 1H), 7.22 (d, J = 10.5 cps), 7.46 (d, J = 10.5 cps).

ah Basis of identification of product: (1) acceptable C and H analyses; (2) absorption at 1600 cm^{-1}; (3) NMR spectrum — τ = 6.47 (s), 4.88 (s, 1H), ~7.47; (4) reaction with Cl$_2$/H$_2$O gave a 58% yield of what was reported to be[ai]

ai Basis of assignment of structure: (1) absorption at 1600 cm^{-1}; (2) NMR spectrum — τ = 5.68 (s), 4.63 (s, 1H), 6.86 (d, J = 10.5 cps), 7.11 (d, J = 10.5 cps); (3) acceptable C and H analyses.

aj Basis of identification of product: (1) acceptable C and H analyses; (2) absorption at 1600 cm^{-1}; (3) reaction with Cl$_2$/H$_2$O yielded 41% of what was reported to be[ai]

ak Basis of identification of product: (1) acceptable C and H analyses; (2) NMR spectrum — τ = 6.44 (q, J = 3.7, 1.8 cps), 4.70 (d, J = 3.7 cps), 6.10 (d, J = 1.8 cps); (3) [530] reaction with CH$_3$CO$_3$H gave >75% of what was reported to be[al]; (4) reaction with Bu^{t}OK gave 99% of what was reported to be[am]

al Basis of assignment of structure: (1) absorption at 1612 cm^{-1}; (2) NMR spectrum — τ = 5.96 (q, J = 3.7, 1.8 cps), 4.73 (d, J = 3.7 cps), 5.69 (d, J = 1.8 cps); (3) sulfone remaining after reaction with Bu^{t}OK was unchanged; (4) reaction with Bu^{t}OK gave 83% of what was reported to be an

am Basis of assignment of structure: (1) acceptable C and H analyses; (2) NMR spectrum — τ = 6.39 (d, J = 6.39 (d, J = 1.9 cps), 5.94 (d, J = 1.9 cps).

an Basis of assignment of structure: (1) acceptable C and H analyses; (2) NMR spectrum — τ = 6.05 (d, J = 1.9 cps), 5.61 (d, J = 1.9 cps); (3) absorption at 1600 and 1620 cm^{-1}.

ao Basis of identification of product: (1) acceptable C and H analyses; (2) NMR spectrum — τ = 7.8 (s), 6.65 (q, J = 3.7, 1.5), 5.02 (d, J = 3.7 cps), 6.38 (d, J = 1.5 cps); (3) [530] reaction with CH$_3$CO$_3$H gave aq ; (4) reaction with Bu^{t}OK gave what was reported to be ap >75% of what was reported to be ap in unspecified yield.

ap Basis of assignment of structure: (1) acceptable H analysis; (2) NMR spectrum — τ = 6.94 (s), 6.14 (q, J = 3.8, 1.6 cps), 4.98 (d, J = 3.8 cps), 5.90 (d, J = 1.6 cps); (3) reaction with Bu^{t}OK gave what was reported to be ar in unspecified yield.

aq Basis of assignment of structure: (1) acceptable C and H analyses; (2) NMR spectrum — τ = 7.77 (s), 6.34 (d, J = 1.9 cps), 5.91 (d, J = 1.9 cps).

ar Basis of assignment of structure: (1) acceptable C and H analyses; (2) NMR spectrum — τ = 7.1 (s), 6.04 (d, J = 2.0 cps), 5.64 (d, J = 2.0 cps).

as Basis of identification of product: (1) acceptable C, H, and S analyses; (2) NMR spectrum (relative areas OK) — τ = 6.63 (q, J = 3.9, 1.8 cps), 5.11 (d, J = 3.9 cps), 6.02 (d, J = 1.8 cps); (3) reaction with 30% H_2O_2 MeOH gave 92% of what was reported to be

at Basis of assignment of structure: (1) acceptable H and S analyses; (2) Isomerization with Bu^tOK at 100° gave material whose n.m.r. spectrum was consistent with the presence of 62% of the original material, considered syn, and 38% of new material, considered anti; (3)   NMR spectrum —   τ = 6.20 (J = 4.0, 1.6 cps), 4.47 (J = 4.0 cps), 5.70 (J = 1.6 cps).

au Basis of identification of product: (1) acceptable C and H analyses; (2) NMR spectrum — τ = 6.57 (q, J = 3.8, 1.6 cps), 5.03 (d, J = 3.8 cps), 6.02 (d, J = 1.6 cps).

av Basis of identification of product: (1) acceptable C, H, and S analyses; (2) NMR spectrum — τ = 6.55 (q, J = 3.8, 1.9 cps), 4.96 (d, J = 3.8 cps), 5.98 (d, J = 1.9 cps).

aw Basis of identification of product: (1) acceptable C and H analyses; (2) NMR spectrum — τ = 6.95 (q, J = 3.7, 1.7 cps), 5.19 (d, J = 3.7 cps), 6.39 (d, J = 1.7 cps).

ax Basis of identification of product: NMR spectrum — τ = 5.83 (q, J = 3.9, 1.9 cps), 4.71 (d, J = 3.9 cps), 6.28 (d, J = 1.9 cps).

ay [530] Basis of identification of product: (1) acceptable C and H analyses; (2) NMR spectrum — τ = 4.97 (q, J = 3.4, 8.6 cps), 5.17 (d, J = 8.6 cps), 5.99 (q, J = 2.1, 3.4 cps), 6.27 (d, J = 2.1 cps).

az [530] Basis of identification of product: (1) acceptable C and H analyses; (2) NMR spectrum — τ = 4.25 (m), 5.85 (m).

ba [530] Basis of identification of product: (1) acceptable C and H analyses; (2) NMR spectrum — τ = 4.48 (d, J = 7.1 cps), 4.71 (q, J = 3.4, 7.1 cps), 6.1 (m).

bb Basis of identification of product: (1) acceptable C and H analyses; (2) reaction with refluxing H_2O_2/HOAc gave 72% of what was reported to be the sulfone. bd

bc Basis of identification of product: (1) acceptable C and H analyses; (2) reaction with refluxing H_2O_2/HOAc gave 75% of what was reported to be the sulfone. be

bd Basis of identification of product: (1) acceptable C and H analyses; (2) NMR spectrum — τ = 4.41 (d, J = 3.9 cps), 5.25 (d, J = 1.2 cps), 5.65 (q, J = 1.2, 3.9 cps).

be Basis of identification of product: (1) acceptable C and H analyses; (2) NMR spectrum — τ = 5.04 (d, J = 3.8 cps), 5.82 (d, J = 1.3 cps), 5.95 (q, J = 1.3, 3.8 cps).

5. Photoreaction of Norbornenes with Oxalyl Chloride

* was reported to give 100% † [517].

No control experiments were reported.

6. HgO-Induced Halogenative Decarboxylation of Acids

With regard to the mechanism of these reactions, the question of whether the immediate precursor of RX is R· is completely open, the only relevant observations being the report [520] that reaction of

BrCH₂—⟨⟩—COOH with Br_2/HgO in refluxing CCl_4 gave 75-85%

BrCH₂—⟨⟩—Br (no allylcarbinyl or cyclopropylcarbinyl products

reported) and the reports of the lack of stereospecificity in the reaction

of exo- and endo- ‡ [283], **,‖ [518],

what were reported to be exo- and endo- **,‖ [518] and

---

* Basis of assignment of structure [519]: (1) prepared in 93%

yield by hydrogenation (10%, Pd/C, EtOH) of ; (2)

melting point same as that of material prepared in 74% yield by reaction of hexachlorocyclopentadiene and ethylene.

† Basis of assignment of structure: (1) acceptable C, H, and Cl analyses; (2) absorption at 1610 (s) cm^{-1}; (3) NMR spectrum - τ = 7.34 (q, J = 2, 10.2 cps), 7.16 (q, J = 1.8, 10 cps), 6.34 (t, J = 1.8 cps)

‡ No control experiments were reported.

** "Some reactions were repeated in the presence of some added product halides to establish that no interconversion of products took place under the reaction conditions."

‖ Products not identified unambiguously.

[518], and what was reported to be

[521] with HgO/Br$_2$ or I$_2$) in refluxing CCl$_4$. It is with this in mind that we mention the generation of "free" norbornenyl radicals by use of this reaction.

Reaction of

with HgO/Br$_2$ in refluxing CCl$_4$ was reported to give

respectively, in yields of 10, 7, 8, and 9%, respectively. ** [518]

---

* No control experiments were reported.

† Products not identified unambiguously.

‡ Basis of assignment of structure: (1) acceptable C and H analyses; (2) NMR spectrum — τ = 6.15 (broad t, J = 1.8, 2.1 cps), 7.12 (q, J = 1.8, 10 cps), 7.37 (q, J = 2.1, 10 cps).

** Basis of assignment of structure: (1) acceptable C and H analyses; (2) absorption at 1610 cm^{-1}; NMR spectrum — τ = 6.87 (d, J = 12 cps), 7.1 (d, J = 12 cps), 8.7 (s).

$^{||}$ Basis of assignment of structure: NMR spectrum — τ = 6.58 (d, J = 8 cps), 7.39 (d, J = 8 cps).

** Basis of assignment of structure: NMR spectrum — τ = 6.13 (d, J = 12 cps), 6.88 (d, J = 12 cps).

C. Energy (Radical Formation from Norbornenyl Precursors)

1. Addition of Radicals to Norbornadienes

Claisse and Davies [437] attempted to determine the relative reactivities of norbornene and norbornadiene toward thiyl radicals. The two olefins were mixed with a quantity of RSH sufficient for <50% reaction and, from the relative fractions of each olefin which remained, a quantity called a "rate ratio" was calculated. For RSH = PhSH and CH_3COSH "rate ratios" (divided by 2) for norbornadiene relative to norbornene of 11 and 0.9, respectively, were reported. The latter provides no indication that nonbornadiene behaves in an unusual way and, in view of the lack of any indication that measures of the relative rates of attack on the two substrates by RS· have been obtained, * the former (corresponding to ~1.4 kcal/mole) cannot be taken to be an indication that the transition state leading to formation of the norbornenyl radical is unusually stable, i.e., that the norbornenyl radical may be bridged [1; II(a)].

The rate constant (divided by 2) for addition of CH_3· to norbornadiene relative to norbornene in isooctane at 65° was reported to be 1.1 [453]. This report would provide no cause to consider the norbornenyl radical to be bridged [1; II(a)].

The first-order disappearance of norbornadiene and norbornene ($PhCl/PhH/CHCl_3$ solvent, 73°, 6 mole % Bz_2O_2)[†] was followed and ($k_{diene}/2k_{monoene}$) was determined to be 1.0 [454]. This result would provide no cause to consider the norbornenyl radical to be bridged [1; II(a)].

---

*No demonstrative control experiments were reported. The following statement was made: " . . . reversibility in the case of phenylthio radicals would result in a build-up of their concentration and formation of PhSSPh, which was not observed as a product . . ."

[†]Reaction of norbornadiene with $CHCl_3$ (PhCl solvent, 80°, 1.1

mole % Bz_2O_2) yielded 60%   .

2. Addition of Radicals to Methylenenorbornenes

As in the first paragraph of Sec. 10; IIC1, for RSH = PhSH and HSCH$_2$CO$_2$Me "rate ratios" for [structure] relative to [structure] of 0.14 and 0.48, respectively, were reported. There is no indication that the radical(s) derived from 5,6-dimethylenenorborn-2-ene would be bridged [1; II(a)]. Similarly, Davies and Parfitt [459], using a quantity of RSH sufficient for 30% consumption of the two olefins, reported data from which one could determine the "relative reactivity" of [structure] relative to ([structure] + [structure]) toward HSCH$_2$COOMe and PhSH to be 1.3 and 1.7 (also, Ref. 437), respectively. There is no indication that the radical(s) derived from 2-methylenenorborn-5-ene would be bridged [1; II(a)]. Using a very similar method, subject to the comments in Sec. 10; IIC1, Huyser and Echegaray [455] reported data from which one could determine the reactivity toward CCl$_4$ (78°, 1.5 mole % Bz$_2$O$_2$) of [structure] relative to ([structure] + [structure]) to be 1.1. This result would provide no cause to consider the intermediate radical(s) to be bridged [1; II(a)].

3. Reaction of Norbornenyl Halides with Organotin Hydrides

The "relative rates of reduction" by Bu$_3$SnH (45°, methylcyclohexane) (determined by allowing an equimolar amount of Bu$_3$SnH to react with two halides simultaneously) of [structure] and [structure] were reported to be 1.0 and 1.7, respectively* [440]. These results would provide no cause to consider the intermediate radical(s) to be bridged [1; II(a)].

---

*. . . unreacted halides did not undergo interconversion under the reaction conditions [norbornenyl with nortricyclyl]." See Sec. 10; IIA5 for information regarding products.

4. Decomposition of Peresters

Rate constants for the decomposition of the t-butyl peresters of the norbornane and norbornene carboxylic acids are given in Table 155. Those of corresponding norbornyl and norbornenyl compounds differ insignificantly and provide no indication that the norbornenyl radical is bridged [1; II(a)].

5. Decomposition of Peroxides

Rate constants for the decomposition of norbornane and norbornene carbonyl peroxides are given in Table 156. Those of corresponding norbornyl and norbornenyl compounds differ insignificantly and provide no indication that the norbornenyl radical is bridged [1; II(a)].

D. Energy (Radical Formation from Nortricyclyl Precursors)

Reaction of Nortricyclyl Halides with Organotin Hydrides

The "relative rates of reduction" by Bu_3SnH (45°, methylcyclohexane) (determined by allowing an equimolar amount of Bu_3SnH to react with two halides simultaneously) of and were reported to be 1.0 and 0.4, respectively*[440]. These results would provide no cause to consider the intermediate radical(s) to be bridged [1; II(a)].

---

*. . . unreacted halides did not undergo interconversion [norbornenyl with nortricyclyl] under the reaction conditions." See Sec. 10; IIA5 for information regarding products.

TABLE 155

$$RCO_3Bu^t \xrightarrow{\text{cumene}}$$

R	Temperature, °C	Concentration, M	Rate constant, sec^{-1}	Ref.
(h)	94.45	0.051[a]	0.448 x 10^{-4}	456
(h)	101.90	0.050[b]	1.18 x 10^{-4}	456
(h)	112.10	0.050[c]	3.97 x 10^{-4}	456
(g)	94.45	0.047[d]	0.605 x 10^{-4}	456
(g)	101.90	0.050[e]	1.53 x 10^{-4}	456
(g)	112.10	0.061[f]	4.72 x 10^{-4}	456

R	Temperature, °C	Concentration, M	Rate constant, sec^{-1}	Ref.
	94.50	0.115	0.311×10^{-4}	456
	101.90	0.115	0.735×10^{-4}	456
	112.00	0.111	2.48×10^{-4}	456
(j)	94.5^i		0.21×10^{-4}	457
(j)	101.90		0.530×10^{-4}	457
(j)	112.0^i		1.8×10^{-4}	457

a A 4.3–fold increase in concentration resulted in a 5% decrease in rate constant.

b A 4.0–fold increase in concentration resulted in an 11% decrease in rate constant.

c A 4.2–fold increase in concentration resulted in a 13% decrease in rate constant.

d A 5.0–fold increase in concentration resulted in a 2% decrease in rate constant.

e A 3.8–fold increase in concentration resulted in a 16% decrease in rate constant.

f A 4.0–fold increase in concentration resulted in an 8% decrease in rate constant.

g The only "major" product was norbornane.

h The only "major" products were norbornene and nortricyclene.

i Interpolated.

j "Kinetics followed . . . for 1 to 3.5 half–lives."

TABLE 156

$$RCO_2)_2 \xrightarrow{\text{b}} \text{a} \quad [458]$$

R	Temperature, °C	Rate constant, $\sec^{-1}$
(structure)	44.5	3.3×10^{-5} c
(structure)	53.9	12.6×10^{-5} c
(structure)	65.9	67×10^{-5} c
(structure)	44.5	5.8×10^{-5} c
(structure)	53.9	17.8×10^{-5} c
(structure)	65.9	93×10^{-5} c
(structure)	44.5	2.22×10^{-5} d
(structure)	53.9	4.52×10^{-5} d
(structure)	65.9	23.7×10^{-5} d
(structure)	44.5	0.91×10^{-5} d
(structure)	53.9	4.33×10^{-5} d
(structure)	65.9	12.8×10^{-5} d

[a] See the original paper [458] for extensive quantitative data regarding the products.

[b] 0.05 N peroxide, 0.2 M styrene in CCl_4.

[c] Average of values determined by "iodometric titration of undecomposed peroxide" and by "disappearance of the peroxide band at 5.65 μ in the infrared."

[d] Determined by "iodometric titration of undecomposed peroxide."

III. BICYCLO[2.2.2] OCTENYL RADICALS

A. Energy (Radical Formation from Allylcarbinyl Precursors)

Addition of Radicals to Methylenebicyclo[2.2.2]octenes

Davies and Parfitt [459] attempted to determine the relative reactivities of bicyclo[2.2.2]octene, methylenebicyclo[2.2.2]octenes, and methylenebicyclo[2.2.2]octanes toward thiyl radicals. Equimolar amounts of two olefins were mixed with a quantity of RSH sufficient for 30% reaction and, from the relative fractions of each olefin which remained, a quantity called a "relative reactivity" was calculated. For RSH = HSCH$_2$COOMe and PhSH data were reported from which one could determine the "relative

reactivity" of  (x) relative to $\left(\text{}\;[†]\; +\; \text{}\;[†††]\right)$

to be 1.0 and 1.1, respectively, and that of 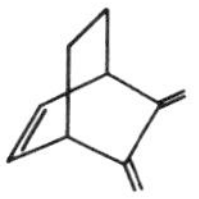 (y) relative to

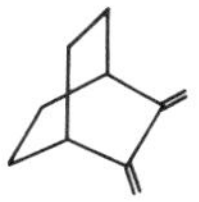

to be 0.62 and 0.98, respectively. There is no indication

that the radical(s) derived from either 2-methylenebicyclo[2.2.2]oct-5-ene or 2,3-dimethylenebicyclo[2.2.2]oct-5-ene is bridged [1;II(a)].

[†] Reaction of  with PhSH at 75° gave a 38% yield of what was

reported to be -SPh [††]

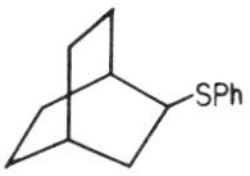

[††] Basis of identification of product: (1) acceptable C and H analyses; (2) n.m.r. spectrum -- τ = 6.63 (q, J = 6, 10 c.p.s.)

[†††]   Reaction of [bicyclo[2.2.2]oct-2-ene structure] with PhSH at 80° gave a 54% yield of

what was reported to be [††††]

[††††]   Basis of identification of product: 1) acceptable C and H analyses; 2) n.m.r. spectrum -- τ = 7.15 (d, J = 7).

[x]   Reaction of [bicyclo diene structure] with PhSH at 80° gave 37%, 28%,

32%, and <7% [xx] of what were reported to be [structure]–CH_2SPh (xxx),

$\left(PhS-[\text{structure}] + PhS-[\text{structure}] \right)$ xxxx , "2:1 adduct", and

$\left(PhS-[\text{structure}] + [\text{structure}]–CH_2SPh \right)$, respectively.

[xx]   Products assumed to be unreactive toward CH_3COSH.

[xxx]   Basis of identification of product: n.m.r. spectrum -- τ = 3.62 (q, J = 3.5, 4 c.p.s.)

[xxxx]   Basis of identification of product: n.m.r. spectrum -- τ = 5.10 (q, J = 4, 2 c.p.s.), 5.30 (q, J = 4, 2 c.p.s.).

[y]   Reaction of [bicyclo diene structure] with PhSH at 14-25° gave ~50%

and 13% of what were reported to be $\left([\text{structure}]–CH_2SPh \ (yy) + [\text{benzene ring with } CH_2SPh \text{ and } CH_3] \right)$

and $\left(PhS-[\text{structure}]–CH_2SPh + PhS-[\text{structure}]–CH_2SPh \right)$ yyyy , respectively.

yy   Basis of identification of product: 1) n.m.r. spectrum --
τ = 3.78, 8.51 (s), 6.52 (s), no exocyclic olefin protons; 2) hydrogena-

tion gave what was reported to be ‑CH₂SPh   (yyy)

yyy   Basis of assignment of structure: 1) n.m.r. spectrum --
no olefin protons, τ = 8.52 (s), 6.52 (s); 2) acceptable C and H analyses.

yyyy   Basis of identification of product: 1) acceptable C and H
analyses; 2) same mixture obtained by reaction of PhSH with what was

reported to be ‑CH₂SPh   (yy)

B.   Structure

γ-Irradiation of [structure] gave material whose ESR spectrum,

attributed to • [structure] on the basis of qualitative consistency with the

structure, was interpreted in terms of α-H, β-H, and β'-H coupling
constants of 20.8, 30.9, and 36.6 G, respectively [512]. These results
would provide no cause to consider the radical to be bridged [1; I(a)].

IV.   3-CYCLOALKENYLCARBINYL RADICALS

A.   Structural Isomers of 3-Cycloalkenylcarbinyl Radicals
Can 3-cycloalkenylcarbinyl radicals exist in both bridged [1; I(a)-(c)]
and unbridged [1; I(a)-(c)] forms?

1.   Reaction of Halides with Organotin Hydrides

Cl‑[structure]   [structure]   [structure]   [structure]
                           Cl

(VI)   (VII)   (VIII)   (IX)

Photoreaction of VI, (VI/VII = 20/80), and (VI/VII = 20/80) with Ph$_3$SnH (0.08, 0.04, and 0.40 M, respectively) in pentane at 15° gave IX/ VIII = 0, 0.01, and 0.04, respectively* [460]. No control experiments were reported. These results, if taken at face value, would be consistent with the coexistence and at-least-one-way interconversion of at least two radicals which give the allylcarbinyl and cyclopropylcarbinyl products in different ratios.

Photoreaction of neat XXXIII and what was reported to be XXXII[†] with an equimolar amount of Bu$_3$SnH at 26° gave, in >95% yield, XXXIV/XXXV/ XXXVI = 1/4.9/0.2 and 1/1.8/0.09, respectively.[‡]

(XXXII)   (XXXIII)

(XXXIV)   (XXXV)   (XXXVI)

---

*. . . when the chloride VII was allowed to stand at room temperature in n-pentane, no rearrangement to VI occurred, even in the presence of large quantities of triphenyltin chloride."

[†] Basis of assignment of structure: 1) Mass spectrum — m/e = 210, 208, 130, 129, 128, 127; 2) n.m.r. spectrum — τ = 9.83 (m, 1H), 9.28 (m, 1H), 8.60-9.12 (m, 1H), 7.10-8.00 (m, 2H), 4.65 (d, J = 2 c.p.s., 1H), 4.25 (d, J = 6.5 cps, 1H), 2.73 (m).

[‡] "Controls showed that a maximum of 7% rearrangement of the [cyclopropylcarbinyl] bromides . . . to [allylcarbinyl] bromide . . . occurred on irradiation at 26° for 4 hr in the presence of tri-n-butyltin bromide."

Similarly, reaction of neat XXXIII (room temperature, hν) and what was reported to be XXXII (room temperature) with an equimolar amount of Ph$_3$SnH gave, in > 95% yield, XXXIV/XXXV/XXXVI = 1/4.0/0.1 and 1/0.5/ 0.06, respectively [509]. These results, if taken at face value, would be consistent with the coexistence and interconversion of at least two radicals which give the allylcarbinyl and cyclopropylcarbinyl products in different ratios.

2. Addition of Thiols to Vinylcyclopropanes

Results are summarized in Table 157. In view of the lack of control experiments and the uncertain identification of some of the products, these results will not be discussed further.

B. 1, 2-Vinyl Migrations

See the first two paragraphs of Sec. 5; IB for a general discussion. Only those products possibly indicative of a 1, 2-vinyl shift are mentioned.

Decomposition of Peresters

Decomposition of ▷—CH$_2$CO$_3$Bu^t $^{(*)}$ at 140° in p-cymene and

PhCCl$_3$ gave 47% of what was reported to be ⬡ $^{(†)}$ and 10%

⬡—Cl [461].

---

* Basis of assignment of structure: (1) acceptable C and H analyses; (2) NMR spectrum — τ = ~4.3 (2H), ~ 7.0 (1H), ~8.7 (s, 9H), 7.4- ~8.7.

† Basis of identification of product: 1) VPC retention time; (2) hydrogenation gave material whose VPC retention time was the same as that of cyclohexane.

TABLE 157

[510]

[CH$_3$SH]/[olefin]	I/II	III/IV	Total % yield	Controls
1.0	0.31	0.18	~60	None
1.5	0.44	0.19	~60	None
2.0	2.0	1.0	~40	None
4.0	5.0	3.6	~40	None

[a] Basis of assignment of structure: (1) prepared in 15% yield by reaction of (cis + trans)- with CH$_3$SK in acetone; (2) acceptable C and H analyses; (3) NMR spectrum — τ = 9.30-9.75, 7.40-8.90, 8.00, 6.67-7.17; (4) absorption at 3060-3070, 3035-3040, 3000, 1020 cm^{-1}; (5) "no C=C absorption"; (6) "the 3040 cm^{-1} CH absorption was enhanced [relative to what was reported to be the trans-isomer[b]], which indicates a cis isomer."

[b] Basis of assignment of structure: (1) as in footnote a(1), expect 40% yield; (2) NMR spectrum — τ = 9.47-9.94, 7.90-8.90, 8.02, 7.10-7.90; (3) absorption at 3070, 3040, 3000, 1025 cm^{-1}; (4) "no C=C absorption"; (5) see footnote a(6).

[c] Basis of assignment of structure of a mixture of cis and trans (identities of individual isomers assumed): (1) acceptable C and H analyses; (2) NMR spectrum — τ = 8.97 (d, J = 7 cps), 8.90 (d, J = 7 cps), 7.58-8.80 (2H), 8.07 (s, 3H), 8.02 (s, 3H), 6.90-7.50 (1H), 6.08-6.48 (1H), 4.30-4.58 (2H); (3) absorption at 3050, 3045, 1600, 1375, 750 cm^{-1}.

C. Energy (Radical Formation from Cyclopropylcarbinyl Precursors)

Chlorination of Norcarane

Results, subject to the qualifications and provisos in Sec. 5; IC2, are summarized in Table 158. When compared to those in Table 69, they would provide no cause to consider to be bridged [1; II(a)].

V. METHYLENECYCLOALKYL RADICALS

Energy (Radical Formation from Methylenecycloalkyl Precursors)

Addition of Radicals to Methylenecycloalkanes

The rate constant (divided by 2) for addition of $CH_3\cdot$ to [figure] relative to [figure] in isooctane at 65° was reported to be 1.6 [453]. This report would provide no cause to consider the 3-methylenecyclobutyl radical to be bridged [1; II(a)].

VI. 3-CYCLOALKENYL RADICALS

Energy (Radical Formation from Cycloalkenyl Precursors)

1. Photoreaction of Cycloalkenes with Cl_2

The "relative reactivity per hydrogen relative to abstraction of a single cyclohexyl hydrogen" at 25° of the "homoallylic" hydrogens of cyclohexene was reported to be 0.72 [462]. This report would provide no cause to consider the 3-cyclohexenyl radical to be bridged [1; II(a)].

2. Reaction of Halocycloalkenes with Organotin Hydrides

The "relative rates of reduction" by Bu_3SnH (45°, methylcyclohexane) (determined by allowing an equimolar amount of Bu_3SnH to react with two

TABLE 158

Photochlorination of 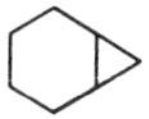 [496]

Reaction conditions	% Yield of reported monochlorinated substrate	Distribution of reported monochlorinated substrates (%)[d] 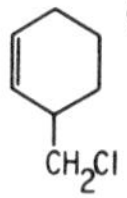	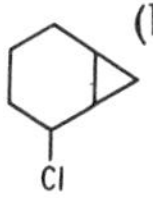	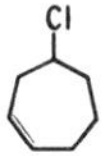	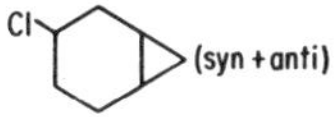	Controls
		(e) CH_2Cl	(b) Cl	(e) Cl	(f) (syn + anti) Cl	
Cl_2, 114°, gas phase	41	57	See footnote a	2[a]	42	None
$(CH_3)_3COCl$, 25°	?	24	See footnote a	45[a]	31	None

[a] What was reported to be gave what was reported to be[c] 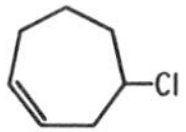 under the conditions of analysis.

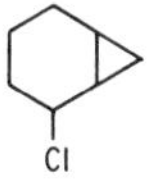

[b] Basis of assignment of structure of isomers: (1) prepared in unspecified yield by reaction of syn- 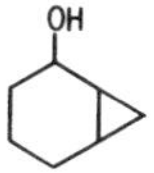 with $SOCl_2$/ether at 0° [NMR spectrum — τ = 5.37 (1H, m), 9.2–9.6 (2H, m); (2) prepared in unspecified yield by reaction of syn- with Ph_3P/CCl_4 [NMR spectrum — τ = 5.41 (1H, m), 9.3–10.0 (2H, m).

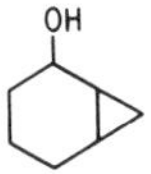

^cBasis of assignment of structure: (1) prepared in unspecified yield by reaction of <u>syn</u>– with excess SOCl$_2$; (2) NMR spectrum — τ = 4.29 (2H, nonet), 6.06 (1H, m), 7.35 (2H, t, J = 6 cps), 7.6–8.9 (6H, m).

^d"Products were identified by spectroscopic and VPC comparison with samples synthesized by other routes" (see footnotes b, c, e, and f).

^eBasis of assignment of structure: (1) prepared in unspecified yield by reaction of with PCl$_5$ in CHCl$_3$; (2) NMR spectrum — τ = 4.24 (2H, m), 6.62 (2H, d, J = 7 cps), 7.0–8.7 (7H, m).

^fBasis of assignment of structure of mixture of <u>syn</u>– and <u>anti</u>-isomers: (1) prepared in unspecified yield by subjection of to "hydroboration-oxidation" followed by reaction with Ph$_3$P/CCl$_4$; (2) NMR spectrum — τ = 6.23 (1H, m), 7.4–8.8 (6H, m), 9.21 (3H, m), 9.90 + 10.16 (1H total, m).

halides simultaneously) of [cyclopentyl]–Br   and   [cyclopentyl]–Br   were reported to be

1.73 and 1.0, respectively [440]. These results would provide no cause
to consider the 3-cyclopentenyl radical to be bridged [1; II(a)].

VII.   7-NORBORNENYL RADICALS

A.   Structure

An ESR spectrum, attributed to [7-norbornenyl-CO₃Buᵗ structure], was obtained by photolysis

of a solution of <u>anti-</u> or (<u>syn</u>/<u>anti</u> = 70/30)- [norbornenyl structure]   in cyclopropane or a

mixture of ethane in cyclopropane [463]. "The spectrum was essentially
unchanged when observed between -98 and -134°. The most prominent
feature of this spectrum [-102°] is a doublet splitting (10.80 G) which is,
no doubt, due to the single H-7. Further splitting into three triplets (2.06,
1.54, and 1.20G) is substantiated by computer simulation . . The tentative
assignments for the hyperfine coupling constants . . . given below are
partly based on analogy with the 7-norbornyl radical (<u>vide infra</u>).

The ESR spectrum [-130°] of the . . . 7-norbornyl radical was obtained

similarly from [7-norbornyl-CO₃Buᵗ structure] . This spectrum consists of a doublet

(16.78 G) of triplets (1.05 G) split further into a pair of quintets (3.53 and
0.72 G). The assignments of these hyperfine coupling constants are as
follows (The assignment of the <u>endo</u> and <u>exo</u> hydrogens is based on the
W-plan formulation)"

(16.78)
(0.72)
(3.53)
(1.05)

If the bridging (?) group is considered to be -CH=CH-, the 7-norbornenyl radical is bridged [1; I(a)]. Unfortunately, the ESR spectrum does not provide a clear answer to the question of whether this bridging [1; I(a)] is accompanied by an unusually short (compared to norbornene or 7-methylenenorbornene) C_7-C_2 (or C_3) distance. The H_5 and H_7 coupling constants are significantly smaller than in the norbornyl radical while the H_2 coupling constant is not exceptionally large, a result not incompatible with either a "short" or a "normal" C_7-C_2 distance.

B. Control of Stereochemistry

Reaction of 7-Norbornenyl Halides with Organotin Hydrides

Reaction of syn- or anti-   Br   with Bu_3SnD gave 7-deuterionorbornene which was reported to be anti/syn = 4.9* [464]. Although bridged

---

* The product was converted [$KMnO_4$(or O_3); $MeOH/H^+$(or CH_2N_2); Na/K; Me_3SiCl; $Bu^tOK/DMSO$], under the assumption that the stereochemistry at C_7 remained undisturbed, to material whose ESR spectrum was considered to be that of a mixture of

(6.5)H D(<0.1)
(2.5) H(2.5)
(2.5) H O•
(2.5) H
H O⁻
(2.5)

and

(1.0)D H(0.4)
(2.5) H(2.5)
(2.5) H O•
(2.5) H
H O⁻
(2.5)

in a ratio of 1:4.9 with the indicated coupling constants and assignments. The assignments, and hence identification of products, were made so as to be consistent with the assignments (shown) made earlier for

(6.5)H H(0.4)
(2.5) H(2.5)
(2.5) H O•
H O⁻
(2.5)

and

(6.5)H H(0.4)
(0.4) D H(2.5)
(0.4) D O•
H O⁻
(2.5)

[1;I(a); see Sec. 10; VIIA], the 7-norbornenyl radical is indicated by this result to be only slightly ($\Delta \Delta G^{\ddagger} < 0.9$ kcal/mole) bridged [1; III(a)].

VIII. γ-VINYLALKYL RADICALS

A. Structure

Photolysis of $CH_2=CHCH_2CH_2CH_2CO_2\overset{+}{)}_2$ in cycloalkane solvent at $-63°$ gave material whose ESR spectrum, attributed to $CH_2=CHCH_2CH_2CH_2\cdot$, was analyzed in terms of $a_\alpha = 22.13$ G, $a_\beta = 28.84$ G, and $a_\gamma = 0.59$ G [431]. These "normal" coupling constants would provide no cause to consider the 4-pentenyl radical to be bridged [1; I(a)-(c)].

B. Energy (Radical Formation from 4-Pentenyl Precursors)

The rate constants for decomposition of $>95\%$ pure, 0.030 M $CH_2=CHCH_2CH_2CH_2CO_2\overset{+}{)}_2$ and $CH_3CH_2CH_2CH_2CH_2CO_2\overset{+}{)}_2$ in toluene at 77° were found to be 1.008 x 10^{-4} and 1.186 x 10^{-4} sec^{-1}, respectively* [471]. "The kinetics data will not support a postulate of neighboring group effect of the double bond . . ."

---

* "The kinetics data for 0.03 M toluene solutions follow the first-order law through two or more half-lives to a high degree of precision. Careful analysis of kinetics of a 0.22 M solution indicates that approximately 3% of the peroxide decomposes by radical induced decomposition at 0.03 M concentration." Products (% yield from 0.11-0.22 M peroxide) (identified from VPC retention time): (1-pentene and 1,4-pentadiene) (80), 1.9-decadiene (16), 4-pentenyl-5-hexenoate (16), 5-hexenoic acid (11), 6-phenyl-1-hexene (22). "Neither cyclopentane, cyclopentene, nor methylcyclobutane was observed . . ."

C. Energy (Radical Formation from Cyclobutylcarbinyl Precursors)

1. Reduction of Ketones by 2-Butanol

See Sec. 10; IE4 and Table 150. The result for the PhCO $-\langle\!\square\rangle\!/$

PhCOCHMe$_2$ system would correspond to a 1.9 kcal/mole difference and is subject to the discussion in Sec. 10; IE4.

2. Decomposition of Azo Compounds

The rate constant for the decomposition (toluene, 80.5°) of 0.02 M

$$\langle\!\square\rangle\!-\underset{\underset{CH_3}{|}}{\overset{\overset{CN}{|}}{C}}-N{=}N-\underset{\underset{CH_3}{|}}{\overset{\overset{CN}{|}}{C}}-\langle\!\square\rangle\!\quad \text{(either stereoisomer) of } 1.51 \times 10^{-4}\ \text{sec}^{-1}$$

[286, 427], when compared to those in Table 152, provides no cause to consider the product radical to be bridged [1; II(a)].

IX.   δ - VINYLALKYL RADICALS

A.   Structure

Photolysis of CH$_2$=CHCH$_2$CH$_2$CH$_2$CO$_2$)$_2$ in cycloalkane solvent at -82° yielded material whose ESR spectrum, attributed to CH$_2$=CHCH$_2$CH$_2$CH$_2$CH$_2\cdot$, was analyzed in terms of a$_\alpha$ = 21.93 G, a$_\beta$ = 28.58 G, and a$_\gamma$ = 0.69 G [418, 431]. These "normal" coupling constants would provide no cause to consider the 5-hexenyl radical to be bridged [1; I(a)-(c)].

An ESR spectrum, attributed to $\langle\!\square\rangle\!-CH_2\cdot$, was obtained by

photolysis in cycloalkane solvent of either CH$_2$=CHCH$_2$CH$_2$CH$_2$CH$_2$CO$_2$)$_2$

at -35° or $\langle\!\square\rangle\!-CH_2CO_2\rangle_2$ at -90° and was analyzed in terms of a$_\alpha$ and a$_\beta$ =

21.3 G [418, 431]. The spectrum is consistent with a strongly bridged [1; I(c)] structure for this δ-vinylbutyl radical.

B.   Structural Isomers of δ-Vinylalkyl Radicals

Can δ-vinylalkyl radicals exist in both bridged [1; I(a)-(c)] and unbridged [1; I(a)-(c)] forms?

See Sec. 10; IXA for relevant ESR data.

1. Reaction of 5-Hexenyl and Cyclopentylcarbinyl Halides with Chromous Salts

See Sec. 5; IC9 for general comments.

Kochi and Powers [147] studied the 1-hexene/methylcyclopentane ratio (total yield $>95\%$) resulting from reaction of $CH_2=CHCH_2CH_2CH_2CH_2Br$ with $Cr^{II}(en)_2^{2+}$ in "83% $DMF-H_2O$ at 25°." A plot of this ratio vs. $[Cr^{II}(en)_2^{2+}]$ was linear with a $(0,0)$ origin. Reaction of ⬠—CH₂Br under the same conditions gave $< 0.01\%$ 1-hexene. These results are consistent with the proposition that $CH_2=CHCH_2CH_2CH_2CH_2Br$ produces a transient species whose sole ultimate fate is either 1-hexene or a different transient species whose sole ultimate fate is methylcyclohexane.

2. Reaction of Halides with Organotin Hydrides

Reaction of ⬠—CH₂CH₂Br with Bu_3SnH [10-20% conversion, 93°/ AIBN or 130°/$(Bu^tO)_2$] was reported [465] to give ⬠—CH₂CH₃ (identified from VPC retention time) and norbornane in a ratio which was strongly dependent on $[Bu_3SnH]$. No control experiments were reported. The data would be consistent with the existence and at-least-one-way interconversion of at least two radicals, one which gives almost exclusively 5-hexenyl product and another which gives a significant amount of cyclopentylcarbinyl product.

The dependence on $[Bu_3SnH]$ of the 1-hexene/methylcyclopentane product ratio resulting from the reaction of $CH_2=CHCH_2CH_2CH_2CH_2Br$ with Bu_3SnH [benzene, 2-10 fold excess halide, 40°/AIBN or 130°/$(Bu^tO)_2$] has been studied [466]. The results at 40° (82-91% total yield) are consistent with the irreversible sequential formation of two radicals, the first whose only fate is to give either 1-hexene or the second and the second whose only fate is methylcyclopentane. "The data at 130° [82-88% total yield], however, show too small a dependence on hydride concentration. They can be fitted if the cyclization . . . [of the first radical to the second] is considered to be reversible, but such a result would predict significant formation of 1-hexene on the reduction of cyclopentyl bromide (36% at . . . [the lower end of the range of hydride concentrations]). Careful examination of this reaction gave methylcyclopentane (77%) as the only C_6

hydrocarbon product, and under 0.1% 1-hexene . . . An alternative hypothesis [which fits the data] is that cyclized product can arise directly from reaction of the uncyclized radical with tin hydride (perhaps by attack of tin on an internal radical- π complex) . . . "

Reaction (91-103% yields) of with Bu$_3$SnH in benzene was reported [467] to give a + +

ratio which is strongly dependent on [Bu$_3$SnH]. No control experiments were reported. These results, if taken at face value, would be consistent with the coexistence and at-least-one-way interconversion of at least two radicals, one which gives mainly 5-hexenyl product and one or more others which lead to cyclopentylcarbinyl and cyclohexyl products.

3. Decomposition of Peroxides

Results of the decomposition of > 95% pure 5-hexenylcarbonyl, cyclopentylcarbinylcarbonyl, and cyclohexylcarbonyl peroxide are given in Table 159. It appears as if three different radical systems are produced (at least in the cage). The cyclohexylcarbonyl peroxide yielded only cyclohexyl products, from which it might be inferred that the intermediate β-vinylbutyl radical is strongly bridged [1; I(c)].

4. Reaction of Hexenyl Halides with Sodium Naphthalenide

The dependence on [sodium naphthalenide] of the 1-hexene/methylcyclopentane product ratio resulting from the reaction of $CH_2=CHCH_2CH_2CH_2CH_2F$ with sodium naphthalenide (dimethoxyethane, 25°) (48% C_6H_{12} products) has been studied [468, 469]. "Reductions of cyclopentylmethyl bromide and chloride yield methylcyclopentane as the only hydrocarbon." The results are consistent with the irreversible sequential formation of two radicals, the first whose only C_6H_{12} fate is either 1-hexene or the second and the second whose only C_6H_{12} fate is methylcyclopentane. *

---

*No cyclohexane, cyclohexene, methylenecyclopentane, 1,5-hexadiene, or C_{12} hydrocarbons were produced. The following statements were made: (1) "No 1-hexene can arise from hydrogen atom transfer reactions with dimethoxyethane because the cyclization of these radicals is much faster than such hydrogen transfer reactions."; (2) "previously reported work shows that . . . the reaction products are stable . . ."

TABLE 159

Decomposition of $RCO_2{\cdot})_2$

Reported Products (% Yield)

R	Reaction conditions	 (methylcyclopentane)	 (methylenecyclopentane)	 (bicyclopentyl)	 (spiro)	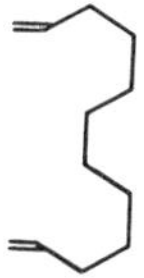 (diene)	1-Hexene + 1,5-hexadiene	 (cyclohexane)	 (cyclohexene)	Ref.
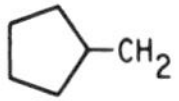 cyclopentyl–CH₂	hν, ~0.05 M, 30°, pentane	33	3	66	0	0	–	–	–	418
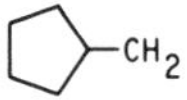 cyclopentyl–CH₂	hν, ~0.05 M, 30°, nujol	27	<2	61	0	0				418
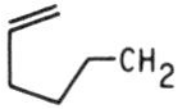 cyclopentyl–CH₂	hν, ~0.05 M, 30°, pentane	33	2	36	3	27				418
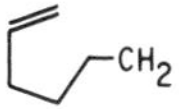 cyclopentyl–CH₂	hν, ~0.05 M, 30°, nujol	34	0	0	0	64	16			418
 cyclohexyl	77°, 0.01 M, toluene	0					0	44[b]	30[b]	470

Structure	Conditions						
cyclopentyl-CH$_2$	77°, 0.093 M, toluene	95[a]		Trace	4[b]	4[b]	470
cyclopentyl-CH$_2$	77°, 0.093 M, toluene, 0.14 M galvinoxyl	2[a]		Trace	4[b]	4[b]	470
cyclopentyl-CH$_2$	77°, 0.093 M, toluene, 0.17 M hydrogalvinoxyl	33[a]		Trace	4[b]	4[b]	470
pentenyl-CH$_2$	77°, 0.093 M, toluene	90[a]	20[a]	16[b]	3[b]	1[b]	470
pentenyl-CH$_2$	77°, 0.102 M, toluene, 0.16 M galvinoxyl	1[a]		13[b]	Trace	Trace	470
pentenyl-CH$_2$	77°, 0.094 M, toluene, 0.16 M hydrogalvinoxyl	1[a]		22[b]	Trace	Trace	470

[a] Identification of product OK.

[b] Basis of identification of product: VPC retention time on two or more columns.

In addition, "ratios . . . as high as 40 were obtained from reactions . . . with very concentrated solutions of sodium naphthalenide . . . This . . . rules out any . . . C=C participation leading to cyclization in the . . . [halogen abstraction] step . . ."

C. Energy (Radical Formation from 5-Hexenyl Precursors)

1. Reaction of 5-Hexenyl Halides with Chromous Salts

See Sec. 5; IC9 for general comments.

"Competition studies show that ω-hexenyl bromide and n-hexyl bromide are reduced at the same rate . . . [by] Cr^{II}(en). Double bond participation, therefore, does not appear to be relevant in the rate-limiting step." [147].

2. Reaction of 5-Hexenyl Halides with Sodium Naphthalenide

The rate constants for reaction of sodium naphthalenide with 5-hexenyl fluoride and n-hexyl fluoride at 25° are $(5.7 \pm 1.6) \times 10^{-4}$ and $(4.3 \pm 1.1) \times 10^{-4}$ M^{-1} sec^{-1}, respectively [468, 469]. This result provides no cause to consider the δ-vinylbutyl radical to be bridged [1; II(a)].

3. Decomposition of Peroxides

In experiments which "are not highly accurate," the half-lives for decomposition of > 95% pure, 0.03 M $CH_2=CHCH_2CH_2CH_2CH_2CO_2)_2$ and $CH_3CH_2CH_2CH_2CH_2CH_2CO_2)_2$ in toluene at 77° were found to be 108 and 93 min, respectively* [470]. These results would provide no cause to consider the δ-vinylbutyl radical to be bridged [1; II(a)].

D. Energy (Radical Formation from Cyclopentylcarbinyl Precursors)

1. Reduction of Ketones by 2-Butanol

See Sec. 10; IE4 and Table 150. The result for the PhCO⟨ ⟩/ PhCOCHMe$_2$ system would correspond to a 0.5 kcal/mole difference and provide no cause to consider the cyclopentylcarbinyl radical to be bridged [1; II(a)].

---

*"All runs obey the first-order rate law through at least two half-lives." See Table 159 for data regarding products.

2. Decomposition of Azo Compounds

The rate constant for the decomposition (toluene, 80.5°) of 0.02 M

(either stereoisomer) of 1.3×10^{-4} sec^{-1} [286, 427], when compared to those in Table 152, provides no cause to consider the product radical to be bridged [1; II(a)].

X. ε-VINYLALKYL RADICALS

Energy (Radical Formation from Cyclohexylcarbinyl Precursors)

1. Reduction of Ketones by 2-Butanol

See Sec. 10; IE4 and Table 150. The results for the PhCO—⟨⟩/

PhCOCHMe$_2$ system would correspond to a 0.1 kcal/mole difference and provide no cause to consider the cyclohexylcarbinyl radical to be bridged [1; II(a)].

2. Decomposition of Azo Compounds

The rate constant for the decomposition (toluene, 80.5°) of 0.02 M

of 2.27×10^{-4} sec^{-1} [286, 427], when compared to those in Table 152, provides no cause to consider the product radical to be bridged [1; II(a)].

3. Decomposition of Peroxides

See Table 149. Cyclohexaneacetyl peroxide decomposed faster ($\Delta\Delta G^{\ddagger} \cong$ 0.9 kcal/mole) than a model compound, $(CH_3)_3CCH_2CH(CH_3)CH_2CO_2)_2$. Given the complexities and uncertainties in the mechanism and the discussion in Sec. 1; VB, it would be premature to conclude from this modest acceleration that cyclohexaneacetyl peroxide decomposes to bridged [1; II(a)] radicals.

ACKNOWLEDGMENTS

This work was supported by the National Institutes of Health and the Petroleum Research Fund, administered by the American Chemical Society.

422

REFERENCES

1. H. L. Goering, P. I. Abell, and B. F. Aycock, J. Am. Chem. Soc., 74, 3588 (1952).

2. T. G. Selin and R. West, J. Am. Chem. Soc., 84, 1860 (1962).

3. J. A. Berson and R. Swidler, J. Am. Chem. Soc., 76, 4060 (1954).

4. F. H. Westheimer, Chem. Revs., 61, 265 (1961).

5. R. P. Bell, Disc. Faraday Soc., 39, 16 (1965).

6. A. V. Willi and M. Wolfsberg, Chem. Ind., 2097 (1964).

7. W. J. Albery, Trans. Faraday Soc., 63, 200 (1967).

8. R. A. More O'Ferrall and J. Kouba, J. Chem. Soc. (B), 985 (1967).

9. R. A. More O'Ferrall, J. Chem. Soc. (B), 786 (1970).

10. M. Katayama, Acta Cryst., 9, 986 (1956).

11. M. Kimura, M. Aoki, and Y. Kurita, Bull. Chem. Soc. Japan, 27, 360 (1954).

12. B. R. Penfold and W. S. Simpson, Acta Cryst., 9, 832 (1956).

13. J. Dajace, Acta Cryst., 8, 851 (1955).

14. M. Yamaha, Bull. Chem. Soc. Japan, 29, 865 (1956).

15. M. Yamaha, Bull. Chem. Soc. Japan, 29, 876 (1956).

16. K. Kuchitsu, Bull. Chem. Soc. Japan, 30, 391 (1957).

17. J. M. Hastings and S. H. Bauer, J. Chem. Phys., 18, 13 (1950).

18. Y. Morino and K. Kuchitsu, J. Chem. Phys., 28, 175 (1958).

19. V. W. Laurie and D. R. Lide, Jr., J. Chem. Phys., 31, 939 (1959).

20. L. Pauling, W. Gordy, and J. H. Saylor, J. Am. Chem. Soc., 64, 1753 (1942).

21. E. Hirota and Y. Morino, Bull. Chem. Soc. Japan, 34, 341 (1961).

22. R. S. Wagner and B. P. Dailey, J. Chem. Phys., 26, 1588 (1957).

23. J. Y. Beach and D. P. Stevenson, J. Am. Chem. Soc., 61, 2643 (1939).

24. R. S. Wagner, B. P. Dailey, and N. Solimene, J. Chem. Phys., 26, 1593 (1957).

25. V. Schomaker and D. P. Stevenson, J. Am. Chem. Soc., 62, 2423 (1940).

26. A. Almenningen, O. Bastiansen, and F. Harshbarger, Acta Chem. Scand., 11, 1059 (1957).

27. J. Donohue, J. Am. Chem. Soc., 72, 2701 (1950).

28. I. M. Dawson and J. M. Robertson, J. Chem. Soc., 1256 (1948).

29. T. Kasuya and T. Oka, J. Phys. Soc. Japan, 14, 980 (1959).

30. R. N. Dixon, Spectrochim. Acta, 9, 59 (1957).

31. C. Flanagan and L. Pierce, J. Chem. Phys., 38, 2963 (1963).

32. R. H. Schwendeman and G. D. Jacobs, J. Chem. Phys., 36, 1245 (1962).

33. D. S. Ashton and J. M. Tedder, J. Chem. Soc. (B), 1031 (1970).

34. H. L. Goering and D. W. Larsen, J. Am. Chem. Soc., 81, 5937 (1959).

35. S. W. Benson, Thermochemical Kinetics, Wiley, N. Y., 1968.

36. J. Dejace, Bull. Soc. Franc. Minér. Crist., 82, 12 (1959).

37. T. Kasuya and T. Oka, J. Phys. Soc. Japan, 15, 296 (1960).

38. D. D. Tanner, D. Darwish, M. W. Mosher, and N. J. Bunce, J. Am. Chem. Soc., 91, 7398 (1969).

39. P. J. Krusic and J. K. Kochi, J. Am. Chem. Soc., 91, 6161 (1969).

40. P. D. Readio and P. S. Skell, J. Org. Chem., 31, 753 (1966).

41. K. Mislow and M. Raban, in Topics in Stereochemistry, (N. L. Allinger and E. L. Eliel, eds.), Vol. 1, Wiley - Interscience, N.Y., 1967, p. 1.

42. J. Horiuti and M. Polanyi, Acta Physicochimica U.R.S.S., 2, 505 (1935).

43. M. H. Mok and J. C. Polanyi, J. Chem. Phys., 51, 1451 (1969).

44. R. A. Ogg, Jr., and M. Polanyi, Trans. Faraday Soc., 31, 604 (1935).

45. R. A. Marcus, J. Phys. Chem., 72, 891 (1968).

46. M. G. Evans and M. Polanyi, Trans. Faraday Soc., 34, 11 (1938).

47. C. N. Hinshelwood, K. J. Laidler, and E. W. Timm, J. Chem. Soc., 848 (1938).

48. J. E. Leffler, Science, 117, 340 (1953).

49. G. S. Hammond, J. Am. Chem. Soc., 77, 334 (1955).

50. G. A. Russell, A. Ito, and D. G. Hendry, J. Am. Chem. Soc., 85, 2976 (1963).

51. M. M. Martin and G. J. Gleicher, J. Am. Chem. Soc., 86, 233, 238 (1964).

52. R. C. Lamb, L. P. Spadafino, R. G. Webb, E. B. Smith, W. E. McNew, and J. G. Pacific, J. Org. Chem., 31, 147 (1966).

53. T. W. Koenig and J. C. Martin, J. Org. Chem., 29, 1520 (1964).

54. D. L. Tuleen, W. G. Bentrude, and J. C. Martin, J. Am. Chem. Soc., 85, 1938 (1963); J. C. Martin, D. L. Tuleen, and W. G. Bentrude, Tet. Letters, 229 (1962).

55. W. T. Dixon, R. O. C. Norman, and A. L. Buley, J. Chem. Soc., 3625 (1964).

56. A. Hudson and R. A. Jackson, Chem. Comm., 1323 (1969).

57. L. H. Sommer, E. Dorfman, G. M. Goldberg, and F. C. Whitmore, J. Am. Chem. Soc., 68, 488 (1946).

58. Y. Nagai, N. Machida, H. Kono, and T. Migita, J. Org. Chem., 32, 1194 (1967).

59. R. A. Benkeser and R. A. Hickner, J. Am. Chem. Soc., 80, 5298 (1958); R. A. Benkeser, M. L. Burrows, L. E. Nelson, and J. V. Swisher, ibid, 83, 4385 (1961).

60. R. Benkeser and R. F. Cunico, J. Organometal. Chem., 4, 284 (1965).

61. R. A. Benkeser, Y. Nagai, J. L. Noe, R. F. Cunico, and P. H. Gund, J. Am. Chem. Soc., 86, 2446 (1964).

62. A. I. Barchukov, T. M. Murina, and A. M. Prokhorov, Opt. i Spekt., 4, 521 (1958).

63. W. R. Cullen and G. E. Styan, J. Organometal. Chem., 6, 117 (1966).

64. H. Sakurai and M. Yamagata, Chem. Comm., 1144 (1970).

65. R. H. Fish, H. G. Kuivila, and I. J. Tyminski, J. Am. Chem. Soc., 89, 5861 (1967).

66. A. E. Fuller and W. J. Hickinbottom, J. Chem. Soc., 3228 (1965).

67. M. Massol, J. Satgé, and Y. Cabadi, Compt. Rend., 268C, 1814 (1969).

68. M. Massol, J. Satgé, and M. Lesbre, Compt. Rend., 262C, 1806 (1966); J. Organometal. Chem., 17, 25 (1969).

69. A. J. Leusink and H. A. Budding, J. Organometal. Chem., 11, 533 (1968).

70. A. J. Leusink, H. A. Budding, and J. W. Marsman, J. Organometal. Chem., 9, 285 (1967).

71. D. Seyferth and L. G. Vaughan, J. Organometal. Chem., 1, 138 (1963).

72. A. J. Leusink, H. A. Budding, and J. G. Noltes, J. Organometal. Chem., 24, 375 (1970).

73. A. L. Buley, R. O. C. Norman, and R. J. Pritchett, J. Chem. Soc. (B), 849 (1966).

74. K. Möbius, K. Hoffmann, and M. Plato, Z. Naturforsch., 23a, 1209 (1968).

75. A. J. Bowles, A. Hudson, and R. A. Jackson, Chem. Phys. Lett., 5, 552 (1970).

76. R. Livingston and H. Zeldes, J. Chem. Phys., 44, 1245 (1966).

77. T. Shiga, A. Boukhors, and P. Douzou, J. Phys. Chem., 71, 3559 (1967).

78. T. Shiga, J. Phys. Chem., 69, 3805 (1965).

79. F. P. Sargent, Can. J. Chem., 46, 1029 (1968).

80. H. L. Benson, Jr., and J. E. Willard, J. Am. Chem. Soc., 88, 5689 (1966); ibid, 83, 4672 (1961).

81. M. Takehisa, G. Levey, and J. E. Willard, J. Am. Chem. Soc., 88, 5694 (1966).

82. R. H. Wiley, W. Miller, C. H. Jarboe, Jr., J. R. Harrell, and D. J. Parish, Radiation Research, 13, 479 (1960).

83. E. B. Dismukes and W. S. Wilcox, Radiation Research, 11, 754 (1959).

84. V. N. Kost, T. T. Sidorova, R.Kh. Freidlina, and A. N. Nesmeyanov, Dokl. Akad. Nauk S.S.S.R., 132, 606 (1960).

85. (a) A. N. Nesmeyanov, R.Kh. Freidlina, and L. I. Zakharkin, Dokl. Akad. Nauk S.S.S.R., 81, 199 (1951); (b) A. N. Nesmeyanov, R.Kh. Freidlina, and L. I. Zakharkin, Quart. Revs., 10, 330 (1956);

(c) R. Kh. Freidlina, in Advances in Free-Radical Chemistry (G. H. Williams, ed.), Logos Press, London, 1, 1965, p. 236; (d) Chem. Abstr., 47, 3789d (1953).

86. R. Kh. Freidlina, M. Ya. Khorlina, and A. N. Nesmeyanov, Izv. Akad. Nauk S. S. S. R., Otdel. Khim. Nauk, 658 (1960).

87. A. N. Nesmeyanov, R. Kh. Freidlina, and A. B. Belyavskii, Izv. Akad. Nauk S. S. S. R., Otdel. Khim. Nauk, 1028 (1959).

88. R. Kh. Freidlina, V. N. Kost, M. Ya. Khorlina, and A. N. Nesmeyanov, Dokl. Akad. Nauk S. S. S. R., 128, 316 (1959).

89. A. N. Nesmeyanov, R. Kh. Freidlina, and V. I. Firstov, Izv. Akad. Nauk S. S. S. R., Otdel. Khim. Nauk, 505 (1951). Descriptions from: Chem. Abstr., 46, 7034b (1952); R. Kh. Freidlina in Advances in Free-Radical Chemistry, (G. H. Williams, ed.), Vol. 1, Logos Press, London, 1965. p. 231.

90. R. Kh. Freidlina, V. N. Kost, M. Ya. Khorlina, and A. N. Nesmeyanov, Dokl. Akad. Nauk S. S. S. R., 137, 341 (1961).

91. M. Ya. Khorlina and V. N. Kost, Dokl. Akad. Nauk S. S. S. R., 137, 1133 (1961).

92. D. N. Hall, J. Org. Chem., 32, 2082 (1966).

93. A. N. Nesmeyanov, R. Kh. Freidlina, R. G. Petrova, and A. B. Terent'eva, Dokl. Akad. Nauk S. S. S. R., 127, 575 (1959); R. Kh. Freidlina, A. B. Terent'ev, R. G. Petrova, and A. N. Nesmeyanov, ibid., 138, 859 (1961).

94. V. N. Kost, T. T. Vasil'eva, and R. Kh. Freidlina, Dokl. Akad. Nauk Belorussk S. S. R., 7, 538 (1963); Chem. Abstr., 59, 13 801e (1963).

95. V. N. Kost, T. T. Vasil'eva, and R. Kh. Freidlina, Izv. Akad. Nauk S. S. S. R., Otdel. Khim. Nauk, 1254 (1962).

96. R. Kh. Freidlina, V. N. Kost, T. T. Vasil'eva, and A. N. Nesmeyanov, Dokl. Akad. Nauk S. S. S. R., 137, 1385 (1961).

97. V. N. Kost, T. T. Vasil'eva, and R. Kh. Freidlina, Dokl. Akad. Nauk Belorussk. S. S. R., 7, 614 (1963); Chem. Abstr., 60, 2475e (1964).

98.   A. N. Nesmeyanov, R.Kh. Freidlina, R. G. Petrova, and A. B. Terent'eva, Dokl. Akad. Nauk S.S.S.R., 127, 575 (1959).

99.   W. H. Urry and J. W. Wilt, J. Am. Chem. Soc., 76, 2594 (1954).

100.   W. H. Urry and J. R. Eiszner, J. Am. Chem. Soc., 74, 5822 (1952).

101.   R. J. Strunk, P. M. DiGiacoma, K. Aso, and H. G. Kuivila, J. Am. Chem. Soc., 92, 2849 (1970).

102.   W. Thaler, J. Am. Chem. Soc., 85, 2607 (1963).

103.   D. D. Tanner and N. Nychka, J. Am. Chem. Soc., 89, 121 (1967).

104.   C. Walling and M. F. Mayahi, J. Am. Chem. Soc., 81, 1485 (1959).

105.   D. S. Ashton and J. M. Tedder, Chem. Comm., 785 (1968).

106.   E. M. Hodnett and P. S. Juneja, J. Org. Chem., 32, 4114 (1967).

107.   P. S. Fredricks and J. M. Tedder, J. Chem. Soc., 144 (1960).

108.   P. S. Fredricks and J. M. Tedder, J. Chem. Soc., 3520 (1961).

109.   N. Colebourne and E. S. Stern, J. Chem. Soc., 3599 (1965).

110.   L. Horner and L. Schläfer, Ann., 635, 31 (1960).

111.   H. C. Brown and A. B. Ash, J. Am. Chem. Soc., 77, 4019 (1955).

112.   F. F. Rust and W. E. Vaughan, J. Org. Chem., 6, 479 (1941).

113.   M. S. Kharasch, W. S. Zimmt, and W. Nudenberg, J. Org. Chem., 20, 1430 (1955).

114.   P. C. Anson, P. S. Fredricks, and J. M. Tedder, J. Chem. Soc., 918 (1959).

115.   J. H. Knox, Chem. Ind., 1631 (1955).

116.   A. Bruylants, M. Tits, C. Dieu, and R. Gauthier, Bull. Soc. Chim. Belg., 61, 366 (1952).

117.   A. Bruylants, M. Tits, and R. Dauby, Bull. Soc. Chim. Belg., 58, 310 (1949).

118.   H. Singh and J. M. Tedder, J. Chem. Soc. (B), 605 (1966); Chem. Comm., 5 (1965).

119.   H. Singh and J. M. Tedder, J. Chem. Soc. (B), 608 (1966).

120.   H. Singh and J. M. Tedder, J. Chem. Soc. (B), 612 (1966).

121.   I. Galiba, J. M. Tedder, and R. A. Watson, J. Chem. Soc., 1321 (1964).

122. H. Singh and J. M. Tedder, J. Chem. Soc., 4737 (1964).

123. H. Magritte and A. Bruylants, Bull. Soc. Chim. Belg., 66, 367
      (1957).

124. H. J. DenHertog, B. DeVries, and J. VanBragt, Rec. Trav.
      Chim., 74, 1561 (1955).

125. H. J. DenHertog and P. Smit, Proc. Chem. Soc., 132 (1959).

126. P. Smit and H. J. DenHertog, Rec. Trav. Chim., 83, 891 (1964).

127. B. Blouri and G. Lanchec, Compt. Rend., 254, 3103 (1962).

128. A. Maillard, A. Deluzarche, and G. Levy, Bull. Soc. Chim.
      France, 1640 (1961).

129. P. Smit and H. J. DenHertog, Rec. Trav. Chim., 77, 73 (1958).

130. A. L. Henne and J. B. Hinkamp, J. Am. Chem. Soc., 67, 1197
      (1945).

131. I. Rosen and J. P. Stallings, Ind. Eng. Chem., 50, 1511 (1958).

132. M. Kosugi, K. Takeuchi, and T. Migita, Bull. Chem. Soc. Japan,
      42, 3242 (1969).

133. P. S. Juneja and E. M. Hodnett, J. Am. Chem. Soc., 89, 5685
      (1967).

134. G. A. Russell, J. Am. Chem. Soc., 80, 4987 (1958).

135. B. Blouri, G. Lanchec, and P. Rumpf, Compt. Rend., 257, 3609
      (1963).

136. J. H. Knox and R. L. Nelson, Trans. Faraday Soc., 55, 937
      (1959).

137. D. D. Tanner and P. B. van Bostelen, J. Org. Chem., 32, 1517
      (1967).

138. C. Walling and B. B. Jacknow, J. Am. Chem. Soc., 82, 6113
      (1960).

139. Y. Shigemitsu, Y. Odaira, and S. Tsutsumi, Bull. Chem. Soc.
      Japan, 38, 1450 (1965).

140. C. Walling and B. B. Jacknow, J. Am. Chem. Soc., 82, 6108
      (1960).

141. C. Walling and P. J. Wagner, J. Am. Chem. Soc., 86, 3368
      (1964).

142. H. Kloosterziel, Rec. Trav. Chim., 82, 508 (1963).

143. J. F. Harris, Jr., J. Org. Chem., 31, 931 (1966).

144. M. S. Kharasch and H. C. Brown, J. Am. Chem. Soc., 61, 2142 (1939).

145. M. S. Kharasch and H. C. Brown, J. Am. Chem. Soc., 62, 925 (1940).

146. C. C. Price, C. D. Beard, and K. Akune, J. Am. Chem. Soc., 92, 5916 (1970).

147. J. K. Kochi and J. W. Powers, J. Am. Chem. Soc., 92, 137 (1970).

148. D. H. R. Barton, N. K. Basu, R. H. Hesse, F. S. Morehouse, and M. M. Pechet, J. Am. Chem. Soc., 88, 3016 (1966).

149. W. C. Kray, Jr., and C. E. Castro, J. Am. Chem. Soc., 86, 4603 (1964).

150. C. E. Castro and W. C. Kray, Jr., J. Am. Chem. Soc., 85, 2768 (1963).

151. D. M. Singleton and J. K. Kochi, J. Am. Chem. Soc., 89, 6547 (1967).

152. J. K. Kochi and D. M. Singleton, J. Am. Chem. Soc., 90, 1582 (1968).

153. J. K. Kochi, D. M. Singleton, and L. J. Andrews, Tetrahedron, 24, 3503 (1968).

154. P. S. Skell, D. L. Tuleen, and P. D. Readio, J. Am. Chem. Soc., 85, 2849 (1963).

155. H. C. Brown, M. S. Kharasch, and T. H. Chao, J. Am. Chem. Soc., 62, 3435 (1940).

156. R. D. Chambers and A. J. Palmer, Tetrahedron, 25, 4217 (1969).

157. M. L. Poutsma, J. Am. Chem. Soc., 87, 2172 (1965).

158. J. T. Pearson, P. Smith, and T. C. Smith, Can. J. Chem., 42, 2022 (1964).

159. G. A. Russell, A. Ito, and R. Konaka, J. Am. Chem. Soc., 85, 2988 (1963).

160. D. S. Ashton and J. M. Tedder, J. Chem. Soc. (B), 1031 (1970).

161. J. C. Little, Y.-L. C. Tong, and J. P. Heeschen, J. Am. Chem. Soc., 91, 7090 (1969); J. C. Little, A. R. Sexton, Y.-L. C. Tong, and T. E. Zurawic, ibid., p. 7098.

162. G. A. Russell and A. Ito, J. Am. Chem. Soc., 85, 2983 (1963).

163. J. L. Brokenshire, A. Nechvatal, and J. M. Tedder, Trans. Faraday Soc., 66, 2029 (1970).

164. Y. Kikuchi, E. Hirota, and Y. Morino, Bull. Chem. Soc. Japan, 34, 348 (1961).

165. M. L. Poutsma, J. Am. Chem. Soc., 87, 4293 (1965).

166. H. Zeldes and R. Livingston, J. Chem. Phys., 45, 1946 (1966).

167. J. E. Bennett, B. Mile, and A. Thomas, J. Chem. Soc. (A), 1399 (1967).

168. M. Kosugi, K. Takeuchi, and T. Migita, Bull. Chem. Soc. Japan, 43, 1535 (1970).

169. J. K. Kochi and D. M. Singleton, J. Org. Chem., 33, 1027 (1968).

170. D. H. Volman, K. A. Maas, and J. Wolstenholme, J. Am. Chem. Soc., 87, 3041 (1965).

171. R. W. Fessenden, and R. H. Schuler, J. Chem. Phys., 39, 2147 (1963).

172. J. K. Kochi and P. J. Krusic, J. Am. Chem. Soc., 92, 4110 (1970).

173. T. S. Zhuravleva and I. A. Misurkin, Zh. Strukt. Khim., 5, 611 (1964).

174. P. J. Krusic and J. K. Kochi, J. Am. Chem. Soc., 90, 7155 (1968).

175. P. I. Abell and L. H. Piette, J. Am. Chem. Soc., 84, 916 (1962); L. H. Piette, in Analysis Instrumentation, Plenum Press, N. Y., 1964.

176. M. C. R. Symons, J. Phys. Chem., 67, 1566 (1963).

177. C. U. Morgan and K. J. White, J. Am. Chem. Soc., 92, 3309 (1970).

178. A. Forchioni and C. Chachaty, Compt. Rend., C264, 637 (1967).

179. D. H. Martin and F. Williams, J. Am. Chem. Soc., 92, 769 (1970).

180. R. J. Neddenriep, and J. E. Willard, J. Phys. Chem., 65, 1206 (1961).

181. D. K. Bakale and H. A. Gillis, J. Phys. Chem., 74, 2074 (1970).

182. W. S. Wilcox, Radiation Research, 10, 112 (1959).

183. P. S. Skell, R. G. Allen, and N. D. Gilmour, J. Am. Chem. Soc.,
      83, 504 (1961).

184. W. O. Haag and E.-A. I. Heiba, Tet. Lett., 3683 (1965).

185. R. J. Field and P. I. Abell, Trans. Faraday Soc., 65, 743 (1969).

186. P. I. Abell, Trans. Faraday Soc., 60, 2214 (1964).

187. R. J. Field and P. I. Abell, J. Am. Chem. Soc., 91, 7226 (1969).

188. R. Barker and A. Maccoll, J. Chem. Soc., 2839 (1963).

189. K. T. Wong and D. A. Armstrong, Can. J. Chem., 47, 4183 (1969).

190. K. T. Wong and D. A. Armstrong, Can. J. Chem., 48, 2426 (1970).

191. S. W. Benson, K. W. Egger, and D. M. Golden, J. Am. Chem.
      Soc., 87, 468 (1965).

192. A. M. North, The Collision Theory of Chemical Reactions in
      Liquids, Wiley, N. Y., 1964, p. 68.

193. L. H. Stein and P. C. Carman, J. S. Afr. Chem. Inst., 9, 55
      (1956).

194. S. J. Cristol, M. A. Imhoff, and D. C. Lewis, J. Org. Chem.,
      35, 1722 (1970).

195. E. F. Ullman and L. Call, J. Am. Chem. Soc., 92, 7210 (1970).

196. J. Kwiatek and J. K. Seyler, J. Organometal. Chem., 3, 421
      (1965).

197. J. Kwiatek and J. K. Seyler, Adv. in Chem. Series, No. 70,
      A. C. S., Wash., D. C., 1968, p. 207.

198. J. Halpern and J. P. Maher, J. Am. Chem. Soc., 87, 5361 (1965).

199. D. R. Lide, Jr., J. Chem. Phys., 30, 37 (1959).

200. C. Flanagan and L. Pierce, J. Chem. Phys., 38, 2963 (1963).

201. K. S. Pitzer, J. Chem. Phys., 12, 310 (1944).

202. J. D. Kemp and C. J. Egan, J. Am. Chem. Soc., 60, 1521 (1938).

203. G. B. Kistakowsky and W. W. Rice, J. Chem. Phys., 8, 610
      (1940).

204. J. G. Aston, R. M. Kennedy, and S. C. Schumann, J. Am. Chem.
      Soc., 62, 2059 (1940).

205. D. R. Lide, Jr. and D. E. Mann, J. Chem. Phys., 29, 914 (1958).

206. J. E. Piercy and M. G. S. Rao, J. Chem. Phys., 46, 3951 (1967).

207.   K. S. Pitzer, J. Chem. Phys., $\underline{5}$, 473 (1937).

208.   K. S. Pitzer, Disc. Faraday Soc., $\underline{10}$, 66 (1951).

209.   J. A. Chen and A. A. Petrauskas, J. Chem. Phys., $\underline{30}$, 304 (1959).

210.   P. H. Verdier and E. B. Wilson, Jr., J. Chem. Phys., $\underline{29}$, 340 (1958).

211.   G. Sage and W. Klemperer, J. Chem. Phys., $\underline{39}$, 371 (1963).

212.   D. R. Herschbach, J. Chem. Phys., $\underline{25}$, 358 (1956).

213.   E. Hirota, J. Chem. Phys., $\underline{37}$, 283 (1962).

214.   J. E. Kilpatrick and K. S. Pitzer, J. Res. Natl. Bur. Std., $\underline{37}$, 163 (1946).

215.   D. R. Lide, Jr., and D. E. Mann, J. Chem. Phys., $\underline{27}$, 868 (1957).

216.   K. D. Möller, A. R. DeMeo, D. R. Smith, and L. H. London, J. Chem. Phys., $\underline{47}$, 2609 (1967).

217.   W. G. Fateley and F. A. Miller, Spectrochim. Acta, $\underline{19}$, 611 (1963).

218.   E. Hirota, J. Chem. Phys., $\underline{45}$, 1984 (1966).

219.   W. G. Fateley and F. A. Miller, Spectrochim. Acta, $\underline{18}$, 977 (1962).

220.   V. W. Laurie, J. Chem. Phys., $\underline{34}$, 1516 (1961).

221.   S. Segal, J. Chem. Phys., $\underline{27}$, 989 (1957).

222.   R. A. Oriani and C. P. Smyth, J. Chem. Phys., $\underline{16}$, 930 (1948).

223.   R. A. Beaudet, J. Chem. Phys., $\underline{37}$, 2398 (1962).

224.   M. L. Unland, V. Weiss, and W. H. Flygare, J. Chem. Phys., $\underline{42}$, 2138 (1965).

225.   L. H. London and K. D. Möller, J. Mol. Struct., $\underline{2}$, 493 (1968).

226.   L. Pierce and J. M. O'Reilly, J. Mol. Spect., $\underline{3}$, 536 (1959).

227.   S. Weiss and G. E. Leroi, J. Chem. Phys., $\underline{48}$, 962 (1968).

228.   J. R. Hoyland, J. Chem. Phys., $\underline{49}$, 1908 (1968).

229.   E. Hirota, C. Matsumura, and Y. Morino, Bull. Chem. Soc. Japan, $\underline{40}$, 1124 (1967).

230.   J. C. Amphlett and E. Whittle, Trans. Faraday Soc., $\underline{62}$, 1662 (1966).

231.   S. W. Benson, J. Chem. Phys., $\underline{43}$, 2044 (1965).

231A. J. H. Sullivan and N. Davidson, J. Chem. Phys., $\underline{19}$, 143 (1951).

232.   S. W. Benson, J. Chem. Phys., $\underline{28}$, 301 (1958).

233.   P. S. Skell, D. L. Tuleen, and P. D. Readio, J. Am. Chem. Soc., $\underline{85}$, 2850 (1963).

234.   F. D. Greene, W. A. Remers, and J. W. Wilson, J. Am. Chem. Soc., 79, 1416 (1957).

235.   P. S. Skell, Organic Rxn. Mechanisms, Special Publication No. 19, The Chemical Society, 1965, p. 131.

236.   W. G. Fateley and F. A. Miller, Spectrochim. Acta, 17, 857 (1961).

237.   R. H. Schwendeman and G. D. Jacobs, J. Chem. Phys., 36, 1245 (1962).

238.   L. J. Altman and B. W. Nelson, J. Am. Chem. Soc., 91, 5163 (1969).

239.   E. Wyn-Jones and W. J. Orville-Thomas, Trans. Faraday Soc., 64, 2907 (1968).

240.   E. Wyn-Jones and W. J. Orville-Thomas, Chemical Society Special Publication No. 20, 209 (1965).

241.   J. H. Andrae, personal communication cited in Ref. [240].

242.   T. Kasuya, J. Phys. Soc. Japan, 15, 1273 (1960).

243.   G. Ya. Kabo and D. N. Andreevskii, Neftekhimiya, 3, 764 (1963).

244.   A. E. Clark and T. A. Litovitz, J. Acoust. Soc. Am., 32, 1221 (1960).

245.   H. Steinmetz and R. M. Noyes, J. Am. Chem. Soc., 74, 4141 (1952).

246.   M. C. Flowers and S. W. Benson, J. Chem. Phys., 38, 882 (1963).

247.   D. M. Golden, R. Walsh, and S. W. Benson, J. Am. Chem. Soc., 87, 4053 (1965).

248.   G. C. Fettis and A. F. Trotman-Dickenson, J. Chem. Soc., 3037 (1961).

249.   M. I. Christie, Proc. Roy. Soc., 244A, 411 (1958).

250.   D. B. Hartley and S. W. Benson, J. Chem. Phys., 39, 132 (1963).

251.   H. C. Anderson and E. R. Van Artsdalen, J. Chem. Phys., 12, 479 (1944).

252.   G. B. Kistiakowsky and E. R. Van Artsdalen, J. Chem. Phys., 12, 469 (1944).

253.   H. Schmitz, H. J. Schumacher, and A. Jager, Z. Phys. Chem. (Frankfurt), B51, 281 (1942).

254. G. A. Oldershaw and R. J. Cvetanović, J. Chem. Phys., 41, 3639 (1964).

255. D. J. Carlsson and K. U. Ingold, J. Am. Chem. Soc., 90, 7047 (1968).

256. S. W. Benson and H. E. O'Neal, Kinetic Data on Gas Phase Unimolecular Reactions, National Standard Reference Data Series, National Bureau of Standards, 21, 611 (1970).

257. J. A. Franklin and G. H. Huybrechts, Int. J. Chem. Kinet., 1, 3 (1969).

258. G. Huybrechts, L. Meyers, and G. Verbeke, Trans. Faraday Soc., 58, 1128 (1962).

259. J. A. Franklin, G. Huybrechts, and C. Cillien, Trans. Faraday Soc., 65, 2094 (1969).

260. P. Goldfinger, G. Huybrechts, and G. Martens, Trans. Faraday Soc., 57, 2210 (1961).

261. P. Goldfinger and G. Martens, Trans. Faraday Soc., 57, 2220 (1961).

262. J. H. Knox and R. G. Musgrave, Trans. Faraday Soc., 63, 2201 (1967).

263. R. K. Boyd, G. W. Downs, J. S. Gow, and C. Horrex, J. Phys. Chem., 67, 719 (1963).

264. C. A. Goy and H. O. Pritchard, J. Phys. Chem., 69, 3040 (1965).

265. J. K. Kochi and P. E. Mocadlo, J. Org. Chem., 30, 1134 (1965).

266. R. A. Beaudet, J. Chem. Phys., 50, 2002 (1969).

267. P. S. Skell and P. K. Freeman, J. Org. Chem., 29, 2524 (1964).

268. N. P. Neureiter and F. G. Bordwell, J. Am. Chem. Soc., 82, 5354 (1960).

269. P. S. Skell and R. G. Allen, J. Am. Chem. Soc., 81, 5383 (1959).

270. O. Maass and C. H. Wright, J. Am. Chem. Soc., 46, 2664 (1924).

271. E. I. Yakovenko, G. B. Sergeev, and M. M. Rakhimov, Dokl. Akad. Nauk S.S.S.R., 171, 890 (1966).

272. N. A. Gac, D. M. Golden, and S. W. Benson, J. Am. Chem. Soc., 91, 3091 (1969).

273. R. Gregory, R. N. Haszeldine, and A. E. Tipping, J. Chem. Soc. (C), 1750 (1970).

274.   M. L. Poutsma and P. A. Ibarbia, J. Org. Chem., 35, 4038 (1970).

275.   P. I. Abell and B. A. Bohm, J. Org. Chem., 26, 252 (1961).

276.   P. I. Abell and C. Chiao, J. Am. Chem. Soc., 82, 3610 (1960).

277.   S.-I. Ohnishi and I. Nitta, J. Chem. Phys., 39, 2848 (1963).

278.   M. D. Sevilla and R. A. Holroyd, J. Phys. Chem., 74, 2459 (1970).

279.   T. Ohmae, S.-I. Ohnishi, K. Kuwata, H. Sakurai, and I. Nitta,
       Bull. Chem. Soc. Japan, 40, 226 (1967).

280.   J. G. Traynham and W. G. Hines, J. Am. Chem. Soc., 90, 5208
       (1968).

281.   P. S. Skell and P. D. Readio, J. Am. Chem. Soc., 86, 3334 (1964).

282.   F. R. Jensen, L. H. Gale, and J. E. Rodgers, J. Am. Chem. Soc.,
       90, 5793 (1968).

283.   S. J. Cristol, J. R. Douglass, W. C. Firth, Jr., and R. E. Krall,
       J. Org. Chem., 27, 2711 (1962).

284.   P. I. Abell, J. Org. Chem., 22, 769 (1957).

285.   S. J. Cristol, L. K. Gaston, and T. Tiedeman, J. Org. Chem.,
       29, 1279 (1964).

286.   D. E. Applequist and A. H. Peterson, J. Am. Chem. Soc., 82,
       2372 (1960).

287.   M. Heintzeler, Ann., 569, 102 (1950).

288.   S. Winstein, Bull. Soc. Chim. France, 70c (1951).

289.   J. M. Riemann, Ph.D. Thesis, Ohio Univ., 1968.

290.   M. Avram, E. Marica, and C. D. Nenitzescu, Acad. rep.
       populare Romîne, Studii cercetari chim., 7, 155 (1959); Chem.
       Abstr., 54, 8664f (1960).

291.   M. Avram, Gh. D. Mateescu, and C. D. Nenitzescu, Rev. Roum.
       Chim., 14, 101 (1969).

292.   R. T. Arnold and P. Morgan, J. Am. Chem. Soc., 70, 4248 (1948).

293.   F. Bell and I. F. B. Smyth, J. Chem. Soc., 2372 (1949).

294.   E. L. Eliel and R. V. Acharya, J. Org. Chem., 24, 151 (1959).

295.   J. D. Berman and C. C. Price, J. Org. Chem., 23, 102 (1958).

296.   D. C. Abbott and C. L. Arcus, J. Chem. Soc., 3195 (1952).

297.   P. Wilder, Jr. and A. Winston, J. Am. Chem. Soc., 75, 5370
       (1953).

298.   R. A. Stein and M. Kayama, J. Org. Chem., 26, 5231 (1961).

299.   R. A. Barnes and R. J. Prochaska, J. Am. Chem. Soc., 72, 3188 (1950).

300.   W. G. Dauben and H. Tilles, J. Am. Chem. Soc., 72, 3185 (1950).

301.   J. Cason and R. L. Way, J. Org. Chem., 14, 31 (1949).

302.   J. W. H. Oldham, J. Chem. Soc., 100 (1950).

303.   W. T. Smith, Jr. and R. L. Hull, J. Am. Chem. Soc., 72, 3309 (1950).

304.   S. J. Cristol and L. K. Gaston, J. Org. Chem., 26, 4672 (1961).

305.   E. R. Buchman and J. C. Conly, J. Am. Chem. Soc., 75, 1990 (1953).

306.   E. L. Sukman, Ph.D. Thesis, Rutgers Univ., 1958.

307.   D. E. Applequist and N. D. Werner, J. Org. Chem., 28, 48 (1963).

308.   H. L. Goering and L. L. Sims, J. Am. Chem. Soc., 77, 3465 (1955).

309.   N. A. LeBel, R. F. Czaja, and A. DeBoer, J. Org. Chem., 34, 3112 (1969).

310.   C. W. Shoppee and R. Lack, J. Chem. Soc., 4864 (1960).

311.   L. H. Gale, J. Org. Chem., 34, 81 (1969).

312.   J. G. Traynham, A. G. Lane, and N. S. Bhacca, J. Org. Chem., 34, 1302 (1969).

313.   N. O. Brace, J. Org. Chem., 28, 3093 (1963).

314.   N. O. Brace, J. Am. Chem. Soc., 86, 665 (1964).

315.   E.-A. I. Heiba and L. C. Anderson, J. Am. Chem. Soc., 79, 4940 (1957).

316.   P. Boldt, L. Schulz, U. Klinsmann, H. Köster, and W. Thielecke, Tetrahedron, 26, 3591 (1970).

317.   J. A. Hirsch, Topics in Stereochemistry (N. L. Allinger and E. L. Eliel, eds), Wiley-Interscience, Vol. 1, 1967, p. 199.

318.   D. M. Tomkinson and H. O. Pritchard, J. Phys. Chem., 70, 1579 (1966).

319.   L. D. Bergel'son and L. P. Badenkova, Izv. Akad. Nauk S.S.S.R., Otdel. Khim. Nauk, 1073 (1960).

320.   G. A. Russell, J. Am. Chem. Soc., 78, 1035 (1956).

321. M. Avram, E. Marica, and C. D. Nenitzescu, Rev. Roum. Chim.,
      4, 253 (1959).

322. M. Avram, E. Marica, and C. D. Nenitzescu, Chem. Ber., 92,
      1088 (1959).

323. D. E. Applequist and A. S. Fox, J. Org. Chem., 22, 1751 (1957).

324. A. J. Fry, W. B. Farnham, B. J. Holstein, M. A. Mitnick, and
      L. C. Riggs, J. Org. Chem., 34, 4195 (1969).

325. J. Kwart and P. J. Chenier, J. Org. Chem., 35, 1562 (1970).

326. H. K. Wart and J. L. Nyce, J. Am. Chem. Soc., 86, 2601 (1964).

327. N. A. LeBel, P. D. Beirne, E. R. Karger, J. C. Powers, and
      P. M. Subramanian, J. Am. Chem. Soc., 85, 3199 (1963).

328. N. A. LeBel, P. D. Beirne, and P. M. Subramanian, J. Am.
      Chem. Soc., 86, 4144 (1964).

329. P. M. Subramanian, M. T. Emerson, and N. A. LeBel, J. Org.
      Chem., 30, 2624 (1965).

330. N. A. LeBel, J. Am. Chem. Soc., 82, 623 (1960).

331. D. I. Davies, L. T. Parfitt, C. K. Alden, and J. A. Claisse,
      J. Chem. Soc. (C), 1585 (1969).

332. M. M. Martin and R. A. Koster, J. Org. Chem., 33, 3428 (1968).

333. D. I. Davies and P. J. Rowley, J. Chem. Soc. (C), 424 (1969).

334. J. A. Claisse, D. I. Davies, and C. K. Alden, J. Chem. Soc. (C),
      1498 (1966).

335. D. I. Davies and L. T. Parfitt, J. Chem. Soc. (C), 2691 (1967).

336. D. I. Davies, J. Chem. Soc., 3669 (1960).

337. A. G. Ludwick and J. C. Martin, J. Org. Chem., 34, 4108 (1969).

338. L. E. Barstow and G. A. Wiley, Tet. Lett., 865 (1968).

339. E. Tobler and D. J. Foster, J. Org. Chem., 29, 2839 (1964).

340. C. L. Osborn, T. V. Van Auken, and D. J. Trecker, J. Am.
      Chem. Soc., 90, 5806 (1968).

341. C. K. Alden and D. I. Davies, J. Chem. Soc. (C), 2007 (1967).

342. C. K. Alden, D. I. Davies, and P. J. Rowley, J. Chem. Soc. (C),
      705 (1968).

343. P. B. Chock and J. Halpern, J. Am. Chem. Soc., 91, 582 (1969).

344. W. S. Trahanovsky and M. P. Doyle, J. Org. Chem., 32, 146
      (1967).

345.   V. I. Smirnova, T. S. Zhuravleva, D. N. Shigorin, E. P.
       Gracheva, and M. F. Shostakovskii, Zh. Fiz. Khim., $\underline{38}$, 469
       (1964).

346.   I. N. Nazarov and L. D. Bergel'son, Izv. Akad. Nauk S.S.S.R.,
       Otdel. Khim. Nauk, 887 (1960).

347.   I. N. Nazarov and L. D. Bergel'son, Izv. Akad. Nauk S.S.S.R.,
       Otdel. Khim. Nauk, 896 (1960).

348.   L. D. Bergel'son, Izv. Akad. Nauk S.S.S.R., Otdel. Khim.
       Nauk, 1066 (1960).

349.   I. N. Nazarov and L. D. Bergel'son, Zhur. Obshch. Khim., $\underline{27}$,
       1540 (1957).

350.   K. R. Kopecky and S. Grover, Can. J. Chem., $\underline{47}$, 3153 (1969).

351.   P. S. Skell and R. G. Allen, J. Am. Chem. Soc., $\underline{86}$, 1559 (1964).

352.   P. S. Skell and R. G. Allen, J. Am. Chem. Soc., $\underline{80}$, 5997 (1958).

353.   L. D. Bergel'son, Izv. Akad. Nauk S.S.S.R., Otdel. Khim. Nauk,
       1235 (1960).

354.   L. D. Bergel'son, Izv. Akad. Nauk S.S.S.R., Otdel. Khim. Nauk,
       1499 (1958).

355.   S. W. Benson, D. M. Golden, and K. W. Egger, J. Chem. Phys.,
       $\underline{42}$, 4265 (1965).

356.   S. W. Benson, J. Chem. Phys., $\underline{38}$, 1945 (1963).

357.   D. M. Golden, S. Furuyama, and S. W. Benson, Int. J. Chem.
       Kinet., $\underline{1}$, 57 (1969).

358.   P. S. Skell and R. R. Pavlis, J. Am. Chem. Soc., $\underline{86}$, 2956 (1964).

359.   S. H. Hastings, J. L. Franklin, J. C. Schiller, and F. A. Matsen,
       J. Am. Chem. Soc., $\underline{75}$, 2900 (1953).

360.   J. G. Traynham and J. R. Olechowski, J. Am. Chem. Soc., $\underline{81}$,
       571 (1959).

361.   J. A. A. Ketelaar and C. van de Stolpe, Rec. Trav. Chim., $\underline{71}$,
       805 (1952).

362.   S. Freed and K. M. Sancier, J. Am. Chem. Soc., $\underline{74}$, 1273 (1952).

363.   F. Fairbrother, J. Chem. Soc., 1051 (1948).

364.   H. McConnell, J. S. Ham, and J. R. Platt, J. Chem. Phys., $\underline{21}$,
       66 (1953).

365. L. J. Andrews and R. M. Keefer, J. Am. Chem. Soc., 74, 458
      (1952).

366. R. Drury and L. Kaplan, unpublished work.

367. J. F. Garst and J. T. Barbas, J. Am. Chem. Soc., 91, 3385
      (1969).

368. J. Q. Adams, J. Am. Chem. Soc., 92, 4535 (1970).

369. A. B. Terent'ev and R. Kh. Freidlina, Dokl. Akad. Nauk S. S. S. R.,
      158, 679 (1964).

370. A. B. Terent'ev and R. G. Petrova, Izv. Akad. Nauk S. S. S. R.,
      Ser. Khim., 2153 (1963).

371. R. Kh. Freidlina and A. B. Terent'ev, Dokl. Akad. Nauk S. S. S. R.,
      152, 637 (1963).

372. R. Kh. Freidlina, A. B. Terent'ev, and R. G. Petrova, Dokl. Akad.
      Nauk S. S. S. R., 151, 866 (1963).

373. R. Kh. Freidlina, A. B. Terent'ev, and R. G. Petrova, Dokl. Akad.
      Nauk S. S. S. R., 149, 860 (1963).

374. R. G. Guy and J. J. Thompson, Chem. Ind., 1499 (1970).

375. R. G. R. Bacon, R. G. Guy, R. S. Irwin, and T. A. Robinson,
      Proc. Chem. Soc., 304 (1959).

376. P. S. Skell and R. G. Allen, J. Am. Chem. Soc., 82, 1511 (1960).

377. P. D. Readio and P. S. Skell, J. Org. Chem., 31, 759 (1966).

378. F. G. Bordwell, P. S. Landis, and G. S. Whitney, J. Org. Chem.,
      30, 3764 (1965).

379. F. G. Bordwell and W. A. Hewett, J. Am. Chem. Soc., 79, 3493
      (1957).

380. J. I. Cunneen, J. Chem. Soc., 134 (1947).

381. C. W. Shoppee, M. I. Akhtar, and R. E. Lack, J. Chem. Soc.,
      877 (1964).

382. R. C. Tweit, F. B. Colton, N. L. McNiven, and W. Klyne, J.
      Org. Chem., 27, 3325 (1962).

383. H. L. Goering, D. I. Relyea, and D. W. Larsen, J. Am. Chem.
      Soc., 78, 348 (1956).

384. H. L. Goering, D. I. Relyea, and K. L. Howe, J. Am. Chem.
      Soc., 79, 2502 (1957).

385. J. F. Ford, R. C. Pitkethly, and V. O. Young, Tetrahedron, 4, 325 (1958).

386. A. A. Oswald, K. Griesbaum, and W. Naegele, J. Am. Chem. Soc., 86, 3791 (1964).

387. S. J. Cristol and R. P. Arganbright, J. Am. Chem. Soc., 79, 6039 (1957).

388. E. P. Blanchard, Jr., and A. Cairncross, J. Am. Chem. Soc., 88, 487 (1966).

389. J. S. Hyde, R. Breslow, and C. DeBoer, J. Am. Chem. Soc., 88, 4763 (1966).

390. E. I. Heiba and R. M. Dessau, J. Org. Chem., 32, 3837 (1967).

391. L. N. Owen and M. U. S. Sultanbawa, J. Chem. Soc., 3109 (1949).

392. Y.-C. Liu and S.-K. Wang, K'o Hsüeh T'ung Pao, 206 (1957); Chem. Abstr., 53, 9115h (1959).

393. M. F. Shostakovskii, E. N. Prilezhaeva, L. V. Tsymbal, and L. G. Stolyarova, Zh. Obshch. Khim., 30, 3143 (1960).

394. W. H. Mueller, J. Org. Chem., 31, 3075 (1966).

395. A. A. Oswald, K. Griesbaum, B. E. Hudson, Jr., and J. M. Bregman, J. Am. Chem. Soc., 86, 2877 (1964).

396. W. E. Truce, H. G. Klein, and R. B. Kruse, J. Am. Chem. Soc., 83, 4636 (1961).

397. J. A. Kampmeier and G. Chen, J. Am. Chem. Soc., 87, 2608 (1965).

398. F. W. Stacey and J. F. Harris, Jr., J. Am. Chem. Soc., 85, 963 (1963).

399. T. L. Jacobs and G. E. Illingworth, Jr., J. Org. Chem., 28, 2692 (1963).

400. J.-M. Surzur, C. Dupuy, M.-P. Crozet, and N. Aimar, Compt. Rend., 269C, 849 (1969).

401. M. F. Shostakovskii, E. N. Prilezhaeva, and L. V. Tsymbal, Trudy po Khim. i Khim. Tekhnol., 4, 198 (1961); Chem. Abstr., 56, 1331b (1962).

402. B. G. Shakhovskoi and A. A. Petrov, Zhur. Obshch. Khim., 37, 1371 (1967).

403.   E. N. Prilezhaeva, I. I. Guseinov, B. V. Lopatin, and M. F. Shostakovskii, Zh. Obshch. Khim., 29, 3227 (1959).

404.   P. S. Skell and J. H. McNamara, J. Am. Chem. Soc., 79, 85 (1957).

405.   S. J. Cristol and J. A. Reeder, J. Org. Chem., 26, 2182 (1961).

406.   W. O. Haag and E.-A. I. Heiba, Tet. Lett., 3679 (1965).

407.   V. W. Laurie, J. Chem. Phys., 31, 1500 (1959).

408.   R. G. Lerner and B. P. Dailey, J. Chem. Phys., 26, 678 (1957).

409.   J. D. Roberts and R. H. Mazur, J. Am. Chem. Soc., 73, 2509 (1951).

410.   E. Renk, P. R. Shafer, W. H. Graham, R. H. Mazur, and J. D. Roberts, J. Am. Chem. Soc., 83, 1987 (1961).

411.   C. Walling and P. S. Fredricks, J. Am. Chem. Soc., 84, 3326 (1962).

412.   L. K. Montgomery and J. W. Matt, J. Am. Chem. Soc., 89, 934 (1967).

413.   L. K. Montgomery and J. W. Matt, J. Am. Chem. Soc., 89, 6556 (1967).

414.   H. K. Hall, Jr., E. P. Blanchard, Jr., S. C. Cherkofsky, J. B. Sieja, and W. A. Sheppard, J. Am. Chem. Soc., 93, 110 (1971).

415.   H. K. Hall, Jr., C. D. Smith, E. P. Blanchard, Jr., S. C. Cherkofsky, and J. B. Sieja, J. Am. Chem. Soc., 93, 121 (1971).

416.   E. S. Huyser and J. D. Taliaferro, J. Org. Chem., 28, 3442 (1963).

417.   T. A. Halgren, M. E. H. Howden, M. E. Medof, and J. D. Roberts, J. Am. Chem. Soc., 89, 3051 (1967).

418.   R. A. Sheldon and J. K. Kochi, J. Am. Chem. Soc., 92, 4395 (1970).

419.   J. Smid, A. Rembaum, and M. Szwarc, J. Am. Chem. Soc., 78, 3315 (1956).

420.   R. C. Lamb, F. F. Rogers, Jr., G. D. Dean, Jr., and F. W. Voight, Jr., J. Am. Chem. Soc., 84, 2635 (1962).

421.   H. Hart and D. P. Wyman, J. Am. Chem. Soc., 81, 4891 (1959).

422.   H. Hart and R. A. Cipriani, J. Am. Chem. Soc., 84, 3697 (1963).

423. T. D. Manly and M. Frost, private communication, cited by R. C. P. Cubbon, Progr. in Rxn. Kinetics, ed. G. Porter, 5, 29 (1970).

424. E. S. Huyser and O. T. Wang, J. Org. Chem., 29, 2720 (1964).

425. D. C. Neckers, A. P. Schaap, and J. Hardy, J. Am. Chem. Soc., 88, 1265 (1966).

426. D. C. Neckers and A. P. Schaap, J. Org. Chem., 32, 22 (1967).

427. C. G. Overberger and M. B. Berenbaum, J. Am. Chem. Soc., 73, 2618 (1951).

428. J. C. Martin and J. W. Timberlake, J. Am. Chem. Soc., 92, 978 (1970).

429. J. C. Martin, J. E. Schultz, and J. W. Timberlake, Tet. Lett., 4629 (1967).

430. J. P. VanHook and A. V. Tobolsky, J. Am. Chem. Soc., 80, 779 (1958).

431. J. K. Kochi and P. J. Krusic, J. Am. Chem. Soc., 91, 3940 (1969).

432. J. K. Kochi, P. J. Krusic, and D. R. Eaton, J. Am. Chem. Soc., 91, 1877 (1969).

433. J. K. Kochi, P. J. Krusic, and D. R. Eaton, J. Am. Chem. Soc., 91, 1879 (1969).

434. L. K. Montgomery, J. W. Matt, and J. R. Webster, J. Am. Chem. Soc., 89, 923 (1967).

435. J. A. Garcia Dominguez and A. F. Trotman-Dickenson, J. Chem. Soc., 940 (1962).

436. J. A. Landgrebe and L. W. Becker, J. Org. Chem., 33, 1173 (1968).

437. J. A. Claisse and D. I. Davies, J. Chem. Soc. (C), 1045 (1966).

438. S. J. Cristol and D. I. Davies, J. Org. Chem., 29, 1282 (1964).

439. V. M. A. Chambers, W. R. Jackson, and G. W. Young, Chem. Comm., 1275 (1970).

440. C. R. Warner, R. J. Strunk, and H. G. Kuivila, J. Org. Chem., 31, 3381 (1966).

441. G. M. Whitesides and J. San Filippo, Jr., J. Am. Chem. Soc., 92, 6611 (1970).

442. E. N. Prilezhaeva, V. A. Azovskaya, and N. P. Petukhova, Zh. Org. Khim., 4, 621 (1968).

443. E. N. Prilezhaeva, V. A. Azovskaya, L. V. Tsymbal, E. N. Gur'yanova, G. Andrianova, and M. F. Shostakovskii, Zh. Obshch. Khim., 35, 39 (1965).

444. A. U. Stepanyants and V. F. Bystrov, Izv. Akad. Nauk S.S.S.R., Ser. Khim., 1003 (1968).

445. V. F. Bystrov and A. U. Stepanyants, J. Mol. Spect., 21, 241 (1966).

446. A. U. Stepanyants and V. F. Bystrov, Zh. Strukt. Khim., 9, 907, 1088 (1968).

447. V. F. Bystrov, O. P. Yablonski, and A. U. Stepanyants, Radiospektrosk. Kvantovokhim. Metody Strukt. Isslid., 177 (1967); Chem. Abstr., 70, 24512q (1969).

448. D. I. Davies, and P. J. Rowley, J. Chem. Soc. (C), 2245 (1967).

449. D. I. Davies, and P. J. Rowley, J. Chem. Soc. (C), 2249 (1967).

450. E. N. Prilezhaeva, V. A. Asovskaya, G. U. Stepanyanz, D. Mondeshka, and R. J. Shekhtman, Tet. Lett., 4909 (1969).

451. C. K. Alden and D. I. Davies, J. Chem. Soc. (C), 1017 (1967).

452. C. K. Alden, J. A. Claisse, and D. I. Davies, J. Chem. Soc. (C), 1540 (1966).

453. J. Gresser, A. Rajbenback, and M. Szwarc, J. Am. Chem. Soc., 83, 3005 (1961).

454. D. J. Trecker and J. P. Henry, J. Am. Chem. Soc., 85, 3204 (1963).

455. E. S. Huyser and G. Echegaray, J. Org. Chem., 27, 429 (1962).

456. M. M. Martin and D. C. DeJongh, J. Am. Chem. Soc., 84, 3526 (1962).

457. P. D. Bartlett and J. M. McBride, J. Am. Chem. Soc., 87, 1727 (1965).

458. H. Hart and F. J. Chloupek, J. Am. Chem. Soc., 85, 1155 (1963).

459. D. I. Davies and L. T. Parfitt, J. Chem. Soc. (C), 1401 (1969).

460. S. J. Cristol and R. V. Barbour, J. Am. Chem. Soc., 90, 2832 (1968).

461. L. H. Slough, J. Am. Chem. Soc., 87, 1522 (1965).

462.   M. L. Poutsma, J. Am. Chem. Soc., $\underline{87}$, 2161 (1965).

463.   P. Bakuzis, J. K. Kochi, and P. J. Krusic, J. Am. Chem. Soc.,
       $\underline{92}$, 1434 (1970).

464.   G. A. Russell and G. W. Holland, J. Am. Chem. Soc., $\underline{91}$, 3968
       (1969).

465.   J. W. Wilt, S. N. Massie, and R. B. Dabek, J. Org. Chem., $\underline{35}$,
       2803 (1970).

466.   C. Walling, J. H. Cooley, A. A. Ponares, and E. J. Racah, J.
       Am. Chem. Soc., $\underline{88}$, 5361 (1966).

467.   D. L. Struble, A. L. J. Beckwith, and G. E. Gream, Tet. Lett.,
       3701 (1968).

468.   J. F. Garst, P. W. Ayers, and R. C. Lamb, J. Am. Chem. Soc.,
       $\underline{88}$, 4260 (1966).

469.   J. F. Garst and F. E. Barton, II, Tet. Lett., 587 (1969).

470.   R. C. Lamb, P. W. Ayers, and M. K. Toney, J. Am. Chem. Soc.,
       $\underline{85}$, 3483 (1963).

471.   R. C. Lamb, W. E. McNew, Jr., J. R. Sanderson, and D. C.
       Lunney, J. Org. Chem., $\underline{36}$, 174 (1971).

472.   A. Hudson and K. D. J. Root, J. Chem. Soc. (B), 656 (1970).

473.   P. J. Krusic and J. K. Kochi, J. Am. Chem. Soc., $\underline{93}$, 0000 (1971).

474.   S. L. Hsu and W. H. Flygare, J. Mol. Spect., $\underline{37}$, 92 (1971).

475.   J. K. Kochi and P. J. Krusic, J. Am. Chem. Soc., $\underline{90}$, 7157
       (1968).

476.   J. K. Kochi and P. J. Krusic, Chem. Society (London) Special
       Publ. No. 24, 147 (1970).

477.   T. Shiga, A. Boukhors, and P. Douzou, J. Phys. Chem., $\underline{71}$,
       3559 (1967).

478.   P. L. Kolker, J. Chem. Soc., 5929 (1964).

479.   D. R. Gee and J. K. S. Wan, Can. J. Chem., $\underline{49}$, 20 (1971).

480.   K. D. King, D. M. Golden, and S. W. Benson, Trans. Faraday
       Soc., $\underline{66}$, 2794 (1970).

481.   W. C. Danen and R. L. Winter, J. Am. Chem. Soc., $\underline{93}$, 716
       (1971).

482.   Yu. A. Ol'dekop and R. V. Kaberdin, Zh. Org. Khim., $\underline{6}$, 1114
       (1970).

483. T. Kawamura, M. Ushio, T. Fujimoto, and T. Yonezawa, J. Am.
      Chem. Soc., 93, 908 (1971).

484. H. Fischer, Z. Naturforsch., 19a, 866 (1964).

485. E. N. Mal'tseva and V. S. Zavgorodnii, Zhur. Obshch. Khim.,
      40, 1773 (1970).

486. B. B. Jarvis, J. Org. Chem., 33, 4075 (1968).

487. B. B. Jarvis, J. P. Govoni, and P. J. Zell, J. Am. Chem. Soc.,
      93, 913 (1971).

488. E. S. Huyser and R. H. C. Feng, J. Org. Chem., 36, 731 (1971).

489. H. Fischer, K.-H. Hellwege, and M. Lehnig, Ber. Bunsenges.
      Phys. Chem., 72, 1166 (1968).

490. H. Zeldes and R. Livingston, J. Am. Chem. Soc., 93, 1082 (1971).

491. R. Wilson, J. Chem. Soc. (B), 84 (1968).

492. A. R. McIntosh and J. K. S. Wan, Can. J. Chem., 49, 812 (1971).

493. R. J. Cvetanović, F. J. Duncan, W. E. Falconer, and W. A.
      Sunder, J. Am. Chem. Soc., 88, 1602 (1966).

494. R. L. Ayres, C. J. Michejda, and E. P. Rack, J. Am. Chem.
      Soc., 93, 1389 (1971).

495. H. D. Roth, J. Am. Chem. Soc., 93, 1527 (1971).

496. R. S. Boikess, M. Mackay, and D. Blithe, Tet. Lett., 401 (1971).

497. P. J. D. Park and E. Wyn-Jones, J. Chem. Soc. (A), 646 (1969).

498. T. H. Thomas, E. Wyn-Jones, and W. J. Orville-Thomas, Trans.
      Faraday Soc., 65, 974 (1969).

499. I. Nitta, T. Ohmae, S. Ohnishi, K. Kuwata, and H. Sakurai,
      Nippon Hoshasen Kobunshi Kenkyu Kyokai Nempo, 6, 295 (1964-5).

500. D. Greatorex and T. J. Kemp, Trans. Faraday Soc., 67, 56 (1971).

501. W. T. Dixon and R. O. C. Norman, J. Chem. Soc., 3119 (1963).

502. J. F. Gibson, M. C. R. Symons, and M. G. Townsend, J. Chem.
      Soc., 269 (1959).

503. R. S. Alger, T. H. Anderson, and L. A. Webb, J. Chem. Phys.,
      30, 695 (1959).

504. B. Fell and L.-H. Kung, Chem. Ber., 98, 2871 (1965).

505. B. Blouri, C. Cerceau, and G. Lanchec, Bull. Soc. Chim.
      France, 304 (1963).

506. P. Smit and H. J. den Hertog, Tet. Lett., 595 (1971).

507. B. A. Krentsel', A. V. Topchiev, and D. E. Il'ina, Dokl. Akad. Nauk S. S. S. R. (Phys. Chem.), $\underline{128}$, 1192 (1959).

508. J. P. Soumillion and A. Bruylants, Bull. Soc. Chim. Belg., $\underline{78}$, 169 (1969).

509. E. C. Friedrich and R. L. Holmstead, J. Org. Chem., $\underline{36}$, 971 (1971).

510. P. K. Freeman, M. F. Grostic, and F. A. Raymond, J. Org. Chem., $\underline{36}$, 905 (1971).

511. R. G. Gasanov, T. T. Vasil'eva, and R. Kh. Freidlina, Dokl. Akad. Nauk S. S. S. R., Otdel. Khim. Nauk, $\underline{193}$, 1058 (1970).

512. R. Marx and L. Bonazzola, Mol. Phys., $\underline{19}$, 899 (1970).

513. J. Rouchaud and A. Bruylants, Bull. Soc. Chim. Belg., $\underline{75}$, 783 (1966).

514. J. Rouchaud and A. Bruylants, Bull. Soc. Chim. Belg., $\underline{76}$, 50 (1967).

515. J. Ph. Soumillion and A. Bruylants, Bull. Soc. Chim. Belg., $\underline{78}$, 435 (1969).

516. G. Lanchec, C. Bejannin, and B. Blouri, Bull. Soc. Chim. France, 4486 (1969).

517. D. I. Davies and P. Mason, J. Chem. Soc. (C), 295 (1971).

518. D. I. Davies and P. Mason, J. Chem. Soc. (C), 288 (1971).

519. C. K. Alden and D. I. Davies, J. Chem. Soc. (C), 700 (1968).

520. K. B. Wiberg and D. S. Connor, J. Am. Chem. Soc., $\underline{88}$, 4437 (1966).

521. G. Snatzke and G. Eckhardt, Chem. Ber., $\underline{101}$, 2010 (1968).

522. M. Ballester, J. Riera, J. Castaner, C. Badía, and J. M. Monsó, J. Am. Chem. Soc., $\underline{93}$, 2215 (1971).

523. K. Eiben and R. W. Fessenden, J. Phys. Chem., $\underline{75}$, 1186 (1971).

524. I. A. Shvarts, M. Ya. Khorlina, and R. Kh. Freidlina, Izv. Akad. Nauk S. S. S. R., Ser. Khim., 2018 (1970).

525. T. A. Foglia and D. Swern, J. Org. Chem., $\underline{31}$, 3625 (1966).

526. K. Schrage, Tet. Lett., 5795 (1966).

527. H. D. Burrows, D. Greatorex, and T. J. Kemp, J. Am. Chem. Soc., $\underline{93}$, 2539 (1971).

528.   H. Sakurai, A. Hosomi, and M. Kumada, J. Org. Chem., <u>34</u>, 1764 (1969).

529.   L. Batt and F. R. Cruickshank, J. Phys. Chem., <u>71</u>, 1836 (1967).

530.   D. I. Davies and P. J. Rowley, J. Chem. Soc. (C), 446 (1971).

531.   R. Alexander and D. I. Davies, J. Chem. Soc. (C), 896 (1971).

532.   G. Lanchec, Chim. Ind., <u>94</u>, 46 (1965).

533.   L. Kaplan, "The Structure of Free Radicals," in <u>Free Radicals</u> (J. K. Kochi, ed.), Wiley, N. Y., 1972.

534.   Yu. I. Porfir'eva, L. A. Vasil'eva, E. S. Turbanova, and A. A. Petrov, Zhur. Org. Khim., <u>5</u>, 591 (1969).

535.   R. Gregory, R. N. Haszeldine, and A. E. Tipping, J. Chem. Soc. (C), 1216 (1971).

ADDENDUM

For readers who wish to use this book as an up-to-date source of references, I have listed below those papers which came to my attention between May 1971 and April 1972.

1. R. A. Bartsch, "The Stereochemistry of Base-Catalyzed β Elimination from 2-Bromobutane," J. Am. Chem. Soc., 93, 3683 (1971).
2. B. L. Hawkins, W. Bremser, S. Borcic, and J. D. Roberts, "Nuclear Magnetic Resonance Spectroscopy. Barriers to Internal Rotation in Some Halogenated Methylbutanes," J. Am. Chem. Soc., 93, 4472 (1971).
3. D. D. Tanner, H. Yabuuchi, and E. V. Blackburn, "Polar Radicals. IV. On the Photobromination of Several Optically Active 1-Substituted 2-Methylbutanes," J. Am. Chem. Soc., 93, 4802 (1971).
4. W. C. Danen, T. J. Tipton, and D. G. Saunders, "Halogen Abstraction Studies. III. Free-Radical Abstraction of Iodine from Bridgehead Iodides," J. Am. Chem. Soc., 93, 5186 (1971).
5. D. D. Tanner, M. W. Mosher, N. C. Das, and E. V. Blackburn, "Polar Radicals. V. The Free-Radical Bromination of Cyclohexyl Bromide and 1-Bromobutane with N-Bromosuccinimide and Molecular Bromine," J. Am. Chem. Soc., 93, 5846 (1971).
6. D. K. Wedegaertner, R. M. Kopchik, and J. A. Kampmeier, "Vinyl Radicals. VII. Stereochemistry of the Free Radical Addition of Ethyl Mercaptan to Ethoxyacetylene. A 1-Alkoxyvinyl Radical," J. Am. Chem. Soc., 93, 6890 (1971).
7. A. R. Lyons and M. C. R. Symons, "Halogen Atom Additions to Olefins. An Electron Spin Resonance Study of the Intermediates," J. Am. Chem. Soc., 93, 7330 (1971).
8. G. B. Watts and K. U. Ingold, "Kinetic Applications of Electron Paramagnetic Resonance Spectroscopy. V. Self-Reactions of Some Group IV Radicals," J. Am. Chem. Soc., 94, 491 (1972).

9. T. Kawamura and J. K. Kochi, "Hyperconjugative and p-d Homocon-jugative Effects of Silicon, Germaniun, and Tin on Alkyl Radicals from Electron Spin Resonance Studies," J. Am. Chem. Soc., 94, 648 (1972).

10. V. A. Brosseau, J. R. Basila, J. F. Smalley, and R. L. Strong, "Halogen Atom Charge-Transfer Complexes in the Vapor Phase," J. Am. Chem. Soc., 94, 716 (1972).

11. T. Kawamura, D. J. Edge, and J. K. Kochi, "p-p Homoconjugation in the β-Chloroethyl Radical," J. Am. Chem. Soc., 94, 1752 (1972).

12. R. L. Hilderbrandt and J. D. Wieser, "Average Structures of t-Butyl Chloride and 9D-t-Butyl Chloride Determined by Gas-Phase Electron Diffraction," J. Chem. Phys., 55, 4648 (1971).

13. R. L. Hilderbrandt and J. D. Wieser, "On a Comparison of the Electron Diffraction and Microwave Spectroscopic Structures for t-Butyl Chloride," J. Chem. Phys., 56, 1143 (1972).

14. D. J. Edge, B. C. Gilbert, R. O. C. Norman, and P. R. West, "Electron Spin Resonance Studies. Part XXVIII. Oxidation of Enols, Enol Ethers, and Related Compounds with the Hydroxyl and the Amino-Radical," J. Chem. Soc. (B), 189 (1971).

15. N. H. Anderson, A. J. Dobbs, D. J. Edge, R. O. C. Norman, and P. R. West, "Electron Spin Resonance Studies. Part XXX. One-electron Reduction of Carbonyl-containing Compounds by the Radicals $\cdot CO_2H$, $\cdot CO_2^-$, and $\cdot CMe_2NH_2$," J. Chem. Soc. (B), 1004 (1971).

16. D. S. Ashton and J. M. Tedder, "Free Radical Substitution in Aliphatic Compounds. Part XXII. The Gas-phase Chlorination of Chlorocyclo-pentane and Methylcyclopentane," J. Chem. Soc. (B), 1719 (1971).

17. D. S. Ashton and J. M. Tedder, "Free Radical Substitution in Aliphatic Compounds. Part XXIII. The Gas Phase Chlorination of Chlorocycloheptane and Fluorocyclohexane," J. Chem. Soc. (B), 1723 (1971).

18. A. J. Bowles, A. Hudson, and R. A. Jackson, "An Electron Resonance Study of Some Reactions involving Silyl Radicals," J. Chem. Soc. (B), 1947 (1971).

19. A. W. P. Jarvie and R. J. Rowley, "Radical Addition Reactions of Alkenylsilanes," J. Chem. Soc. (B), 2439 (1971).

20. V. M. A. Chambers, W. R. Jackson, and G. W. Young, "The Stereochemistry of Organometallic Compounds. Part X. Further Evidence as to the Mechanism of Borohydride Reduction of Organomercurials from a Study of Photochemical and Organotin Hydride Reductions," J. Chem. Soc. (C), 2075 (1971).

21. R. Alexander, D. I. Davies, D. H. Hey, and J. N. Done, "Free-radical Reactions of Halogenated Bridged Polycyclic Compounds. Part XIV. The Cobalt (II) Chloride-catalysed Reaction of Methylmagnesium Iodide with 5-endo-Bromo-1,2,3,4,7,7-hexachloronorborn-2-ene and Related Compounds," J. Chem. Soc. (C), 2367 (1971).

22. W. E. Truce and G. C. Wolf, "Adducts of Sulfonyl Iodides with Acetylenes," J. Org. Chem., 36, 1727 (1971).

23. S. J. Cristol and R. Kellman, "Bridged Polycyclic Compounds. LXX. Rearrangements Accompanying Free-Radical Addition of Thiophenol to 3-Methylenenortricyclene," J. Org. Chem., 36, 1866 (1971).

24. H. G. Kuivila, J. D. Kennedy, R. Y. Tien, I. J. Tyminski, F. L. Pelczar, and O. R. Khan, "Addition of Trimethyltin Hydride and Methylhalotin Hydrides to Norbornadiene," J. Org. Chem., 36, 2083 (1971).

25. H. H. Szmant and J. J. Rigau, "Nonstereospecific Oxidative Addition of Benzenethiol to Indene," J. Org. Chem., 37, 447 (1972).

26. W. Adam and J. Arce, "Stereospecific Dehalogenation of vic-Dibromides by Sodium Naphthalenide," J. Org. Chem., 37, 507 (1972).

27. A. L. J. Beckwith and G. Phillipou, "Stereoelectronic Effects in Radical Fragmentation: Rearrangement of 3β, 5-Cyclocholestan-6-yl Radical," Chem. Comm., 658 (1971).

28. J. E. Anderson and H. Pearson, "Effect of Alkyl Groups on the Barrier to Rotation in Substituted Ethanes," Chem. Comm., 871 (1971).

29. A. R. Lyons and M. C. R. Symons, "Electron Spin Resonance Evidence for Greatly Enhanced Hyperconjugation in Radicals Containing Tin, Phosphorus, and Arsenic," Chem. Comm., 1068 (1971).

30. R. G. Pews and T. E. Evans, "Photochemical Addition of Sulphonyl Cyanides to Olefins," Chem. Comm., 1397 (1971).

31. D. S. Ashton, J. M. Tedder, and J. C. Walton, "Relative Selectivities in Bromination Reactions - A Note of Warning," Chem. Comm., 1487 (1971).

32. H. Paul and H. Fischer, "Electron Spin Resonance of Transient Radicals during Photoreactions of Aliphatic Ketones," Chem. Comm., 1038 (1971).

33. H.-P. Löffler, "Radical Rearrangements in some Bicyclic Systems," Chem. Ber., 104, 1981 (1971).

34. K. Herwig and C. Rüchardt, "On the Mechanism of the Hunsdiecker Reaction," Chem. Ber., 105, 363 (1972).

35. I. Tabushi, Y. Tamaru, and Z.-i. Yoshida, "The Chlorination of Sulfones with Sulfuryl Chloride," Tet. Lett., 3893 (1971).

36. D. Touchard and J. Lessard, "The Addition to Olefins of Acylamino Radicals Generated by Photochemical Decomposition of Halogenated N-Chloro- and N-Bromoacetamides," Tet. Lett., 4425 (1971).

37. M. C. R. Symons, "Directive Effects of Substituents CH_2X in Aromatic Electrophilic Substitutions," Tet. Lett., 4919 (1971).

38. A. Ohno and Y. Ohnishi, "Effect of Heteroatoms in Free Radicals. VII. Transannular Participation of Sulfur Atom in Thiacyclohexyl Free Radicals," Tet. Lett., 339 (1972).

39. C. Ronneau, J.-P. Soumillion, P. Dejaifve, and A. Bruylants, "Photobromination of N-Butyl Bromide Labelled with [82]Br. On the Intermediacy of Bridged Bromoalkyl Radicals," Tet. Lett., 317 (1972).

40. P. Neta and R. W. Fessenden, "Electron Spin Resonance Study of Radicals Produced in Irradiated Aqueous Solutions of Thiols," J. Phys. Chem., 75, 2277 (1971).

41. K. W. Watkins and L. A. O'Deen, "Isomerization of Vibrationally Excited 3-Methyl-1-buten-1-yl Radicals via Hydrogen Atom Migration. Quantum Statistical Weight Effect," J. Phys. Chem., 75, 2665 (1971).

42. P. J. Krusic, P. Meakin, and J. P. Jesson, "Electron Spin Resonance Studies of Conformations and Hindered Internal Rotation in Transient Free Radicals," J. Phys. Chem., 75, 3438 (1971).

43. D. Greatorex and T. J. Kemp, "Electron Spin Resonance Studies of Photo-Oxidation by Metal Ions in Rigid Media at Low Temperatures. Part 2.- Ce(IV) Photo-Oxidation of Carboxylic Acids," Trans. Faraday Soc., 67, 1576 (1971).

44. H. G. Benson, A. J. Bowles, A. Hudson, and R. A. Jackson, "Electron Resonance Spectra of some Radicals derived from Acetylenes and Nitriles," Mol. Phys., 20, 713 (1971).

45. A. R. McIntosh and J. K. S. Wan, "Electron Spin Resonance of Some Triphenylsiloxydiphenylmethyl Radicals," Mol. Phys., 22, 183 (1971).

46. C. L. Norris and W. H. Flygare, "The Microwave Spectrum, Barrier to Internal Rotation of the Methyl Group, and ^{14}N Nuclear Quadrupole Coupling Constants in Methacrylonitrile," J. Mol. Spect., 40, 40 (1971).

47. W. Good, R. J. Conan, Jr., A. Bauder, and Hs. H. Günthard, "Microwave Spectra of Deuterium Substituted 2-Chloropropenes and the Conformation of the Methyl Group," J. Mol. Spect., 41, 381 (1972).

48. H. Sakurai, A. Hosomi, and M. Kumada, "Abstraction of Methyl Hydrogen of Organosilicon Compounds by t-Butoxy Radicals," Bull. Chem. Soc. Japan, 44, 568 (1971).

49. D. F. McMillen, D. M. Golden, and S. W. Benson, "Kinetics of the Gas-Phase Reaction of Cyclopropylcarbinyl Iodide and Hydrogen Iodide. Heat of Formation and Stabilization Energy of the Cyclopropyl-carbinyl Radical," Int. J. Chem. Kinetics, 3, 359 (1971).

50. P. C. Reeves, "Inductive Effects in the Chlorination of 1-Chloro-butane," J. Chem. Educ., 48, 636 (1971).

51. T. Kawamura and J. K. Kochi, "Electron Spin Resonance Study of the Adduct of Trialkylstannyl Radical to Butadiene," J. Organometal. Chem., 30, C8 (1971).

52. J. Ph. Soumillion, P. Gouverneur, T. Burton, and A. Bruylants, "Studies of Directed Chlorination. XVI. Mesomeric Effect and Reactivity in Radical Chlorination Reactions," Bull. Soc. Chim. Belges, 80, 233 (1971).

53. A. L. J. Beckwith and P.K. Tindal, "Free-Radical Acetoxy Group Migration: An E.P.R. Study," Aust. J. Chem., 24, 2099 (1971).

54. T. A. Halgren, J. L. Firkins, T. A. Fujimoto, H. H. Suzukawa, and J. D. Roberts, "Evidence for Hydrogen Abstraction by Classical Radicals in the Norbornenyl-Nortricyclyl System," Proc. Nat. Acad. Sci. USA, 68, 3216 (1971).

55. N. V. Komarov and O. G. Yarosh, "Ethynylsilanes in the Radical Hydrobromination Reaction," Izv. Akad. Nauk SSSR, Ser. Khim., 1573 (1971).

56. V. A. Azovskaya, N. P. Petukhova, A. U. Stepanyants, D. Mondeshka, and E. N. Prilezhaeva, "Stereochemistry of Bridged Bicyclic Systems. III. The Configurational Stability of Chlorine-Containing Norbornene Sulfones in the Process of their Preparation by Oxidation or by Diene Condensation," Zhur. Org. Khim., 6, 2451 (1970).

57. V. A. Azovskaya, A. U. Stepanyants, D. Mondeshka, R. I. Shekhtman, I. Koshel'skaya, and E. N. Prilezhaeva, "Stereochemistry of Bridged Bicyclic Systems. IV. Free-Radical Addition of t-Butyl Mercaptan to Hexamethylnorbornadiene," Zhur. Org. Khim., 6, 2458 (1970).

58. C. Descoins, M. Julia, and Huynh V. S., "Radical Cyclizations. XIII. Cyclization onto an Ethylenic Double Bond of a Cyclopropyl Radical Formed during the Reduction of a Geminal Dihalide by an Organotin Hydride," Bull. Soc. Chim. France, 4087 (1971).

59. E. Körös, "Relation between Iodine Exchange and Iodination Reactions," Acta Chim. Acad. Scient. Hung., 67, 195 (1971).

60. R. Kh. Freidlina, A. B. Terent'ev, M. Ya. Khorlina, and S. N. Aminov, "Rearrangement of Radicals during Telomerization of Ethylene by Oxygen-Containing Telogens," Zhur. Vses. Khim. Obshch. im. Mendeleeva, 11, 211 (1966).

61. Yu. A. Ol'dekop and R. V. Kaberdin, "Reaction of Acetyl Peroxide with 1,1,1-Trichloroethane and 1,1,1,2-Tetrachloroethane, Homolytic Rearrangement of Polychloroalkyl Radicals with the 1,2-Migration of Chlorine," Vestsi Akad. Navuk Belarus. SSR, Ser. Khim. Navuk, 82 (1971); Chem. Abstr., 75, 140184m (1971).

62. R. Perrey, Thesis, Univ. of Paris, 1969, cited by M. Julia, "Free-Radical Cyclizations," Accts. Chem. Res., 4, 386 (1971).

AUTHOR INDEX

Numbers in brackets are reference numbers and indicate that an author's work is referred to although his name is not cited in the text. Underlined numbers give the page on which the complete reference is listed.

A

Abbott, D.C., 240[296], 436
Abell, P.I., 2[1], 4[1], 187[175],
  188[175, 185, 186, 187],
  236[175], 237[275, 276],
  238[175], 240[284], 242[284],
  249[175], 259[175], 260[276],
  264[276], 285[175], 423, 431
  432, 436
Acharya, R.V., 240[294], 436
Adam, W., 451
Adams, J.Q., 306[368], 440
Aimar, N., 338[400], 441
Akhtar, M.I., 320[381], 321[382],
  440
Akune, K., 117[146], 430
Albery, W.J., 2[7], 423
Alden, C.K., 269[331], 273[334,
  342], 274[341], 332[331],
  381[452], 382[334, 452],
  383[334], 384[451], 438,
  444, 447
Alexander, R., 448, 451
Alger, R.S., 39[503], 446
Allen, R.G., 193[183], 226[269],
  292[351], 293[352], 432, 435,
  439, 440
Almenningen, A., 3[26], 424
Altman, L.J., 210[238], 434
Aminov, S.N., 454
Amphlett, J.C., 138[230],
  229[230],
  433

Anderson, H.C., 138[251], 434
Anderson, J.E., 451
Anderson, L.C., 253[315], 437
Anderson, N.H., 450
Anderson, T.H.. 39[503], 446
Andrae, J.H., 134[241], 135[241],
  434
Andreevskii, D.N., 134[243], 434
Andrews, L.J., 125[153],
  127[153], 128[153], 299[365],
  430, 440
Andrianova, G., 380-383[443],
  444
Anson, P.C., 61[114], 63[114],
  71[114], 428
Aoki, M., 4[11], 414[11], 423
Applequist, D.E., 240[286],
  241[307], 246[307],
  264[323], 373[286], 415[286],
  421[286], 436, 437
Arce, J., 451
Arcus, C.L., 240[296], 436
Arganbright, R.P., 331[387],
  332[387], 441
Armstrong, D.A., 188[189],
  195[189], 216[190], 223[190],
  432
Arnold, R.T., 240[292], 436
Ash, A.B., 78[111], 115[111],
  117[111], 428
Ashton, D.S., 5[33], 85[105],
  152[105, 160], 153[160], 155[160],
  156[160], 161[105], 424, 428, 430,
  450, 452

455

Colton, F. B. , 321[382], <u>440</u>
Conan, R. J. , Jr. , <u>453</u>
Conly, J. C. , 240[305], 241[305],
    <u>437</u>
Connor, D. S. , 394[520], <u>447</u>
Cooley, J. H. , 416[466], <u>445</u>
Cristol, S. J. , 240[283, 285, 304],
    241[283, 285, 304], 331[387],
    332[387], 394[283], 406[460],
    <u>432</u>, <u>436</u>, <u>437</u>, <u>441</u>-<u>444</u>, <u>451</u>
Crozet, M.-P. , 338[400], <u>441</u>
Cruickshank, F. R. , 229[529], <u>448</u>
Cubbon, R. C. P. , 367[423], <u>443</u>
Cullen, W. R. 21[63], 28[63],
    <u>425</u>
Cunico, R. F. , 22[60, 61], <u>425</u>
Cunneen, J. I. , 314[380], <u>440</u>
Cvetanović, R. J. , 216[254],
    299[493], <u>435</u>, <u>446</u>
Czaja, R. F. , 250[309],
    314-320[309], <u>437</u>

D

Dabek, R. B. , 416[465], <u>445</u>
Dailey, B. P. , 3[24], 4[22],
    414[22], <u>423</u>, <u>442</u>
Dajace, J. , 3[36], 4[13], <u>423</u>,
    <u>424</u>
Danen, W. C. , <u>445</u>, <u>449</u>
Darwish, D. , 197[38], 200[38],
    239[38], <u>424</u>
Das, N. C. , <u>449</u>
Dauben, W. G. , 240[300], <u>437</u>
Dauby, R. , 70[117], 71[117],
    78[117], 79[117], <u>428</u>
Davidson, N. , 138[231A],
    229[231A], <u>433</u>
Davies, D. I. , 269[331], 270[333],
    273[333-336, 342], 274[333,
    335, 336, 341], 332[331],
    376[437], 380[449], 381[449,
    452], 382[334, 452], 383[334],
    384[448, 451], 385[448],
    394[517, 518], 395[518],

396[437], 397[459], 403[459],
    <u>438</u>, <u>443</u>, <u>444</u>, <u>448</u>
    <u>451</u>
Dawson, I. M. , 3[28], <u>424</u>
Dean, G. D. , Jr. , 366[420], <u>442</u>
DeBoer, A. , 250[309], 314-
    320[309], <u>437</u>
DeBoer, C. , 334[389], <u>441</u>
DeJaifve, P. , <u>452</u>
DeJongh, D. C. , 399[456], <u>444</u>
Deluzarche, A. , 71-73[128],
    <u>429</u>
DeMeo, A. R. , 134-136[216],
    <u>433</u>
den Hertog, H. J. , 70[125, 126,
    129], 78[124], 79[124-126],
    85[125, 126, 129, 506],
    86[506], 100[126], <u>429</u>, <u>447</u>
Descoins, C. , <u>454</u>
Dessau, R. M. , <u>441</u>
DeVries, B. , 78[124], 79[124],
    <u>429</u>
Dieu, C. , 69-71[116], 78[116],
    79[116], 89[116], 90[116],
    93[116], <u>428</u>
DiGiacoma, P. M., 58[101],
    59[101], 215[101], <u>428</u>
Dismukes, E. B. , 43[83], <u>426</u>
Dixon, R. N. , 3[30], 4[30], <u>424</u>
Dixon, W. T. , 19[55], 38[55, 501],
    39[501], 170[55], 370[55],
    <u>425</u>, <u>446</u>
Dobbs, A. J. , <u>450</u>
Done, J. N. , <u>451</u>
Donohue, J. , 3[27], <u>424</u>
Dorfman, E. , 16[57], <u>425</u>
Douzou, P. , 38[77], 39[77], <u>426</u>
    <u>445</u>
Downs, G. W. , 42[263], <u>435</u>
Doyle, M. P., 285[344], <u>438</u>
Drury, R. , 300[366], 301[366],
    <u>440</u>
Duncan, F. J. , 299[493], <u>446</u>
Dupuy, C. , 338[400],
    <u>441</u>

[For a radical of particular structure, see the Table of Contents]